# Is the Universe a Hologram?

# Is the Universe a Hologram?

Scientists Answer the Most Provocative Questions

Adolfo Plasencia

foreword by Tim O'Reilly

The MIT Press
Cambridge, Massachusetts
London, England

This book was set in ITC Stone Sans Std and ITC Stone Serif Std by Toppan Best-set Premedia Limited. Printed and bound in the United States of America.

Library of Congress Cataloging-in-Publication Data

Names: Plasencia, Adolfo, editor.
Title: Is the universe a hologram? scientists answer the most provocative
    questions / [interviews by] Adolfo Plasencia ; foreword by Tim O'Reilly.
Description: Cambridge, MA : The MIT Press, [2017] | Includes bibliographical
    references and index.
Identifiers: LCCN 2016038291 | ISBN 9780262036016 (hardcover : alk. paper)
Subjects: LCSH: Science--Miscellanea. | Science--Popular works.
Classification: LCC Q173 .I8 2017 | DDC 500--dc23 LC record available at
    https://lccn.loc.gov/2016038291

10  9  8  7  6  5  4  3  2  1

To my mother, Rafaela, and my father, Ángel, who did everything in their power to make me a better person. To my sister, María Jesús, and my nephew, Ángel, a prodigy of nature. To my grandmother, Concha, an extraordinary woman who lived for 102 years, spanning two centuries, and who continues to guide my heart and investigations.

To my grandfather, Adolfo, to whom I owe my name, but whom the Spanish Civil War prevented me from ever knowing. From him, apart from my name, I must have inherited some mysterious force that inexplicably allows me to face up to the most difficult moments of my life.

Plato's Academy. Mosaic, House of T. Siminius Stephanus, Pompeii, 110–80 BCE.
Museo Archeologico Nazionale di Napoli, Italy. Photograph by Adolfo Plasencia.

*Though the gods were far away, he visited their region of the sky, in his mind, and what nature denied to human vision he enjoyed with his inner eye. When he had considered every subject, through concentrated thought, he communicated it widely in public, teaching the silent crowds, who listened in wonder to his words, concerning the origin of the vast universe, and of the causes of things; and what the physical world is; what the gods are; where the snows arise; what the origin of lightning is; whether Jupiter, or the storm-winds, thunder from colliding clouds; what shakes the earth; by what laws the stars move; and whatever else is hidden.*

—(Ovid describes Pythagoras) *Metamorphoses XV*

# Contents

# Foreword

Tim O'Reilly

The future is not something that "happens." It is something we create.

Yes, there are elements and influences beyond us. The laws of nature, and our own nature, constrain our choices. And there are great catastrophes, earthquakes, plagues, and floods that shape events. But increasingly, we humans are the source of our own destiny, our own greatness, and our own failure.

All the more reason, then, to reflect on the future, and on the choices we make.

In his essay "Imagination as Value," found in the collection *The Necessary Angel*, the poet Wallace Stevens wrote, "The truth seems to be that we live in concepts of the imagination before the reason has established them. If this is true, then reason is simply the methodizer of the imagination." The future is the result of countless creative acts, visions of what can be that are made real through persuasion and effort. The computing pioneer Alan Kay echoed this thought when he said, "It is easier to invent the future than it is to predict it."

But each invention, each new idea, each new social or political moment, only becomes real when it is shared. The spark leaps from mind to mind, sometimes slowly, sometimes in a conflagration. But without that leap, the spark dies.

A simplistic version of history focuses on single individuals and single moments: the defining battle, the great man or woman, the momentous discovery. But the truth is that while some moments matter more than others, and some individual choices or discoveries do seem to send the world careening off on a new path, no discovery, no new idea, and no momentous choice exists in isolation. Rather, they seem to spring from a thousand conversations, a stumbling together toward a shared consensus. The fashions of the moment, and what endures or is rediscovered (sometimes too late for the creator's benefit), have always been crowdsourced.

Now, though, social media accelerate the process, with viral storms of discovery and sharing. There is so much we don't know about how this new cultural transmission vector will work over the long run, but already we can see the following at work:

• The pathways of attention are found not in ephemeral "Likes" but in the deeper persistence of search engines, which echo the way our brains themselves preserve memories, by laying down repeated tracks, growing stronger and stronger over time, so that some things take precedence over others. As some narratives become dominant, others are forgotten.

• As in the brain, memories fade over time, constantly overwritten by what is new. What was once popular becomes a curiosity, perhaps even fades from view. A website is taken offline, a document disappears and the link is redirected.

But is this really different (except in speed, scale, and the electronic means of production) from what went before? I remember standing over my father's grave, my aunt, herself quite learned, lamenting, "So much learning. So much knowledge now gone." Now, thirty years on, my aged aunt herself is a repository of knowledge and memories about to go "404."

In the old cultural order, works considered worthy of note were preserved in libraries. Now, apart from the Internet Archive and the accidental archiving provided by search engine caches, there is little formal preservation. This may well turn out to be one tragedy of our age.

That is why what Adolfo Plasencia has done in the dialogues reproduced in this book is so important. He has gathered a series of important conversations, the transmission of ideas from mind to mind, debates that shaped the future, important concepts that once were new and controversial, that were perhaps at first ignored, then argued over, and only then finally adopted widely enough to subside into that sea of the present that we call "common knowledge," eventually to sink below the waves and become history.

# Acknowledgments

Without a doubt, I can safely say that writing this book would have been an impossible task without the help of a multitude of people, in some cases premeditated and constant in their role as accomplices, in other cases the help offered unconsciously or altruistically.

First among all those who have helped and guided me is Douglas Morgenstern, who embodies an amazing combination of wisdom and modesty, utopian idealism, and love of the truth, while always keeping his feet firmly on the ground. A few hours after meeting me, Douglas took me to a meeting that changed my life. The meeting was with William J. Mitchell, at that time dean of the MIT School of Architecture and Planning, in his office in Lobby 7, under the grand dome of the Rogers Building. That encounter opened a marvelous door into MIT. Since then, everything has changed in my life, and this book is just one of the consequences. For that reason, my most sincere thanks go to Douglas, who, besides providing that opening, collaborated with me on an extraordinary and exciting initiative that we founded together, the MITUPV Exchange, a language and cultural exchange program that ran for the next twelve years at MIT.

My deep gratitude goes to all the participants in these dialogues for their ideas and generosity, and especially to Tim O'Reilly, who somehow found the time, when none existed in an intense year for him, to write the introduction to this book. I am grateful to Gita Devi Manaktala and Susan Buckley, my editors at the MIT Press, for their patient, efficient, and enthusiastic support, their extremely helpful suggestions and their indispensable guidance; and to Deborah Cantor-Adams and Marjorie Pannell for their valuable help in editing and production.

Although there is insufficient space here to mention everyone who has helped me, I wish to name at least a few. First, I am truly grateful for the revisions and advice on the text provided by José Manuel Gironés. I am grateful to my lawyer, Luis Sáez Mariscal, for his enormous help and ability to achieve the most difficult of things, and to María Jesús Plasencia and Ana Gómez for their ongoing support. I also sincerely appreciate the support provided by Francisco Mora, president of the Polytechnic University of Valencia, and José E. Capilla, the Vice-Rector for Research, and my colleagues Roberto Aparici and Sara Osuna of the Universidad Nacional de Educación a Distancia, Spain, for their support. I am especially grateful for the help of Israel Ruiz, Michail Bletsas, Manuel Ramírez, Tom Burns Marañón, Mark McCreary, Rafael de Luis, George Mattingley, Javier Benedicto, Ricardo Baeza-Yates, Avelino Corma, Justo Nieto, Miguel Ángel Sánchez, Rodney Cullen, Rosa Martínez, Elisa Cuenca, and Jaime Gómez. I also wish to express my gratitude to Vinton Cerf for his kindness and help. I acknowledge the diverse help of Natalia Navas, Ramón Diago, Elena Benito, Miles Roddis, Enrique Dans, Antonio Córdoba, Miquel Ramis, Norberto M. Ibáñez, Juncal Iglesias, Mercedes Gómez-Ferrer, Gabriela Ruiz Begué, and Juan Quemada. I also appreciate the support of my colleagues from Innovadores, Rafael Navarro, Eugenio Mallol and María Climent, and the unconditional support of Juan Reig, Antonio J. Araque, Juan Tatay, Pilar Roig, and Fernando Brusola. Finally, my heartfelt gratitude goes to all those friends and acquaintances who at different moments encouraged me to continue striving to meet the challenges involved in writing this book.

# How This Book Came About

*"And what is the use of a book," thought Alice, "without pictures or conversations?"*

The first of many apparently naïve questions that Alice poses in *Alice's Adventures in Wonderland* is really a veiled criticism of the type of teaching common during Lewis Carroll's time. The methods Carroll opposed ignored the example of great teachers such as Plato and Rousseau, who considered dialogue to be essential for a sound education. This book takes seriously Alice's desired formula as foundational for education and ultimately for science. It brings dialogue and images together to explore the frontiers of thought as practiced by some of the leading researchers at work today.

Today's scientific landscape teems with conversations. The cutting edge of new knowledge is the product of collaboration across traditional disciplinary boundaries. It emerges from places where researchers from diverse backgrounds come together to solve problems. Knowledge and its practical applications arise from intense dialogue *across* fields and the formation of new intersections *among* them.

This book offers a brief, subjective, and far from comprehensive inventory of what these collaborations are achieving. The answers come from practitioners in fields ranging from physics to the arts, computing, and biology. The book tries to parse some of the conversations going on in the humanities and sciences today and to convey the still contested and competing views that are emerging.

## How do new and transformative ideas arise?

Recently I visited an astronomical observatory to learn about the Eagle Nebula, home to the singular gas formations that have been called the Pillars of Creation.[1] Today, thanks to several famous images obtained by the Hubble Space Telescope, these gas clouds form part of the general iconography of the universe. With its generative gas and dust activity, the Eagle Nebula is now understood to be a major birthplace for new stars. As of now, we don't know why the giant clouds of the Eagle Nebula produce so many new stars, only that they do.

The creation of new stars is a useful metaphor to call on when discussing how new ideas arise. Where do they come from? How are they created, and why? Who will be capable of bringing them to light?

Bill Aulet, managing director of the Martin Trust Center for MIT Entrepreneurship, told me that entrepreneurship is not an algorithm, and neither, apparently, is success. With this in mind, I questioned the computer scientist Ricardo Baeza-Yates about creative process mechanisms in his field. He replied that it was impossible to say, as we cannot imagine how something new and previously unimagined arises from what we already know. There is no single method or mechanism for the production of new knowledge.

It is clear that certain people are capable of innovating thanks to a comprehensive vision that allows them to connect disparate ideas and subjects. This type of vision is far from universal; not everyone has it. Ricardo Baeza-Yates cites the example of artistic creation: the artist makes something new by bringing a singular vision to bear on her medium and realizing this vision through exceptional craft and skill, precisely because others did not see or execute it in the same way before.

My method in this book has been to establish certain connections between the different dialogues presented. What the scientists I conversed with share is this comprehensive vision and the craft of invention. They are alchemists of new knowledge, each exceptional in his or her field and each in different circumstances. For this reason, I have not attempted to label, group, or divide the texts in this book according to some canonical classification that would capture all the twists and nuances, or even to provide a framework into which all the different disciplinary *quadrivia* would fit.[2] As Ricardo says, complexity arises from diversity. My hope is that the heterodox diversity of the creators' visions will itself stimulate and generate new thought.

**What is the book about?**

Every scientist, creator, or inventor who makes a significant advance in his or her field has struggled to come up with the right questions. Following Plato, for whom good questions were always much more valuable than answers, I have attempted to structure this book around key questions and ideas, a list of which appears after the prologue. The thinkers conversed with for this book offer specific observations on these questions. They also engage the wider frameworks of thought that inform these subjects.

Pablo Picasso noted that technique and technology are no match for the grand questions of the human condition. The artist who subverted the art of the twentieth century recognized that understanding has its limits as well as its possibilities. This book similarly shows that creative discoveries, especially those with a high degree of subversion, do not produce greater certainties but greater uncertainties. From those uncertainties more questions arise, and it is precisely such questions that drive further inquiry. Just as young stars emerge from the Eagle Nebula, unexpected ideas shed new light on the universe we thought we knew, subverting old beliefs and revealing new avenues of inquiry.

The way of creating new knowledge is changing, especially in science, where nothing lasts forever. We can see this without looking further than CERN (the European Organization for Nuclear Research), the largest scientific laboratory ever built by man, whose philosophy, vision, and human and technical machinery José Bernabéu describes in this book. A recently published paper offers CERN's findings concerning a new type of particle, the pentaquark.[3] This paper credits 724 authors. A more recent publication on research work at CERN also exists, which attempted to make a more precise estimation of the Higgs boson mass. It was the first joint publication by the two teams operating ATLAS and CMS, the two huge detectors in the Large Hadron Collider (LHC) at CERN. It was published in *Physical Review Letters* in May 2016 and broke the record for the number of researchers participating in the same publication to date. The paper was signed by a total of 5,154 authors![4] Nowadays, in these cases the nationality of these individual researchers is irrelevant. What matters is that they got together at the giant Large Hadron Collider tunnel to collaborate on the largest machine ever built. It is safe to say that their discovery would not have been realized without a collaboration at this scale. The image of the lonely scientist working away on his or her own in a laboratory is increasingly distant from the reality of scientific discovery. Research today is fast-moving, intellectually hybrid, and scientifically promiscuous, producing findings that often can be shared instantly.

The conversations in this book can be read in any order. The researchers profiled here share many concerns, questions, and methods of analysis. Despite a lack of disciplinary orthodoxy in their work, and in contemporary science more generally, common themes emerge. I leave it to the reader to draw out the full implications of these overlaps and connections.

My aim is stimulate thought and, if possible, to provide some synthesis along the way.

**How did the content and diversity of this book originate?**

These dialogues (and many more that I could not include) took place over quite a few years as part of my professional life. They include a number of conversations with researchers and professors at MIT. They also include talks with scientists, technologists, and humanists who share their country of origin with me; I was born and live in Spain. Many of these researchers have changed their nationalities as a consequence of where they currently live and work. Language and nationality were not the determining factors in my choice to include them in the book. Their scientific achievements or success in the humanities provided the rationale for their inclusion.

That said, it did help to share country of origin with some of these participants, whose scientific careers I follow closely. I have spent years writing science and technology articles for magazines and newspapers in Spanish and once directed a television program on science and technology that was broadcast and seen throughout the Spanish-speaking world. All of that, plus attendance at international conferences, gave me access to technologists and scientists from all over the world, particularly those of interest to a global, Spanish-speaking audience.

In 2000, with my colleague Douglas Morgenstern, I cofounded a pioneering project called MITUPV Exchange, which operated for twelve years in Spanish. Thousands of MIT and Spanish university students from the Polytechnic University of Valencia (UPV), as well as students from several Latin American universities, participated.[5] That project required me to make annual visits to MIT to collaborate in classes and meetings and greatly informed my knowledge of the MIT ecosystem, along with the wider sphere of university scientific research in Cambridge, Massachusetts. I have been fortunate to view the U.S. research environment, which includes scientists and technologists from all over the world, through this exceptional window. The conversations in this book were selected to provide a similar window on the ideas, visions, and questions that inform current science. Such a selection can never be comprehensive; at best, it can evoke a dynamic landscape at a particular moment in time.

As with any birth, the genesis of the book was neither simple nor easy. Through questions and answers, my collaborators and I have tried to describe what the physicist Bernabéu points out in his conversation: the desperately slow expansion of our island of knowledge within the vast ocean of our ignorance.

The order in which the dialogues appear is intended only to provide a general framework for reading and enjoying them. The book begins with "The Physical World," followed by sections titled "Information," "Intelligence," and a final epilogue that closes with a critical reflection on the connections between science, technology, and the humanities through art. Preceding each section is a brief introduction to the specific themes and ideas that arise in the immediately following conversations.

### What is explored in this book?

Achieving this final arrangement of the book has been both easy and difficult: easy because many of the participants would be content to have their dialogue in any section. The work that each does transcends the simple labels applied to these sections. The questions I posed were designed to elicit connections among their dialogues, despite the diversity of the respondents. Progress in science is difficult, in part because each step forward by a research community raises the level of complexity for everyone else. Complexity comes precisely from increasing diversity, which helps to explain some of the uncertainty we live with today.

Progress also requires an assertive and positive vision of the future that awaits us, as Hiroshi Ishii asserts in his conversation. "I really savor divergence or difference of viewpoints about the future. For me, it's entirely natural. I also think it's healthy that there should be variety in the different versions of the future that people predict."

The book seeks to challenge a number of highly qualified minds from science, the humanities, and technology to weigh in on topics outside their own particular field. Many participants accepted the challenge and crossed over the line of their own disciplinary specialization.

In case the era of specialization in science might seem to have ended, I offer this anecdote from a recent conference. I had the good fortune to hear an outstanding paper by a scientist, whose name I won't mention. As I hadn't foreseen talking to him and had not prepared any question, I asked, "What scientific subject can I ask you about?" Without hesitation, he replied, "Ask me about chromosome 22. I've dedicated my life to chromosome 22. I can speak for months about it. Ask me anything you like about it, but don't ask me anything else."

Specialization remains because it works, despite what we may say about the need to contemplate the vast panoply of human knowledge as a whole.

The advance of new knowledge demands robust collaboration among specialists from different realms, so as to advance toward the horizon of discovery in a more informed and accelerated way. On the other hand, the complexity that gives rise to those advances comes from growing diversity. The new and sophisticated instruments created by super-specialization allow us to access scenarios that we have neither explored nor observed before, whether these are the deep universe and its exoplanets or the frontier of the nanometer. In the space between the most immense and the most diminutive things we begin to glimpse our reality, that nature is infinitely more complex than we ever imagined.

**What is this book like inside?**

In a certain sense, this set of texts is indebted to the vision of the literary agent John Brockman and his book, *The Third Culture: Beyond the Scientific Revolution*. Like that book, this one aims to take a modest step away from the two cultures that C. P. Snow famously described.[6]

Nearly fifty years later, we are faced with a persistent paradox: in a world of unprecedented access to information, a world propelled by the urgency for the new, many of the most decisive questions remain, year after year, untouched. These seem oblivious to the obsolescence that seems to affect everything else as a result of digital development and its Moore's law. The list of questions that opens this book suggests the many important problems and issues are far from being resolved. Rather than becoming obsolete, they have come to appear timeless amid the acceleration of everything else around us. Not all such problems can be covered here, but I believe the questions my respondents have engaged are significant ones, worthy of investment and attention.

**Why this dialogue model?**

Almost all cutting-edge scientific disciplines today are hybrids. Their hybridization has accelerated the process of knowledge discovery and creation. A number of these conquests of knowledge will endure as genuine and timeless advances. If I have been able to document any of these timeless ideas, it is because I have had the remarkable fortune to converse with— if not everyone I would have liked to meet—many relevant scientists. In this book, I have tried to present something like the ideal dinner party—a rich seam of ideas that interact with one another.

A goal of this book is to provoke some cross-fertilization between different lines of thought. One strategy was to use parts of the replies from some dialogue questions for other respondents. Another was to ask those inter-viewed to give opinions on matters outside the scope of their specialty. The result, I believe, is something unusual: thinkers in a particular field have been drawn into reflecting on the same concept or powerful idea from very different perspectives. To the extent that this can be done in a mutually enriching way, the result is a panorama of ideas on questions such as, for example, what is intelligence. Replies come from the cutting edge of neuroscience, neurophysiology, computation, artificial intelligence, and the humanities and converge to illuminate the question from multiple angles.

Before ending, I would like to express my gratitude to all the respon-dents for devoting so much of their precious time and many ideas to this book, and for their all-important endeavor to help me achieve my main aim, which is simply to provide a worthwhile and fascinating experience for the reader.

## Notes

1. An image of the Eagle Nebula (catalogued as Messier 16 or M16, and as NGC 6611), taken from the La Silla Observatory, Chile, is available on Wikipedia (https:// en.wikipedia.org/wiki/Eagle_Nebula#/media/File:Eagle_Nebula_from_ESO.jpg).

2. The *quadrivium* (plural: *quadrivia*) comprised arithmetic, geometry, music, and astronomy, advanced subjects taught from the Classical period through the medi-eval period.

3. A pentaquark is a subatomic particle consisting of four quarks and one antiquark bound together.

4. G. Aad et al. (ATLAS and CMS Collaboration), "Combined Measurement of the Higgs Boson Mass in pp Collisions at $s\sqrt{}=7$ and 8 TeV with the ATLAS and CMS Experiments," Physical Review Letters 114, no. 19 (May 14, 2015), doi: http://dx .doi.org/10.1103/PhysRevLett.114.191803.

5. Douglas Morgenstern, Adolfo Plasencia, and Rafael Seiz, "Students as Designers and Content Creators: An Online Multimedia Exchange between the U.S. and Spain," *Campus Technology,* September 29, 2003, http://campustechnology.com/ articles/2003/09/students-as-designers-and-content-creators-an-online-multimedia -exchange-between-the-us-and-spain.aspx.

6. John Brockman, *The Third Culture* (New York: Simon & Schuster, 1995); C. P. Snow, *The Two Cultures and the Scientific Revolution* (New York: Cambridge University Press, 1959).

# Powerful Ideas Dealt with in the Book

*"Computers are useless. They can only give you answers."*
—*Pablo Picasso*

The approach to the dialogues in this book focuses on this principle: the reality is much more complex and less certain than before, with more possibilities. It can bring about more questions than answers. In other words, the best responses are those that generate more questions. These new questions turn out to be more important than the answers.

Some powerful and thought-provoking ideas dealt with in the book are as follows:

• What is intelligence, how does it work, where does it reside, and how is it measured?

Alvaro Pascual-Leone / José Hernández-Orallo / Ricardo Baeza-Yates / Javier Echeverria

• What will intelligent machines be like? Will there be nonbiological intelligence (not based on *Homo sapiens*)?

Michail Bletsas / José Hernández-Orallo / Sara Seager

• What will happen with global warming?

Mario J. Molina / Avelino Corma

• How far should our commitment go to economizing energy and minimizing its use?

Avelino Corma / Mario J. Molina / Alejandro W. Rodriguez

• Has Moore's law come to an end? Is graphene the answer?

Pablo Jarillo-Herrero / Ignacio Cirac / Anne Margulies

• How does the brain really work? Where is the "I"?

Alvaro Pascual-Leone / José Hernández-Orallo

• Where does consciousness reside and how does it emerge?

Alvaro Pascual-Leone / Javier Echeverria

• Are we deterministic and are we determined? Is human behavior determined or do we have free will?

Ignacio Cirac / Javier Echeverria / Alvaro Pascual-Leone / Ricardo Baeza-Yates

• Is the universe a hologram? What happens to dark matter?

Javier Echeverria / José Bernabéu

• What causes behavior in humans? In robots?

Alvaro Pascual-Leone / Michail Bletsas / Rosalind W. Picard

• What is going to happen with learning and universities, and what role will they play in society? What will the best model be? What values will they have? What should an efficient university be like nowadays?

Hal Abelson / Israel Ruiz / Anne Margulies

• Is convergence technology also a matter of culture?

Henry Jenkins

• How should technology be integrated into learning?

Anne Margulies / Hal Abelson / Henry Jenkins

• Are we going to become bionic? To what extent?

Jose M. Carmena

• Is technology modifying human perception? How should it engage with our cognitive system?

Hiroshi Ishii / Michail Bletsas / Javier Echeverria

• Should we use technology only as manufacturers wish? Should technology be designed to force us to do so?

Richard Stallman

• Is our relationship with technology changing the "settings" of our senses?

Hiroshi Ishii / Michail Bletsas / Rosalind W. Picard / Howard Rheingold

- Is the breaking of the symmetry between a movement and its reversal-in-time observed in particle physics?

José Bernabéu

- Can ethics combined with open knowledge create a sustainable economy of ideas?

John Perry Barlow / Michail Bletsas / Hal Abelson

- Have we come to see the final unknown particle?

José Bernabéu

- Why is nature quantum, yet our logic won't accept it? Is computing going to become quantum? Will we see quantum computers as a normal part of our lives?

Ignacio Cirac / Pablo Jarillo-Herrero / Alejandro W. Rodriguez

- Why does mass exist? Does anything lie beyond the Higgs boson?

José Bernabéu

- Is it possible to govern uncertainty and live with their stochastic effects? Can we plan the impossible?

Javier Benedicto

- Will there ever be a "completed work"?

Paul Osterman

- Is the semantic web going to make everything that is implicit clearly explicit? How is rendering knowledge on the Internet going to evolve?

Bernardo Cuenca / José Hernández-Orallo / Ricardo Baeza-Yates

- Are search technologies now allowing us to remember the future?

Ricardo Baeza-Yates

- Is it important to have a free and open Internet?

Bebo White

- Has the impact of the Web come to maturity? Is Web 3.0 the mobile Internet, the Semantic Web? The Internet of Things? The Internet of Everything? Will there be a definitive Web?

Tim O'Reilly / Ricardo Baeza-Yates / Bebo White

• Could the universe of the great "connected brain" of the Web finally turn out to be the "cement" that leads us back to a period of understanding human knowledge as a whole?

Tim O'Reilly

• Is the change brought about by quantum physics much greater than that brought about by the theory of relativity?

Ignacio Cirac / Pablo Jarillo-Herrero

• Encryption and decryption: should the "right to encode" personal communication be viewed as a human right?

David Casacuberta / Michail Bletsas / José Hernández-Orallo

• Is reality much more open than the mathematics that tries to explain it?

Ignacio Cirac

• Is the expression "affective computing" an oxymoron?

Rosalind W. Picard

• Will nanotechnology change our operational frameworks in a way that we are unable to imagine?

Pablo Jarillo-Herrero

• Will we be able to discover another earth different from our own during our lifetime?

Sara Seager

• Beauty ≠ truth? Can we contradict John Keats?

José María Yturralde

• In art, can we go back to the past and change it?

José María Yturralde

• Is it possible to paint the void?

José María Yturralde

# I  The Physical World

Exploration of what we once called the physical world is not what it used to be. Even the word "explore" has changed in meaning. The pages of this book bear witness to that.

In his book *Ideas and Opinions*, published in 1954, Albert Einstein positioned himself on the side of pure mathematics and logic when it came to advancing knowledge of the "objective world" or the world of "things."[1] He contrasted his own alignment with abstract concepts and ideas to the prevailing scientific approach, which favored the raw evidence our sensory impressions can provide. It should not come as any surprise to us, Einstein argued, that Plato placed greater importance on the reality of ideas than on empirically proven things. He called Plato's vision of knowledge the "aristocratic" position, considering it unlimited, as opposed to the plebian view of naïve realism in science, in which the knowable world is confined to what we perceive through our senses. Obviously, Einstein was not being totally objective here. He was defending his own approach in the context of a debate that was taking place in the scientific world at the time. In those days, science, based on direct sensory perception, was linked in many cases to the *realia* of the daily lives of people or animals, including Darwin's theory of evolution, developed from his observations on the Galápagos Islands. That was a decisive period for all of the sciences, particularly the natural sciences, which set out to discover and demonstrate, through the scientific method, the realities of the physical world—and ultimately the larger universe.

The philosopher and mathematician Bertrand Russell shared Einstein's view. In *An Inquiry into Meaning and Truth*, Russell wrote:

We all start from "naive realism," i.e., the doctrine that things are what they seem. We think that grass is green, that stones are hard, and that snow is cold. But physics assures us that the greenness of grass, the hardness of stones, and the coldness of snow are not the greenness, hardness, and coldness that we know in our own experience, but something very different. The observer, when he seems to himself to be observing a stone, is really, if physics is to be believed, observing the effects of the stone upon himself. Thus science seems to be at war with itself: when it most means to be objective, it finds itself plunged into subjectivity against its will. Naïve realism leads to physics, and physics, if true, shows the naïve realism is false. Therefore naïve realism, if true, is false; therefore it is false.[2]

And let's bear in mind this is not a physicist talking here but a philosopher who is also a mathematician. Like other human beings, scientists cannot escape subjectivity. There are cases of very expressive scientists who, 'looking at the same thing,' saw different things. They saw things that others did not. They saw the same thing but arrived at a different conclusion. The debates between Wallace and Darwin, or between Ramón y Cajal and Camillo Golgi, also take up this question of naïve realism in science. Their conclusion? Direct observation is necessary to discovery but it is rarely sufficient.

Almost a century has gone by since Einstein made his first advances in science, and the tools that are now used for observation have changed enormously. The Hubble Space Telescope and the Atomic Force Microscope (AFM), among other instruments, now provide us with a very different and wider vision of the physical world compared with the traditional view based only on our senses. Perhaps such tools have merely given us a new naïve realism for the twenty-first century, based on information we can gather at a distance and simulate with abstract digital data (zeros and ones) rather than on what we directly perceive with our senses. However, before reaching that conclusion, we would have to clarify whether the observation enabled by these sophisticated instruments is in any sense equivalent to the direct perception and realism that Einstein and Russell described.

Marshall McLuhan argued that all technologies, even the most powerful, are extensions of human ability. They extend our bodies and our minds. If we accept McLuhan's argument that the Hubble Space Telescope and the AFM extend human observation rather than fundamentally changing it, we can perhaps infer that direct apprehension of the world is still possible in contemporary science, at least to some extent. Other ideas that McLuhan proposed, however, now seem less likely to be considered valid: "We become what we behold" and "We shape our tools and afterwards our tools shape us."[3] That would certainly have to be more clearly explained in today's world.

What does seem clear is that we are far from direct perception when faced with something like the Hubble or the AFM. To transcend immense distances, in the case of the Hubble, or to perceive single atoms, in the case of the AFM, is a fundamentally different way of seeing and knowing than we had before. The information provided by these instruments is qualitatively different from the information provided by our senses, not least because it is a digital representation of the physical world.

The following conversations discuss many fascinating aspects of the physical world, including how the quantum bit functions; the primordial cosmology of the universe; what exoplanets are like; quantum and thermal fluctuations related to the design of black bodies; the finest materials that have ever existed, exist now, or will exist; the wisdom that hewn stones offer us from past centuries; the challenge of global warming; why the farther we look into the universe, the earlier we see; and how best to combine atoms and bits into a whole.

In short, what you are about to read is a series of conversations overflowing with ideas. I hope that reading them will tempt you to reflect on whether the naïve realism of the senses that Einstein and Russell criticized should be declared officially obsolete or whether, as McLuhan claims, it still remains valid.

## Notes

1. Albert Einstein, *Ideas and Opinions by Albert Einstein* (New York: Crown Publishers, 1954), 20.

2. Bertrand Russell, *An Inquiry into Meaning and Truth* (New York: W. W. Norton, 1940), 14–15.

3. Marshall McLuhan, *Understanding Media: The Extensions of Man* (Cambridge, MA: MIT Press, 1994; first published 1964), 21.

# 1  Quantum Physics Takes Free Will into Account

**Ignacio Cirac and Adolfo Plasencia**

Ignacio Cirac. Photograph by Adolfo Plasencia.

*Quantum physics gives you a new vision of nature, a new vision that perhaps has both philosophical and physical repercussions. It tells us, in a way, that the properties of the objects are not defined, and we are defining them when we observe them.*

*The change brought about by quantum physics is much greater than that brought about by the theory of relativity.*

—*Ignacio Cirac*

Juan Ignacio Cirac Sasturain is Professor at the Institute for Theoretical Physics in Innsbruck, Austria, and Director of the Theoretical Division of the Max Planck Institute for Quantum Optics, Garching, Germany. A Spanish physicist, he is renowned for his research in quantum computing and quantum optics, encompassed within quantum theory and theoretical physics.

Cirac received a degree in theoretical physics from the Complutense University of Madrid, where he also earned a doctorate in optics. His research focuses on quantum information theory. He has developed applications that prove the viability of his tenets and has shown how one could carry out calculations that are impossible with current systems. According to his theories, the quantum computer, which he developed, will revolutionize the information society. Cirac has published more than two hundred articles and is one of the most cited authors in his field.

Adolfo Plasencia:    Ignacio, thank you for having me.

Ignacio Cirac:    My pleasure. Thanks for coming.

A.P.:    Ignacio, you're well aware that great expectations have been aroused by quantum physics. Some people think that Moore's law is seeing its end days and that other alternatives have to be sought to continue our progress in computer science and information technology (IT). The ultimate alternative in this sense is quantum physics. But in addition to this, quantum physics combines with philosophy; it deals with who we are and why we exist. There is a big debate surrounding all this. For instance, in a recently published debate, some intellectuals have associated quantum mechanics with such things as human freedom, free will; with the kind of criteria more linked to metaphysics than to physics. Some people have linked quantum physics with freedom of choice and suggested they are not compatible, thereby leading to determinism.

The physicist Carlo Rovelli, leader of the Quantum Gravity Group (Équipe de gravité quantique) of the Centre de Physique Theorique de Luminy, refuted these statements in a text published by *Edge*: "Free Will, Determinism, Quantum Theory and Statistical Fluctuations: A Physicist's Take."[1] There he reminded us that Democritus assumes that the movement of an atom is deterministic; that is, a different future does not happen without a different present.

Over the past century, Newton's equations have been replaced by your and your colleagues' equations and math—I mean those of quantum theory—which include an element of uncertainty governed by highly rigorous probabilistic dynamics. You have so many references and so much precise information that is difficult to refute. Your equations do not determine what is going to happen, but they strictly determine the probability of what is going to happen. In opposition to those who have established a link between the two, Rovelli says that free will has nothing to do with quantum physics because we are highly unpredictable as human beings, as is the case with most macroscopic systems, and there is no incompatibility between free will and microscopic determinism. In other words, people's freedom of choice does not contradict your quantum mechanics.

Rovelli states that our idea of being free is right, but this is only a way of saying that we are ignorant about why we make decisions.

What do you think of Rovelli's statements?

I.C.:   This is a very interesting and deep discussion. I don't think it can be summarized in just a few words, but, even so, a few things should be highlighted. First, quantum physics gives you a new vision of nature, a new vision that perhaps has both philosophical and physical repercussions. It tells us, in a way, that the properties of objects are not defined, and we are defining them when we observe them. It's really quite strange, a rather odd and striking theory. It's startling that nature behaves this way.

A.P.:   Well, *strange* if we view it in terms of the framework that has governed physics since Newton. What do you think?

I.C.:   It's strange from the point of view of what we are used to seeing. If somebody explains to you the principles of quantum physics, it looks like something extraordinary, a really bizarre and almost unbelievable thing. Thus, some people are looking for ways to keep our vision of nature unchanged, to preserve the vision we used to have.

And perhaps a way not to change this previously proposed vision, which I won't go into, is to state that we do not have free will. But what would happen, you think, if we did not have free will? Could we then salvage some of nature's properties? For instance, if, every time I do an experiment, the results conform to or don't contradict quantum physics, it might just be that it was already programmed to turn out that way; that is, I have no power of decision and there is nothing I can do or decide about it, which would put any theory in a position of vulnerability. A few papers have been published in the last few years dealing with this issue, but I am not an expert on that.

A.P.:   However, does it make you feel obliged to be even more rigorous, to "demonstrate" and think even more?

I.C.:   These are things that cannot be demonstrated. I mean, if a robot is programmed, it's likely that it won't itself realize that it has been programmed, but I think very few people think about this. It is just one opinion, which is a small part of a huge range of opinions that exist, and is a very extreme one indeed.

But there are other, much more consistent possibilities, or at least they appear much more reasonable to us than the one you've mentioned. Quantum physics in a way takes free will into account. In itself, although maybe not in its hypotheses, it makes the assumption that we are capable of choosing and deciding how, for example, to make measurements. This is what gives rise to all the experiments and experimental arrangements that we have. But what happens with quantum physics, which is also very interesting, is that it is really different from other previous theories, which were supposed to include a description of ourselves as well. In other words, after Newton put forward his laws and Maxwell also developed his laws and equations, people thought that these principles applied to nature as a whole, which includes us, because we are made of atoms and matter. In other words, we ourselves should also follow the laws of nature. And in a way, this is what led people to think about determinism—that is, if we follow Newton's laws and they are deterministic, that means we are determined. But some people said no, that's not right: what is completely outside Newton's laws and doesn't follow them is our conscience, or whatever you want to call it. This is a real possibility until someone proves the opposite.

Quantum physics is something else. On the one hand, it states what happens with everything else, yet on the other hand it cannot define itself, which is very strange indeed. In fact, a small problem called the measurement problem arises with quantum physics precisely on this point. Why is it that this branch of physics cannot describe what we do but can describe everything else? It's a fascinating subject. At present, various options are open to us, and none can be discarded as false until further research is undertaken.

A.P.:   The famous MIT professor Walter Lewin says in his book *For the Love of Physics* that the most important thing about measurement in physics is, precisely, accuracy and precision.[2] As Lewin says in his book, and as he also told his students, "Something which all university text books on physics always leave out when taking measurements, concerns the issue of imprecision in the measurements."

He also kept telling his students, "Any measurement you make without knowing its imprecision is completely meaningless."

This gives you an idea of the importance that precision has in measurements in physics. Yet, Ignacio, you quantum physicists actually make measurements that are so precise they are almost irrefutable, and everybody agrees on that.

I.C.:   Yes, what we have in quantum physics, especially in what is called quantum electrodynamics, is that its predictions are highly precise. So you can measure a physical property to twelve digits of precision. This is something that nobody imagined could be measured, but nevertheless it is measured. Therefore, it's a very strong and sturdy theory, one that has been highly tested. Nevertheless, you always have to say that this is not the final theory. But if we go along with this line of thought, we'll never have a final theory; there will always be experiments that we haven't done that might have produced results leading to a different theory.

A.P.: Have quantum physicists, like you, observed any resistance from within your field of work to your breakthroughs in the world of physics?

I.C.: No. That happened a bit during the 1930s and 1940s, when quantum physics was under development and naturally these strange properties were found, features of nature that were so different from the classical way of thinking that there was some reluctance.

There are some examples in which there were some difficulties, but the mind of a physicist is very open, and all that most of them wanted to do was to carry out experiments and see if things were like that. As soon as the experiments were done, things started to open up. Now it is difficult to find anybody, any physicist, who does not believe in quantum physics.

A.P.: Let's talk a little about some of that historical opposition. Albert Einstein wrote in a letter to Max Born in 1926, "Quantum mechanics is a very serious matter, but I heard an interior voice saying, this is not the way."[3] According to Roger Penrose, Einstein didn't like the probabilistic side of quantum mechanics. He said that this side to it was not acceptable for him because Einstein was convinced that there must be a physical world, objective in itself, even on the minute scales of quantum phenomena, which is the environment in which quantum physicists thrive.

You mentioned in an interview, "Usually, when we observe something, we see that it exists and is well defined. Whenever we see a yellow object, we think that this is an 'objective' property the object has, which doesn't depend on me. That is, when I am not watching it, the object still remains yellow."[4] Now, quantum physics, according to you, says no; it says that some properties of the microscopic objects in movement are not defined when they aren't being observed and only become defined when we watch them.

If I understood properly, what you have said moves away from this intrinsic "objectivity" of matter, which Einstein preferred.

Do you think it has been difficult for quantum physics to firmly contradict someone as great as Einstein?

I.C.: I don't think it was that difficult. In Einstein's time, people discussed and debated at great length. Because, of course, when he says, "This theory cannot be right," somebody wonders, what is wrong about the theory? Tell me why it's wrong. He wasn't able to say what was wrong. He tried to find contradictions, but he couldn't find any. But I think there has been a process to this. On the one hand, many scientists thankfully said, "Well, it's a strange thing, but we are going keep moving forward." They kept working on particle theory and developed the Standard Model without worrying themselves about the issue. On the other hand, another group of physicists said, "We are going to do experiments to find out whether this is true or false." These experiments took place, and moved forward, especially in the 1980s, and today the evidence clearly shows that nature is like this. And when you get used to it, well, I think if Einstein were alive today and had become used to it, he wouldn't be too surprised to discover that's how things were.

A.P.:   I don't think he would find it strange because, in fact, I believe he did the same thing. When he'd made a number of discoveries about his theory of relativity, in a way, and in certain fields, he questioned Newton's mechanics, which had been upheld for centuries. So could we say that you quantum physicists have done the same to him as he did to Newton?

I.C.:   Yes, in fact, they did. We haven't done much, but they did. However, I think that the change brought about by quantum physics is much greater than that brought about by the theory of relativity. Relativity, of course, is extraordinary; it has made a huge change. But quantum physics in addition to that gives us a new vision of nature, which is not just a question of ensuring that some specific laws are observed or not, or the fact that things move and time changes when you move, and so on. Yes, it is really strange, but it tells us something else. It's telling us that reality is stranger than what we thought. When we speak about reality, the reality of objects, it's much more than that.

A.P.:   Much more complex?

I.C.:   Yes, reality is much more complex, has more possibilities, is more uncertain, and is beginning to provide us with more questions than answers. Well-known physicists such as Richard Feynman say that nobody understands quantum physics. Even if you try hard to think about it there is no way to relate it to any other analogy that you can find in the ordinary world. Whereas I do think that you can imagine it, that it's easier for the imagination to grasp.

A.P.:   Ignacio, as you have put it, the change in quantum mechanics is just starting, and what you are doing is probably just the start of a huge change.

Do you have any hypothesis about which changes of scale the applications of quantum information theory might involve for our present world, which is a highly computerized and technological world, having a global network shared by more than one-third of the world's population and with more cell phones than people?

What do you think would change if digital information being used now became quantum and networks became quantum networks?

I.C.:   We are just starting to scratch the surface of this world of quantum physics and we have just begun to be aware of the first applications, but, as happens whenever we have access to new laws of physics, the most important applications are yet undiscovered, and most likely any forecast I could make now about quantum physics applications would have nothing to do with anything that happened during the next thirty years. However, what we know now is that if we can have access to these laws of quantum physics, we will be able to build systems capable of processing and transferring information in a very different way. This allows us to envisage much faster computing, maybe not for every type of calculation or computing, but for some of them. We might also have more efficient and safer forms of communication, transmitted in such a way that nobody will be able to hack our communication. I don't know what impact this may have, for instance, on cell phones. Smart phones already cover most of our present-day needs.

A.P.:   But also, as you know, there are already clichés about your science—something can be in two different places at the same time, the cat may be alive and dead at the same time[5]—things that make you imagine something even stranger than you can imagine. And very often our imagination is wrong, and that is why we have your experiments. Of course, after watching *Star Trek*, people think that bodies can be teletransported to another galaxy.

Now, we know this cannot be true, but we can't prevent people from imagining things that you physicists have never said, which are truly impossible, even for quantum physics.

I.C.:   Yes, that's right. Sometimes we physicists use some unfortunate language to name phenomena, such as teletransportation—that word has a very clear meaning. Quantum teletransportation (quantum teleportation) means in a way that it is information and not matter that disappears from one place and appears in another. And this is true. But as soon as we have access, and as soon as communication systems based on quantum physics can be built into the computer, that is when ideas about how to use them will emerge, as, for instance, a huge computation power. Don't you agree?

We already know today that such computing power might be used in drug design and new materials design. This computing is done today by supercomputers. But I guess that when we achieve it, somebody will find out what it can be used for. And the same happens with communication. We know that quantum communication is safe. As you quite rightly said, information may disappear from one place and reappear in another without going through anything on the way, and that means nobody can read it. That is an application. There may also be some talk of quantum credit cards, which no one can copy, so the information is unique and nobody can use it in your name. We already know about some of these applications, but there must be many that have not yet been exploited because we need young people with good ideas, not us, the scientists formulating and developing these phenomena, but people having ideas about how these new laws of nature should be developed.

A.P.: That's right, but this doesn't mean that anything is possible. The imagination of a scripwriter may generate hypotheses that are and will always be impossible. They do, however, associate them with names as strange as the ones you have discovered, don't they?

I.C.: Yes, we must be careful because sometimes quantum physics seems to hide mysteries. It is even used in a wrong way, such as to insist, "This cannot happen, that can happen"—there are many instances on record. You mentioned *Star Trek* and physical teletransportation. We do not know how to do this today. We do not even know whether the laws of physics will allow it, but probably the answer is no. Another thing that people hear is that they can influence their future simply by—

A.P.: By traveling into the past and changing it?

I.C.: Well, that is another thing. I was referring to the so-called quantum superposition, that is, you can do two things at the same time, or I can use my mind to cause or make something happen. This has nothing to do with quantum physics. For this reason, there are misconceptions about the ideas of quantum physics because some people who speak about them don't have a clear concept or understanding, or they are speaking about something completely different from quantum physics.

A.P.: I saw you being interviewed on TV, and you said that for you there is a before and an after. Until 1994 people, even those connected to science, thought that applying quantum theory in practice was not going to be possible. But in 1995, you and the Austrian theoricist Peter Zoller together presented the first theoretical description of a quantum computer architecture.[6] It was based on ion traps in which electrically charged atoms, cooled almost to absolute zero, were trapped by electrical fields and manipulated by lasers. Could you describe this architecture? Is there any equivalence between the description of a quantum computer architecture and that of present IT? *Wired* magazine referred to the computer you described in that paper as the holy grail of computer science, which has been sought by scientists since 1980.[7] Do you find this a bit exaggerated?

But first, how is this architecture you described?

I.C.: If we can use quantum physics to transfer information, then the first thing quantum physics says is that instead of storing and processing data in terms of bits, zeros and ones, it must be done in terms of quantum bits, or qubits, which means they have to be physical systems, such as zero and one, and also have the property of quantum + superposition. We know that this happens at an atomic level, so the only thing one has to be able to do is choose a series of atoms, in this case ions, and manipulate the properties of the electrons composing these atoms, specifically the property called electron spin, with lasers, in such a way that the electrons change from zero to one and from one to zero, and which, in addition, can also have quantum superpositions and interact with each other, in order to carry out quantum computing .

A.P.: And in a controlled way? With an aim, I mean.

I.C.: That's right, with an aim. In the same way as ordinary computers handle zeros and ones in terms of logic gates, qubits can also be used to make the appropriate calculations in terms of logic gates, which we call quantum gates. And handling these quantum logic gates takes place by means of lasers—that is, by using lasers aimed at these ions, which send a little amount of light to each of them. The intensity of light sent and the time during which the light pulse is sent depend on how the program is made, based on what you want the ions to do, that is, which logic gates you want to be "executed." This is simply what a quantum computer like those we have today does. Today's are prototypes. They are very small, but they prove that all this works.

A.P.: I am also very interested in the human aspect. What you were in search of, as *Wired* magazine says, is the holy grail of tech research, right? And for a long time, many important scientists said this was impossible. Now, all of a sudden, you present a paper and say: It is not impossible! It is possible, and that's it. Amazing! How was that moment? Was it difficult for you and Peter?

I.C.:   Well, yes. This is a strange story. We were working in quantum physics, basically on how to cool atoms, how to cool ions, how to make them stop, and observing those strange properties produced by quantum physics, but we had hardly heard of quantum computing. We had heard about some other different things. Then, at a conference in Colorado in 1994, it was mentioned—in an abstract way, in theory—that these quantum computers might exist, but we still didn't know whether they could be built. And as we were working on cooling these ions, we thought, perhaps this can be a way to build them, because these were the ones that we understood better, up to that moment, from the quantum point of view. So, based on this, we started to work. We had several ideas, and three months later, we concluded that, in fact, the answer was yes, this would be possible, after taking a series of steps. As we weren't from the field of quantum computing, we wanted to know whether what we were doing was right. So we took a train to Torino to attend a conference on quantum computing theory and present our proposal before physicists working on quantum computing.[8]

A.P.:   What was the reaction? Were there any big surprises?

I.C.:   It was funny. Of course, people trusted us, more or less, because we had already predicted some experiments, which had already been tried and tested. On the other hand, their background was completely different, and they didn't have sufficient knowledge about ions.

I particularly remember someone who was at that time working in quantum computing saying to me when I finished my speech, "This is impossible!," and I thought: Why? "Because there is a theorem that says this is impossible!," he said.

And I thought, but how is that possible if, mathematically, everything is correct! After the lecture I went to talk to him and I started to explain to him that we were using the qubits with two internal levels, and in order to make the logic gate we were using another internal level.

And then he told me, "No, no, but this is forbidden! You only have two levels!," and I told him, "No, atoms have many more levels!" Then I realized what had happened. He had developed his theorems thinking that there was not another level. What we saw was that it couldn't be done unless you added another level, so, as far as we knew, as far as we had observed, it could be done since atoms also have more possible levels.

A.P.:   So with energy, you can make an electron jump from one level to another, and, depending on the energy, it can jump to a different level, and he thought there were no intermediate levels. Is that right?

I.C.: Yes, understanding that is a bit like understanding the difference between mathematicians and physicists. His structure was of the kind: "If I have that and that, then this is possible and this is not." The physicist instead says: "Well, if this is impossible, what I should do is add things until it is possible, shouldn't I?" This is a little like the way I see it. It happened to me several times during my career, experiencing the fact that we physicists try to do everything possible—in some ways breaching the a priori approaches and considerations of mathematicians—to transform things so that they stop being certain and and immovable when we want to achieve something.

A.P.: That means not only once but many times you have come across surprised faces in the audience, haven't you?

I.C.: Yes, and not only with things done by me but also things done by other physicists, for example the time when some theorists predicted that Bose-Einstein condensation was practically impossible with some particular atoms. Proof of the phenomenon received the Nobel Prize in Physics in 2001; the result had been experimentally produced in 1995.[9]

The experiments were done in Colorado, where I was living at the time. I remember asking one of the experimenters, "But if it has proved impossible, why do you keep carrying out the experiment?," and the experimenter answered, "Because I don't follow what they mean!" Well, finally the experiment eventually worked out! And the expermentalist in question, Eric A. Cornell, was given the Nobel prize.

A.P.: Do you mean that reality is much more open than the mathematics that claims to represent it?

I.C.: It depends. These kinds of things have happened many times, although the opposite has also happened. Many people doing a certain experiment take a long time, and then a mathematician arrives and says, "No, you do it this way, the other way is impossible," and if that is understood, a lot of money can be saved in research because you can see that this is not possible. So these are the two sides, but in this particular case, reality surprised them.

A.P.: Let's talk about another case that is happening now. In copper semiconductor technology, it is now calculated that by 2016 there will be chips with a technology of 12 nanometers. If they continue like this, physically, the specialists say, there will come a time when, for a pure physical reason, the electrical charges may jump from a copper microprofile of the chips to another, and this method will be exhausted because of the extreme reduction of its scale. From your point of view as a researcher in quantum mechanics, is it your understanding that a truly functional quantum computer will arrive before Moore's law is exhausted?

I.C.:   I don't think so. The first demonstration of a quantum computer, which was a basic demonstration with just one of the qubits, took place in 1995, just after we published a paper in which one of the basic parts was shown in one of the experiments. Then in 1997, experiments were done with two of the qubits. In 2000 there were four qubits; in 2004 there were eight. Today we have fifteen or sixteen, and there are people who say they have probably reached thirty. Now, if this is extrapolated to 10,000, which is what is needed, or to 100,000 or 1,000,000, which would be the optimum, there are still many years ahead.

A.P.:   What's better right now, 10,000 or 1,000,000 qubits?

I.C.:   We know that with about 1,000 qubits, we could do some interesting calculations. The problem is that there are likely to be errors, but if we had 1,000 qubits that were perfect, we could carry out some really interesting calculations. The problem is that qubits are not perfect, and we have to perform error correction, or "debugging." That means that the number of qubits that you have to use in practice is at least one hundred times higher than the original. Therefore, to do computing in the presence of errors, we would need about 100,000 to 1,000,000 qubits. However, to reach this figure, an important technological development is still needed, and this might come in five years or fifty. We do not know.

A.P.:   In formulating the next question, I had to consult a friend who is a wonderful physicist in the field of condensed matter, Pablo Jarillo-Herrero, and pose the question to him first. Another of your contributions, yours and Peter Zoller's, is the quantum simulator.[10] When you hear the word "simulator," you imagine a flight simulator that simulates the outside world, with you at the controls and everything you see behaves interactively. That's what I thought a simulator was. Pablo explained the following to me. Yours is a simulator for artificial matter with real atoms. With a simulator like this, you could work out how conducting materials behave at high temperatures, for example. In other words, it is actually a simulator of real material made artificially, but which allows the possibility of understanding how materials behave with real atoms in certain environments that are very difficult to see in nature. Is that right?

I.C.:   That's right. The idea is somewhat similar to what I mentioned before, that it will probably take a long time for quantum computers to be constructed, and you might stop to think: Well, what do we want a quantum computer for? What applications do quantum computers have? And one of the applications, perhaps the most important one, is the one that would be capable of solving scientific problems that we cannot figure out with normal computers, problems related to materials design, perhaps with chemical reactions, the chemical composition of some materials, and so forth

So, from this analysis, we get the idea that maybe it isn't necessary to build a quantum computer to solve these problems; maybe we can actually do it using an analogical computer in which we choose a totally different system, a system of atoms, for example, where they are made to interact in such a way that they behave like the material you want to simulate. And if you take measurements in this atomic system, you might be able to make predictions about what is going to happen with the material. We explicitly proposed that something like this should be built, and the first experiments were carried out in 2002. Today the first quantum simulations have been carried out with this equipment—simulations that we are unable to describe with normal computers. In other words, the first quantum simulator that specifically works faster than a normal computer has already been built. The problem is that there isn't any scientific interest in this simulation, which means it is an "artificial problem."

A.P.:   Where was this done? In Europe?

I.C.:   Yes, in Europe. In fact, the first experiment was carried out at the Max-Planck-Institut, Munich, by Immanuel Bloch's group.[11] And now many experimenters are trying to replicate these simulations, and I believe that we'll soon start to see people who have worked out problems that we couldn't solve before with normal computers using these quantum simulators.

A.P.:   But what you're saying isn't only physics, it's also chemistry, isn't it? In another conversation in this book I was told by the renowned chemist Avelino Corma, that when you work on a scale below 10 nanometers, physicists and chemists work in the same setting doing practically the same thing.[12] Would it be right to say you work in the field of physico-chemistry?

I.C.:   Yes, it would. I fully agree with Avelino on this account, to the extent that we understand each other and use the same language when we speak under these conditions.

A.P.:   There's something else I'd like your opinion on. I've been asking you a lot of things about where this experiment is. You tell me "at our Institute in Germany, close to Munich," and that it is cutting-edge experimentation. I'm interested in knowing whether the approaches have any geographic relationship, or perhaps some scientific cultural nuance. That's why I'm asking you if there is a vision of science that characterizes and differentiates European science from what may exist in other "sciences," for example, the United States', even though I'm aware that science is now global, with people from all over the world working in teams. But in your opinion, is there anything that characterizes European science with respect to other scientific views?

I.C.:   I believe that European science in general is more conservative than in North America. The Americans are much more intrepid. Young people in the United States have ideas they want to put into action as soon as possible, whereas in Europe, it's more step by step; of course, there are numerous exceptions to this rule.

A.P.:   But surely you're not talking about scientific semantics because European science is great at inventing words and has nothing to fear from America, right?

I.C.:   No, semantically it doesn't. But then again, it's simply something you can see. For instance, there's more help for young people in America. There is an idea that when you get a PhD and do a bit of postdoctoral study, what you need to do is step aside and let the ideas flow, which will probably lead to the best ideas. It's not the same in Europe, in general. When you finish a postdoctoral course, it is possible that you might have some independence, but you always depend on somebody because they want to focus research on certain issues, which are the ones they want to be solved within many years. It is different, though. The American way is likely to be more successful on the applied research side and the European perhaps in the more theoretical domains. In fact, at the moment I think that my area of research, which is quantum computing, is on the same level or even higher than in North America.

A.P.:   I heard a scientist who was defending the European version saying it isn't that we are more conservative, we are just more rigorous! We are more cautious about presenting results that haven't been fully substantiated. That was how he defended the European vision. I don't know if you agree.

I.C.:   Well, yes, I do, but the European system has its advantages and disadvantages. There are people who might think that so much rigor—as I said in the example I gave before—means it isn't possible to do something, when perhaps they should be saying, "If it isn't possible, let's make it possible!" It was good to change experimental conditions, but even so, I think it's difficult to differentiate European and American science except in very general terms.

A.P.:   From what you've told me, mathematicians are even more conservative than physicists, right?

I.C.:   Yes. The thing is that within mathematics there is also originality, people who not only figure out the problem but who also realize that some problems are more important than others, and there's a lot of originality and art in that. There are people who find new formulas, new ways of solving problems that aren't conservative but break away from all concepts. People used to think that to solve a problem, you had to follow a set of guidelines along the way; then someone else comes along on a completely different path that actually solves the problem in a simpler way.

A.P.:   Ignacio, thanks very much. It has been a pleasure, and I hope to see you again soon. Maybe you'll have a quantum computer up and running by then. Thanks!

I.C.:   Great! Thanks to you too!

## Notes

1. Carlo Rovelli, "Free will, determinism, quantum theory and statistical fluctuations: A physicist's take," *Edge*, May 24, 2014, http://edge.org/conversation/free-wil l-determinism-quantum-theory-and-statistical-fluctuations-a-physicists-take.

2. Walter Lewin, *For the Love of Physics: From the End of the Rainbow to the Edge of Time* (New York: Free Press, 2011).

3. Albert Einstein, in "The Born-Einstein Letters," Internet Archive, http://archive .org/stream/TheBornEinsteinLetters/Born-TheBornEinsteinLetters_djvu.txt.

4. R. Corcho, "Juan Ignacio Cirac: 'La física cuántica requiere un cambio drástico de nuestra visión de la naturaleza,'" *La Tercera Culture*, November 19 , 2008, http:// www.terceracultura.net/tc/?p=584.

5. The reference is to Schrödinger's cat. "The cat may be both alive and dead ... as the result of being linked to a random subatomic event that may or may not occur" (https://en.wikipedia.org/?title=Schr%C3%B6dinger%27s_cat). This is an example of quantum superposition, which comes up later in the conversation.

6. J. I. Cirac and P. Zoller, "Quantum Computations with Cold Trapped Ions," *Physical Review Letters* 74, no. 20 (May 15, 1995), doi:http://dx.doi.org/10.1103/ PhysRevLett.74.4091.

7. Cade Metz, "Physicists Foretell Quantum Computer with Single-Atom Transistor," *Wired*, February 20, 2012, http://www.wired.com/2012/02/sa-transistor. See also Eric Smalley, "D-Wave defies world of critics with 'first quantum cloud,'" *Wired*, February 22, 2012, http://wrd.cm/1TPMtXP.

8. "Quantum Computations with Cold Trapped Ions," J. I. Cirac and P. Zoller, paper presented at the "Workshop on Quantum Computation," Torino, Italy, June 1995.

9. The 2001 Nobel Prize in Physics 2001 was awarded jointly to Eric A. Cornell, (University of Colorado, Boulder) and Wolfgang Ketterle (MIT) "for the achievement of Bose-Einstein condensation in dilute gases of alkali atoms, and for early fundamental studies of the properties of the condensates" (http://www.nobelprize.org/nobel_prizes/physics/laureates/2001). In 1995 the first gaseous condensate was produced by Eric A. Cornell and Carl Wieman at the University of Colorado, Boulder, NIST-JILA lab using a gas of rubidium atoms cooled to 170 nanokelvin ($1.7 \times 10^{-7}$ K) (http://en.wikipedia.org/wiki/Bose%E2%80%93Einstein_condensate). M. H. Anderson, J. R. Ensher, M. R. Matthews, C. E. Wieman and E. A. Cornell, "Observation of Bose-Einstein Condensation in a Dilute Atomic Vapor," *Science*, July 14, 1995, 198–201.

10. J. Ignacio Cirac and Peter Zoller, "Goals and Opportunities in Quantum Simulation," *Nature Physics* 8 (April 2012): 264–266, doi:10.1038/nphys2275.

11. Professor Immanuel Bloch is with the Quantum Optics Group, Ludwig-Maximilians-Universität, Munich, and the Max-Planck-Institut for Quantum Optics, Munich.

12. Avelino Corma intervenes in dialogue 7.

## 2 Unifying Particle Physics with the Cosmology of the Primordial Universe

José Bernabéu and Adolfo Plasencia

José Bernabéu. Photograph by Adolfo Plasencia.

*Over the last decade we have discovered that 95 percent of the matter and energy of the universe is unknown to us: it is dark matter and dark energy.*

*The discovery of the Higgs boson, announced at CERN in the first week of July 2012, will always remain in the annals of science as a great scientific landmark.*
*—José Bernabéu*

José Bernabéu is a Professor in the Theoretical Physics Department and the Instituto de Física Corpuscular, the University of Valencia-CSIC, Spain. After earning a doctorate in physics (with Extraordinary Prize) at the University of Valencia, he went to the European Organization for Nuclear Physics (CERN), in Geneva. Subsequently he was appointed Chair of Theoretical Physics at the University of Barcelona and then at the University of Valencia.

His research work has mostly been devoted to elementary particle physics, in the field of unified electroweak interactions within and beyond the Standard Theory. His results on the nondecoupling effects associated with the spontaneous breaking of the gauge symmetry responsible for the origin of mass and the Higgs boson have been influential. Another area in which his work has achieved international recognition is neutrino physics.

His awards and honors include the Order "Alfonso X the Wise"; Academician of the National Academy of Exact, Physical and Natural Sciences of Argentina, of the Royal Academy of Sciences of Spain, and of the Royal Academy of Medicine of the Valencia Community; the Prize King Jaime I in Basic Research (2008); and the Medal of the Spanish Physical Society-Prize BBVA (2011).

Adolfo Plasencia:   José, thanks very much for finding the time to see me.

José Bernabéu:   My pleasure.

A.P.:   Christopher Llewellyn Smith as director general of CERN once said, "Clearly, our Project has a spiritual dimension, something to do with our feelings, with the question of what our place is in the universe and what we are made of."

Does finding the Higgs boson have anything to do with a spiritual feeling?

J.B.:   Absolutely. The human being has always been interested in the grand questions of existence. We can, with well-posed questions, understand nature. Modern science poses these types of question, just as ancient Greek culture, the basis of Western civilization, did. It's what's called the theory of knowledge, epistemology. Religion too tries to come up with answers to these questions. I believe curiosity has been the driving force behind the development of thought and the reason why, five hundred years ago, modern science appeared. Science seeks to combine theory with experiment so as to discover nature's secrets and reduce them to certain laws, to certain regular behaviors. This is what the advance of knowledge is based on, and so yes, there is a religious or philosophical component to it. That component is associated with human curiosity about the grand questions, which have always been pending and which are now standardized through modern science, or what we call the scientific system.

The discovery of the Higgs boson is quite clearly one of the great landmarks in the advance of modern science. In particle physics, our aim is to reveal the behavior of the elementary constituents in the makeup of matter. Recent decades have witnessed spectacular developments, but an essential piece was missing because the advances that had been achieved to describe fundamental interactions were only understandable in a situation of massless particles, against all experimental evidence.

What I mean is that fundamental physics had a problem with this mystical question: What is the origin of mass? The Brout-Englert-Higgs mechanism, in modern theory, suggested how knowledge could be systemized, how all the results that we had could be taken into account and, at the same time, provide a mechanism for understanding how mass arises from the properties of the vacuum state. In this sense, I would go so far as to say that this has been the most important issue in recent decades that we needed to respond to in order to push forward the frontiers of knowledge rather than remain in ignorance. The frontier we are talking about, of course, is that surrounding the question, what is the origin of matter?

The discovery of the Higgs boson is being confirmed by recent results from the analyses carried out by the different groups participating in the experiments at CERN. They confirm that that particle does indeed have the properties corresponding to the Higgs boson. This particle remains as a signal, a remnant through which we may understand how matter originates in the behavior of the elementary constituents existing in nature.

A.P.:   If we know how mass originates, then we should know why the universe continues to expand. So what is missing? What is it that we still don't see, so that the equations that confirm that expansion, that extension of the universe that is observed, make sense?

J.B.:   On the one hand, we know how it originates, but that does not say that we now have all the information on what all the constituents of the universe are. That is a fascinating question for future generations. The discovery of the Higgs boson is not the final point in understanding. On the contrary, it's the starting point. Having a mechanism available that can provide us with information on how matter originates does not tell us what the content of matter and the energy of the universe are. I would like to make a comparison between this and the first Copernican revolution, when the human being was indeed at the center of the universe and planet Earth was the center for describing the movement of all objects in the cosmos. That revolution in modern-day cosmology meant that today we know the universe has no center. It is no longer centered on the human being, on planet Earth, the Solar System, the Milky Way. No, there is no center to the universe. Over the last decade we have discovered that 95 percent of the matter and energy of the universe is unknown to us: it is dark matter and dark energy. So it seems we are made up of a kind of constituents that only appear in 5 percent of the total contents of the universe that we know of.

Over the past few years we have been undergoing a second Copernican revolution: not only are we not the center of anything, but also the type of material that we are made of represents only 5 percent of the total content of the universe that we know of.

A.P.:   It seems we will have to accept someday that the questions about the universe will never end and that there will be more and more facts to be discovered.

J.B.:   Yes. That's a very interesting thought that is directly linked to the idea of epistemology or the theory of knowledge. What does advancing knowledge mean? I don't see it as a linear advance at all, rather, as an analogy of how that advance takes place, I would say it's something like the shoreline or the contour of an island that advances into an ocean of ignorance. In that analogy we can see that the more we know, the more frontier there is. That is, there are more and more questions.

This is case with the CERN laboratory and with the results that are being obtained in the detectors that record the results of proton-proton collisions in the Large Hadron Collider (LHC). As important as or even more so than the answers we now have to questions that we have been asking ourselves over recent decades is that the results of the LHC experiments will allow us to formulate new questions. As we will know which questions to ask, this will lead to developments for the next generation of experiments to be undertaken. What I want to say is that it's the advance of knowledge itself that generates questions, which is something that is never going to end in science. It is the advance of knowledge that is creating the new questions we are posing for each of the stages.

A.P.:   In the end, Picasso will be proved right. He dampened the spirits of the first computer enthusiasts by saying, "Computers are useless. They can only give you answers."[1] What is important are good questions rather than the answers.

J.B.:   Exactly. The questions, as far as I am concerned, are the most important thing. When I say "new questions," these should be put in a more important category: faced with new answers or new questions, I put the latter first. By the time we have a definite criterion on how to formulate a question to nature, we have already advanced a long way toward finding the correct answer.

A.P.:   José, you have dedicated a lot of time and effort and thought to getting thousands of physicists to agree, to convincing dozens of countries and thousands of leading politicians to build the biggest machine that humankind has ever built, and to bringing together thousands of the best physicists and engineers to work together here, in a Europe that not very long ago was at war with itself. Those nations are now reunited at CERN.

How do you feel about the fact that it is physics that has led the way from war to peace in Europe, from world wars to the twenty-first century with the Europe of CERN?

J.B.:   I think it can safely be said that science has been the pioneer in building Europe. It was scientists and science politicians in the 1950s who, after the tearing apart that took place between European nations in World War II, realized that Europe would never be able to compete as an advanced society either with the United States or with the developments of the former USSR if we didn't take a step forward and build a united Europe.

CERN was the way of giving form to this idea. The European particle physics laboratory was created in 1954. I should remind you that 1954 was even earlier than the first European treaties on coal and steel. It's not just that there weren't treaties of an economic nature in Europe (let alone political ones, which in my opinion we still don't have), there were not even commercial ones! Yet CERN came into existence in 1954, first with five countries, and then other countries joined in. CERN has consolidated the idea of Europe from the scientific point of view and is also encouraging it to take steps.

For example, CERN is now opening up its frontiers to go beyond Europe, inspired by the vision that science is a way of joining human forces with the aim of furthering universal human civilization. Besides the CERN member countries at the moment, there are also associate member countries (non-European countries) that attend advisory sessions and participate in CERN experiments, not only as external members but collaborating on building the detectors, analyzing the results with data obtained from them, and working together—American, Asian, and European physicists— as equals. CERN thus not only embodies the idea of Europe, it embraces science from all round the world, all over the whole planet, our planet.

I think that when circumstances like these arise, then automatically we should not only support science as such but also support the concept through which an advanced society can be built, in which the scientific component will play an important part. It's for that reason that there is a consensus among all European countries that CERN is not only the most important laboratory in the world but also the most visible flagship of an advanced society, one we should all be proud of. Science is culture, understanding for understanding's sake, the most worthy and sublime expression of humans.

Furthermore, CERN produces economic and social benefits. I'll give you just one example, because it's something that represents a revolution in modern times. Where was the Web invented? Tim Berners-Lee invented it back in 1989. And on April 30, 1993, CERN announced that the components of Web software would enter the public domain, so allowing them to be used, duplicated, modified, or distributed. That software, made at CERN, changed the world. It's revealing to think that one has access to information without paying a European cent or dollar for it. It was invented at CERN, and at CERN patents are very unusual; the content is free to use, while the intellectual property right is maintained. All the development work is returned to society.

A.P.:   CERN has opened up society. I'm not sure whether you are aware of something that many people are not. Something I was told not long ago by the only Spaniard among the cofounding group of Arduino. The group wanted their Arduino Diecimila to have a universal, open source hardware license.[2] The group turned to the legal department at CERN, who got their attorneys working on drawing up an open, universal license for everyone involved in the world of hardware, and within a few months they had legally created the CERN Open Hardware License (CERN OHL), which can be used by everybody.[3] It's a good example.

The flagship that CERN represents, as you say, is the flagship of grand science. Can we understand grand science as a catalyst for civilization?

Does more science mean more civilization?

J.B.:   It's clear that in modern societies, science plays a fundamental role in development, but also in coexistence. What I mean is that science consolidates the advances that are taking place in society. Science, of course, may be used for ends that may not be the most appropriate ones, but one cannot attack science for that. That's the fault of certain individuals using scientific results.

A.P.:   We often talk about grand science, but CERN is not just words. It is a huge mechanism, an enormous machine, and a fact that really functions. I think that the CERN mechanism may be seen as a real instrument for "catalyzing civilization," don't you?

J.B.:  Yes. I'm 100 percent in agreement with that idea because CERN is a meeting point not only for scientists but also for people interested in culture and the development of society, who wish to take CERN as an example of how to effectively achieve communication and collaboration. I'd just like to add one thing that contributes to this idea. Now, using the results of the Higgs boson, there are two different experiments going on at two different intersection points of the proton beams at the LHC. Clearly, if there are two experiments, it is because in science, it is essential to compete. Competition arises from the fact that the results of one group have to be compared with another. It has to be like that, but at the same time the two groups collaborate. In the commercial world, competition normally means the opposite of collaboration. They are usually two opposing terms. It's not like that in science. In science you can compete at the same time as you collaborate. They are complementary terms.

A.P.:  Your life seems to be closely linked to CERN even during your holidays.

J.B.:  Not only my scientific life but also my personal life. When I return to CERN, I feel at home. I don't consider it a foreign laboratory. Recently I was at SLAC, the laboratory of the University of Stanford concerned with the results of broken time-reversal symmetry. Although scientific atmosphere is universal and friendship without frontiers is a much-appreciated value among scientists, I did feel that I was abroad. However, when I go to CERN, because I have a strong link with the laboratory and its surroundings, for me it's like being at home.

A.P.:  What are the great challenges of CERN?

J.B.:  The first great discovery at CERN took place in 1973. I have always said that it was the first experimental result leading to the Standard Model of particle physics. The paper with the observation of a new weak force was "Discovery of Weak Neutral Currents at CERN."[4] This was a great catalyst for later theoretical and experimental developments, which took place quickly and in great depth. When the ideas are there, there's immediately a hurry, and everything accelerates. That's what I was saying before. When we know how to formulate the questions, the answers come quickly and everything advances at great speed. That is something that Carlo Rubbia, the Nobel Prize winner in 1984 and later director general of CERN, said at the beginning of the 1980s. Only ten years after the discovery of the weak neutral currents there are data to confirm what I was telling you earlier about competition versus collaboration. Since then there have been more American physicists working at CERN than European physicists working in the United States. What I mean is that the advance that a laboratory like CERN represents for Europe is considerable. Nobody today disagrees that it is the number one world laboratory, and that should make Europe proud because it's true it is the catalyst for great developments that are taking place in scientific advances.

A.P.:   What for you would be the second greatest landmark? The Higgs boson?

J.B.:   The discovery of the Higgs boson, announced at CERN in the first week of July 2012, will always remain a landmark in the annals of science.

But that is not the end of the story in the advance of fundamental knowledge. On the contrary, both from the consistency of the theoretical scheme of the Standard Model and from unequivocal experimental signals, we are convinced that a new type of physics will have to arise that is capable of explaining pending problems, such as the mass of neutrinos, dark matter, and dark energy, or why at present the universe contains hardly any antimatter in natural form. Moreover, the present experimental results will provoke new questions that we still do not know of as yet.

A.P.:   In Shakespeare's *Hamlet*, act 2, scene 2, the prince says:

"I could be bounded in a nutshell
and count myself a king of infinite space."

This verse was the inspiration for the title of Stephen Hawking's recent book, *The Universe in a Nutshell*.

Did you feel like the king of infinite space at the LHC once the tunnel and machine had been built in that gigantic subterranean space at CERN?

What kind of emotions do you feel there?

J.B.:   Well, you feel satisfaction and, obviously, excitement, because the fact that humans are capable of building a machine that can provide answers to the secrets of nature in this way, a fantastic way, is very positive and exciting. But that's true not only with the experiments being done at CERN but also with the experiments that are being undertaken from satellites for observing the fossil trace that remains from the beginning of the universe. The advance on the two frontiers of physics from the smallest to the biggest is fantastic. Moreover, there is a sense of unity and—talking about emotions—a great human feeling that we are all involved in trying to reduce phenomena that appear very distinct to a single great law.

Today, the study of the conditions of particle physics and of what the physics associated with the primordial universe consists of is all coming together. It's a wonderful manifestation of the unity of physics and well expressed in that quotation from *Hamlet*. From the observation of the most intimate detail we are answering the question of how the universe behaves overall and why the universe has evolved as it has up till now in accordance with several conditions that now are being recreated in the laboratories of particle physics.

So now it's not just a question of advancing knowledge. It's the unity of science, the knowledge that we are capable of working together to understand what is happening from observations of the smallest things to observations of the biggest things. When we talk about the smallest and the biggest, we have to bear in mind that, from our metric scale to the constitution of the atom, there are ten orders of magnitude in length, to use our jargon. This means we have to multiply the length of an atom by one followed by ten zeros to reach one meter. And there are eight other orders of magnitude down, to reach from the atom to what is now being explored at the smallest distances. On the other hand, we have also to go to the greatest distances. But then we are asked, "Why go to the greatest distances if what you want to know is the primordial universe?" This is another wonderful aspect of that analogy of the nutshell—that we are capable from there of arriving at the limits of the universe, and not only that but to find out from there what the universe was like in the past, because signals are transmitted at a finite speed and, therefore, when I am observing what is happening out there. ...

A.P.:   In the universe, the further out you look, the earlier in time you are looking.

J.B.:   You see an earlier time, exactly. That is what is taking place with the Hubble Space Telescope and other telescopes.

A.P.:   That is: the further, the earlier.

J.B.:   Earlier, yes. An alternative is the observation of the background radiation we have in the present universe, because it has remained like a fossil from that primitive period. These are the two ways through which we have access to all that. That quotation from *Hamlet* is wonderfully appropriate because if there is something today that is a manifestation of the unity of science, it is that convergence between the physics of the smallest thing and the physics of the largest thing, the connection between particle physics and the physics of the primordial universe.

A.P.:   Physics seeks to unify natural laws that describe different phenomena in a common dynamics. That dynamics provides us with evolution over time and it is linked to symmetries of matter. But that doesn't happen with time itself. So I have the following questions:

Why is time asymmetric?

What does the "time reversal" that you speak about in your research and publications mean?

J.B.:   The great scientific advances that have been made, and with them the knowledge attained, have gobbled up a considerable part of the ocean of ignorance by unifying what were two apparently distinct phenomena. The names of Newton, Maxwell, Einstein, and Bohr, or the recent unification of forces, electromagnetism, responsible for the formation of the atom, and the "weak" interaction, responsible for the generation of energy by our Sun, are associated with that unifying dynamical laws as a consequence of symmetries.

Time is a concept that runs in a single direction. For that reason we speak of the "arrow of time." We see that complex systems have to follow the dictates of the second law of thermodynamics, always evolving (if they are isolated) toward an increase of so-called "entropy," an increase of disorder! If a glass falls and breaks into a thousand pieces it is impossible that, spontaneously, the pieces will get together again and the glass reconstruct itself. The falling of the glass is irreversible! That is how Eddington almost a hundred years ago explained, with the increase of entropy, the meaning of the "arrow of time" concept.

A.P.:   That's clear, but what about that expression you use, "time reversal," for a possible symmetry of the physical laws?

J.B.:   That arrow of time does not eliminate the question of whether the dynamics of the fundamental laws for elementary particles, for those that we observe to have reversible processes in time, are able to describe both the direct process and the reverse process in time. Perhaps the expression "time reversal" is not the best one because people think it means reversal *of* time, time running backward. That makes no sense. What is reversed is the movement. So, in the same year, 2012, a few months after the discovery of the Higgs boson, it was experimentally established that several processes governed by weak force are asymmetrical under time reversal. Although reversible, the two processes occurring in one sense of time evolution and in the reverse sense do not show up with the same frequency. This discovery was made in the BaBar experiment installed in the SLAC laboratory, in which IFIC scientists participate and play an essential role, both in the theoretical proposal and in the analysis of the results.

A.P.:   But how did you arrive at that idea?

J.B.:   It was known that, in certain processes resulting from weak interactions, there is an asymmetry between the behavior of matter and antimatter. It was natural to ask whether, in these processes, there is a breaking of time-reversal symmetry too. However, these are processes in which the particles decay. If the particle disappears, the process is irreversible; you cannot study the reverse process; it is like the arrow of time discussed before for a falling glass. We proposed a bypass to this no-go argument using some spectacular properties of quantum mechanics, able to transfer the information from the decaying particle to a partner that is still alive! And you do the experiment with the partner. Thus the difficulty of this experiment was in knowing what one had to measure. The concept and the method are explained in my article, "Time-Reversal Violation with Quantum-Entangled B Mesons."[5]

Today we do know that, in certain processes in which there is a breaking of the symmetry between matter and antimatter, there is an asymmetry under time reversal, too.

A.P.:   José, thanks for speaking to us and for providing such fascinating conversation.

J.B.:   You're welcome. Thanks for coming. It's a pleasure.

## Notes

1. On Pablo Picasso's statement, "Computers are useless. They can only give you answers," the earliest known appearance is in Herman Feshbach, "Reflections on the Microprocessor Revolution: A Physicist's Viewpoint," in *Man and Technology*, ed. Bruce M. Adkins (Cambridge: Cambridge University Press, 1983), where the attribution is described as "rumoured" (http://en.wikiquote.org/wiki/Pablo_Picasso).

2. Wikipedia, "The Arduino Diecimila, Another Popular and Early Open Source Hardware Design," https://en.wikipedia.org/wiki/Open-source_hardware#/media/File:Arduino_Diecimila.jpg.

3. Open Hardware Repository, "CERN Open Hardware License," http://www.ohwr.org/projects/cernohl/wiki.

4. The Standard Model of particle physics is a theory from the 1970s that describes three of the four fundamental interactions (strong nuclear, weak nuclear, electromagnetic, and gravitational) among the known elemental particles comprising all material in the universe. It also describes a classification of all known subatomic particles. It is consistent with the special theory of relativity and with quantum mechanics. The theory is considered to be incomplete, however, as it does not

provide a coherent explanation for the origin of gravity—the fourth of the known forces—or for energy and dark matter.

5. J. Bernabéu and F. Martínez-Vidal, "Time-Reversal Violation with Quantum-Entangled *B* Mesons," *Review of Modern Physics* 87 (February 23, 2012): 165, http://journals.aps.org/rmp/abstract/10.1103/RevModPhys.87.165.

6. For a thirty-year review of this work, see "The Discovery of the Weak Neutral Currents," *CERN Courier,* October 4, 2004, http://cerncourier.com/cws/article/cern/29168.

## 3   For Exoplanets, Anything Is Possible

Sara Seager and Adolfo Plasencia

Sara Seager. Photograph by Adolfo Plasencia.

*Yes, exoplanets found me before I found them.*

*My personal opinion about life that could traverse the galaxy or travel to a distant star system is that it probably has to be nonbiological.*
—*Sara Seager*

Sara Seager is Professor of Planetary Science and Physics, astrophysicist, and planetary scientist in MIT's Department of Earth, Atmospheric, and Planetary Sciences. She is known for her work on extrasolar planets and their atmospheres.

Seager earned a doctorate in astronomy at Harvard University, at a time when the first reports of exoplanets around Sun-like stars began appearing, and she studied the atmospheres of these so-called hot Jupiter planets. She then joined the cadre of postdoctoral fellows at the Institute for Advanced

Study at Princeton University, followed by a stint on the senior research staff at the Carnegie Institution of Washington before making her way back to Harvard.

Her research focuses on the theory, computation, and data analysis of exoplanets. She is Co-Investigator on the MIT-led TESS, a NASA Explorer Mission to be launched in 2017, and chairs the NASA Science and Technology Definition Team for a Probe-class Starshade and telescope system for direct imaging discovery and characterization of Earth analogues. She is a 2013 MacArthur Fellow, the 2012 recipient of the Raymond and Beverly Sackler Prize in the Physical Sciences, and the 2007 recipient of the American Astronomical Society's Helen B. Warner Prize.

Adolfo Plasencia:   Thank you for having me, Sara.

To start with, I have two questions. The first one is about your education. I read that when you were a child, someone put you in front of a telescope and you looked at the Moon through it for the first time.

Was it a turning point in your life?

Do you remember?

Sara Seager:   Adolfo, I just want to tell you a story. I was teaching astrophysics at a winter school in Guatemala, and many of these students come from all sorts of backgrounds, but they're not as privileged as we are in North America. I told them a personal story about the Moon. It was a slightly different story but it was amazing because it resonated with all those students, many of whom had had a very similar experience, so the very fascinating thing is that such experience with the Moon is common with people who love astronomy.

Yes, at first I felt like I had nothing in common with the students. It was more the story of the Moon following me and I couldn't figure out why. And every student just had a huge smile because they realized that the MIT professor from far away is really just like them.

A.P.:   You earned a BSc in mathematics and physics and later a PhD in astronomy. What is your relationship to mathematics? Is it a rational or a logical thing, or is it more related to emotions and passion?

Is it the same with physics?

S.S.:   Well, math for me is just a tool. I use it just as an application to help solve problems; although I know there's an aesthetic beauty, it's a challenge for me to reach that level. Physics, on the other hand, is fascinating to me, very phenomenal, because it describes the world around us, and although it's somewhat disappointing that you have to make many approximations, physics I see as both a tool and a beautiful way to describe the universe.

A.P.: After completing your undergraduate degree in mathematics and physics, you earned a PhD in astronomy from Harvard University. At that time, exoplanet astronomy was developing very quickly: in 1995 researchers found the first known planet orbiting a Sun-like star. Named 51 Pegasi b, it was about as massive as Jupiter but orbited its star so closely that its surface temperature must have been almost 2,000° Fahrenheit. In 1996 Geoff Marcy, an astronomer at the University of California, Berkeley, along with his collaborator, Paul Butler, discovered six more exoplanets, three of which were also big and broiling. Humanity finally had hard proof that the universe is full of other solar systems, something that until then had been an act of faith in science fiction. You followed these discoveries very closely. Apparently, you—and you later devoted your life to finding them—were found by them, by the exoplanets.

Was it like that? As far as I know, you were doing your PhD work when the first exoplanet was discovered. So exoplanets entered your life even before you made the decision to find them. Is that right?

S.S.: Yes, that's a fascinating way of looking at it. In fact, nobody has ever described it in that way. You're very right: the exoplanets found me. It's a little bit more like you're hiking in the forest and you come across a huge mountain. Do you climb the mountain or not? So exoplanets presented me with an opportunity, which I took. Yes, they did indeed interrupt my life, but I had to make a conscious choice to follow the path of exoplanets.

A.P.: Let's talk about the "what." Which features must a celestial object have for us to call it an exoplanet? Can this concept be understood by schoolchildren? I understand there are "gas dwarfs" (mini-Neptunes), up to "super-Earth" planets, with a large mass—ones that predominantly consist of hydrogen and helium, like exoplanet Gliese 581 c. But the term *gas dwarves* has also been used for planets smaller than gas giants, with hydrogen or helium atmospheres. It sounds complicated, doesn't it?

S.S.: The simple way to look at it is that every star in the sky is a sun, and the planets of our Solar System orbit our Sun, just as exoplanets orbit other stars or other suns. We could leave it at that, but I think that what you're trying to explain is that exoplanets come in all masses and all sizes and there's literally a continuum from a small rocky planet to a bigger rocky planet, to a small planet with gas, to a bigger planet with gas—and it is, I agree, somewhat complicated. We are struggling a little bit to decide how to define planets, but I think it's just better to see the bigger picture; that there are planets of all sizes and masses and that nature forms many kinds of planets, and we just simply call an exoplanet a planet that orbits a star other than our Sun.

A.P.:   According to the May 2014 *Smithsonian Magazine,*

Kepler space telescope scientists announced the discovery of 715 new planets orbiting other stars; the current total is 1,693. (In the 4,000 years from the emergence of Mesopotamian astronomy until the 1990s, scientists found a total of three new planets—two if you are a Grinch and don't count Pluto.) There may be tens of billions of Earth-sized worlds in our galaxy alone. NASA recently approved TESS, the Transiting Exoplanet Survey Satellite, to identify other worlds around the nearest stars. You [Sara] are involved as a project scientist.[1]

All this really seemed unimaginable not too long ago.

How can you convince people responsible for budgets that really remote things like exoplanets are relevant to all of us?

S.S.:   It's not hard to convince anyone in the world at any level how exciting and compelling the search for rocky exoplanets is. In fact, people hardly need convincing to understand how exciting the search for exoplanets actually is. When it comes to a mission like TESS, there's a competition that unfolds over many years, and one has to have a very compelling scientific case and an airtight technical case, so that's why TESS was selected by the higher officials at NASA. In general, though, the astronomy community gets together and decides what its priorities are. It's sort of like a wave passing through matter: more and more astronomers want to work on exoplanets. And it's a kind of democratic process at some level, deciding which science of which mission gets chosen. However, I think the science of exoplanets is very special because, unlike most fields in astronomy or in science, it generally captures the world's fascination—and that includes the general public and the higher-level officials managing the budget—like almost no other topic we've ever seen before.

A.P.:   Sara, there are many assumptions about the universe: some argue there is just one universe, while others say there are several universes connected to each other.[2] The last theory I read describes the universe as being some sort of depiction or a huge hologram. What do you think?

S.S.:   Adolfo, exoplanets used to be like science fiction, but now they're scientific fact, and their study seems almost practical in comparison to thinking about multiple universes. So my personal opinion is that the multiple-universe study and that of extra dimensions are really more math and philosophy rather than actual astronomy. I think it will be some time, if ever, before we have a way to find real evidence for any universe beyond our own or even what our universe really is.

A.P.: Is this type of deliberation really worthwhile since it all seems so unreal and impossible?

S.S.: Yes, absolutely; it is something that should be discussed. I think a great analogy is inflation: we believe that our universe initially had to rapidly expand by something called inflation, which for some was just an idea, a concept worked out mathematically through physics, and recently we've begun to think there's actual evidence in the echoes of the Big Bang. Therefore, some idea today that seems a little crazy needs to be worked through because whatever we think of, nature is always smarter than we are and it may actually have implemented something that is at the limits of our creativity. This said, there may be a time downstream in the future when we can actually prove it so. Yes, definitely, it must be worked on.

A.P.: And now let's talk about the "how." The way some exoplanets were found is incredible. I don't know if this is right, but it seems that some exoplanets were found without our actually being able to see them. They are celestial bodies with no light and therefore difficult to see with a telescope, yet they were detected thanks to the apparent small oscillation of some star resulting from that planet's gravitational force. Is this a way used by science to find them?

Can a remote planet that can't be seen be found?

S.S.: Well, Adolfo, most planets are seen indirectly. A planet is so small, so less in mass and so faint compared to the big star that it is right next to that it's nearly impossible to see a planet directly. So, as for most planets that have been discovered, we only see them indirectly, by their effects on their host star. We've kind of accepted this to such a degree that we barely even talk about it, so I'm really glad that you asked about that.

A.P.: You said (please correct me if I'm wrong): "I'm dedicated to finding another Earth, other Earth-like planets or planets that can support life." Based on this, you are convinced there is life elsewhere in the universe. I have a few questions on this:

What type of life can there be far away from us in the universe?

Michail Bletsas told me he is convinced that in this century there will be nonbiological intelligence, intelligence not based on *Homo sapiens*.[3]

Could life on remote planets already be intelligent life? Would it compare to human life as we know it? Would it be biological or nonbiological life?

S.S.:   That's such a great question and concept. I have to start out by saying that in astronomy, we're more focused on what we can see and find, and remotely, all we can see are chemicals; we can see the atmospheres of other planets. With future space telescopes we'll be able to look at the atmospheres of planets the size of Earth and see what chemicals are there; here on Earth, plants and photosynthetic bacteria produce oxygen, and oxygen is a very reactive gas and shouldn't be in our atmosphere at all, so if we can see oxygen on a planet far away we have a clue that there may be biological life that uses chemistry, that uses chemical reactions to release and store energy. From astronomy we can only search for life that makes some kind of by-product that is chemical, so I think in astronomy we can only look for biological life. My personal opinion about life that could traverse the galaxy, if we are now talking about life that could come to Earth, or in the future, if we're able to travel to a distant star system, it probably has to be nonbiological because space is very harmful for people. We can barely survive on Earth, if you think about it, and Earth is a very safe, well-designed place for us, or rather we are adapted to our environment. So I think for us initially as human beings to find life elsewhere, it's bound to be biological, since that's all we can see; it's all we know how to do. But if we ever think of traveling through the galaxy or of alien life coming here, then I believe on a personal level that it will be nonbiological.

A.P.:   I'd like to talk to you about a long-term matter. Whatever the case, humankind's mission to find life is really a very long-term goal. You said it could take "more than a generation, as happened with the construction of the Great Wall of China or Europe's great cathedrals." You also said that "we are now fully prepared for it."[4]

I suppose you meant scientists and human beings, because some pessimistic voices said that NASA no longer has a long-term mission, a mission that can live up to the challenges that made it necessary.

And there is another corollary: your projects are obviously long-term projects. Young people today need to have instant satisfaction.

Do we live in a world of immediacy? Are today's young people patient enough to be astronomers?

S.S.:   I'll answer that question in a couple of different ways. The first is that we have a wonderful new thing happening in the United States. We call it the "private commercial spaceflight world"; there are companies like SpaceX and several others that are going off and doing their own thing, since NASA, as you mentioned, is big, engages in big projects, and has a lot of bureaucracy and risk aversion, so these smaller companies are making access to space cheaper and easier. They want to reduce the cost of going to space by a factor of 100 or 1,000, and that will enable even NASA, for example, to launch much larger telescopes to carry out surveys to find other Earths. So it may actually be easier to go into space sooner than it has been in the last couple of decades.

But now let's turn to the notion that finding signs of life on another planet could be a project lasting over many generations. I think that we are all hard-wired in a certain way, and even though young people may seem less patient, I believe that there are enough talented, dedicated people out there to carry on the search for other Earths. The way I actually describe this to my students goes like this—and I'm sure it's true for yourself and others. You have many things going on; it's the pie chart of investment. Some of them give short-term satisfaction; we all must have these or we can't get up every day and do our job. Then you have much longer-term tasks that you know are going to take a very long time, so I think that as long as these younger people have a better balanced way of doing things, it will definitely be possible for them to continue.

A.P.:   I've been involved in a twelve-year project with 4,600 students. That's why I see young people this way.

S.S.:   Hopefully they can balance that immediacy. I'm going to have to think about that too, because it's incredible, really. I mean, I have two children of my own, aged nine and eleven. I got each of my two boys an iPhone because I leave them alone now for short periods of time. I was surprised at the direction things took. For example, they maintain friendships with adults now; they mostly see my friends or my students. When you're nine, do you have a series of adult friends you can message with? I'm just trying to put that in context. When you think of moving to a technological, biological life—you sort of see that happening now—in one or two generations, kids are practically born with their brains connected, they're with their friends in constant communication. We have to think about that more. … I'll be thinking about these questions after you leave.

We see through people like Elon Musk and the SpaceX Corporation that the younger generation thinks differently. It may seem negative but it has huge positives, and they're able to get things done in a way that prior generations were not, so perhaps their sense for immediacy may fold over into getting things done more quickly. It's certainly possible.

A.P.:   It takes less and less time to discover a new exoplanet. Are some of them more exciting than others as far as the possibility of finding life is concerned?

I have been told Kepler-62f is a good candidate in this respect, as it ranks among the top twenty. At 1,200 light-years from Earth, it is one of five planets orbiting the same star. Are you excited about such a remote possibility?

S.S.:   I'd like to backtrack a little and say that, initially, I and others used to know every single exoplanet by its name and features when there were only ten of them; we knew them all. Then, as the years went by, there were too many to remember. If you have one or two children, you remember them, but perhaps if you have so many great-grandchildren it's as though there are just too many of them to recall. So it's true that now we have less emotional connection. As for these new planets that are found and are supposedly habitable—we actually don't know for sure whether they're habitable—we only know their size and the amount of energy they're receiving at the upper limit of their atmosphere. We don't really know what they're like, we don't have enough information on them to know if they're habitable; that's probably why I wouldn't get emotional. On the whole, it's not so much the distance; it's the idea that our universe and our galaxy are full of small rocky planets. We know it for a fact, they are so common— every star must have a handful of them—and it's emotional to think that every time you look in the sky those stars probably have a rocky planet, whether or not it's habitable. Some of them actually will be, so the very fact that we're finding a few of them makes it exciting to think how many of them are actually out there.

A.P.:   You are known for your determination, among many other positive things. I read a very nice story about you in the *Smithsonian Magazine* article by Corey S. Powell. It was about your very unconventional birthday party thrown at MIT, at the new, elegant extension of the MIT Media L ab, Building E15. You invited a few dozen colleagues, including a prestigious former astronaut and the director of the Space Telescope Science Institute.

You told them you did not want any presents. Well, just one: you asked them to respond to a challenge: to help you "plot a winning strategy to find another Earth," and do it within your lifetime.

A huge challenge, isn't it? What a present!

What was the party like? And what was that challenge like? I suppose it is the challenge of your life.

S.S.: The occasion was my fortieth birthday, which I kind of see as the halfway point in my life, although I do expect to live to be a hundred or older. I did invite all my so-called famous friends. And I asked them to be bold, not like at a regular conference. I gave them a short amount of time to say something important, and I would say that not a single specific thing came out of the conference. It was more a question of momentum building; you had all these people stand up there and say how important it is to find a planet, how they believed that we need to work harder, that we need to focus our efforts. I think it made an important mark on the world, and I had each talk videotaped. You didn't see the part afterward where we celebrated! That part was not filmed, but the actual talks are filmed so people can go back and look at them. So yes, I have had a growing sense that it is the biggest challenge of my life, you're correct, to try to find another Earth with signs of life on it. However, I already know that all these people are with me. It wasn't just the scientists who came and talked at the meeting, it's also about people like you who are writing about us, the viewers who are listening to us, everybody in the world who wants us to succeed. So although it's a challenge, we believe it's one that we can meet and for which we have the help of the entire world.

A.P.: Sara, thank you for your words.

S.S.: Thanks, Adolfo, for this great conversation.

## Notes

1. Corey S. Powell, "Life in the Cosmos: Sara Seager's Tenacious Drive to Discover Another Earth," Special Report, *Smithsonian Magazine*, May 2014, http://bit.ly/1fpU9cE.

2. For more on multiverses and the debate over universe versus multiverses, see Steven Weinberg, "Anthropic Bound on the Cosmological Constant," *Physical Review Letters*, November 30, 1987, http://bit.ly/1qNETvR; Raphael Bousso and Joseph Polchinski, "Quantization of Four-Form Fluxes and Dynamical Neutralization of the Cosmological Constant," Arxiv.org, June 26, 2000, http://arxiv.org/pdf/hep-th/0004134.pdf; Delia Schwartz-Perlov and Alexander Vilenkin, "Probabilities in the Bousso-Polchinski Multiverse,"Arxiv.org, January 20, 2006, http://arxiv.org/abs/hep-th/0601162; and Natalie Wolchover, "New Physics Complications Lend Support to Multiverse Hypothesis," *Quanta Magazine* (Scientific American), June

1, 2013, http://www.scientificamerican.com/article/new-physics-complications-len d-support-to-multiverse-hypothesis.

3. Michail Bletsas participates in dialogue 16.

4. EFE, "Sara Seager, astrofísica del MIT: 'He decidido dedicar mi vida a encontrar vida en otro planeta'" ("Sara Seager, MIT astrophysicist: 'I decided to dedicate my life to searching for life on another planet'"), *20 Minutos*, March 30, 2014, http:// www.20minutos.es/noticia/2100152/0/sara-seager/vida-extraterrestre/mit.

# 4   From Casimir Forces to Black-Body Radiation: Quantum and Thermal Fluctuations

Alejandro W. Rodriguez and Adolfo Plasencia

Alejandro W. Rodriguez. Photograph courtesy of A.W.R.

*The problems we are working on right now do have the potential to completely change the way that we think about energy.*

*The idea of achieving higher energy efficiency, of using every drop of energy at our disposal, is definitely revolutionary, though the world is moving in that direction. Greater energy efficiency will revolutionize the way we view ourselves and our relationship to energy, technology, and the environment.*

*—Alejandro W. Rodriguez*

Alejandro W. Rodriguez is Assistant Professor of Electrical Engineering, Department of Electrical Engineering, Princeton University. He received his doctoral degree in physics from MIT in 2010. Prior to joining Princeton University, he held joint postdoctoral positions with the School of Engineering and Applied Sciences at Harvard University and the Department of Mathematics at MIT.

He has received several awards, including the National Science Foundation Early Career Award, the MIT Infinite Kilometer Award, and the MIT Orloff Award for Service in Physics. He is also a National Academy of Sciences Kavli Fellow (2014) and was named a World Economic Forum Global Shaper (2011–2013).

About himself, he says:

I was born in Havana, Cuba, a by-product of loud rumbas, a family of physics enthusiasts, and Afro-Cuban folklore. At age twelve I emigrated to the United States. Although my last name is Rodriguez-Wong, a reflection of my dual Cuban and Chinese ancestry, I generally publish under the name Alejandro W. Rodriguez. When I am not thinking about photons, I am either dancing salsa, watching old films, listening to Cuban music, or playing video games.

## Part I: At the MIT Physics Department

Adolfo Plasencia:    Alejandro, thanks for meeting me.

Alejandro W. Rodriguez:    Thank you for coming.

A.P.:    Today, we are in the Physics Department at MIT. You were, however, born in Cuba.

A.W.R.:    Yes, I'm Cuban.

A.P.:    How does one get from Cuba to the Physics Department at MIT?

A.W.R.:    Well, to be honest, my family instilled in me a love of science. My dad studied physics. My mother studied astrophysics. My stepfather was a physics professor at the University of Havana. They never forced anything on me; to like physics you have to get to know it, try out a physics class. They did, however, instill in me a love of science in general. After arriving in the United States, from high school and even before that, in preschool in Cuba, I fell in love with physics and I firmly decided already in my second year at high school that I was going to come to MIT.

A.P.:    Is there anyone, any professor, you have met who has been decisive for your getting involved in this field of physics?

A.W.R.:    I did my undergraduate studies and started research here at MIT as well. During those three years I interacted a lot with professors whom I work with today, and I wasn't absolutely sure what I was going to like most. I knew I liked theoretical physics, but I ended up working in a new kind of physics, partly computational and partly theoretical. The convergence of these two areas is what made me decide in the end; obviously, the love I felt for the intersection of physics and computing is related to the relationship with my adviser at MIT, Steven G. Johnson, and also with John D. Joannopoulos. They are two professors who had a big impact on me and started me out on this path.

A.P.:   Why did you move toward quantum physics on a nanoscale and not toward the physics of the universe, for instance? Why the infinitely small universe and not the big one? Why between 0 and 200 nanometers?

A.W.R.:   There are two reasons. When I started my degree I was in love with general relativity, which you know is related to astrophysics, to the macroscopic world, and to hugeness, but at MIT there's quite a practical environment. There are many theorists; theory is very strong at MIT. But then there are people around you who are always thinking about applications and how to help society by exploiting technology, by thinking about specific applications. This ideas approach to science is what made me view quantum mechanics or the nanoscale world as a source, where I felt that my work could have a more direct impact on everyday technology. It isn't that astrophysics and general relativity are not important for society or technology but that quantum physics has a much broader connection to everyday things.

A.P.:   Let's speak about the Casimir forces. I remember a news item reported on the MIT website under the title "Mysterious quantum forces unraveled," which cited your work.[1] What mysterious forces had you unraveled?

A.W.R.:   Casimir forces were discovered a long time ago. In 1948 Hendrik Casimir, a physicist, discovered that if you have two neutral objects—that is, they have no electrical charge—placed a few nanometers apart, they will attract one another. In classical physics, the absence of an external current also means that no field exists between the two objects, and theoretically there would be no force between them. There would be no reason for any interaction at this scale. We're talking about a mysterious attraction in the sense that it cannot be easily explained by classical laws. It occurs because electromagnetic waves, or light, permeate the vacuum. The vacuum is not empty.

A.P.:   The vacuum is not nothing. ...

A.W.R.:   Exactly. It is not nothing.

A.P.:   And is there also energy within a vacuum?

A.W.R.:   There is energy in the form of "virtual photons."

A.P.:   But can photons be virtual?

A.W.R.:   Yes, of course. Then again, they're not so "virtual" when they have so many physical ramifications, when you can observe their existence. They are called virtual because they come from quantum mechanical processes, from Heisenberg's uncertainty principle, which reveals quantum features that aren't explained by Newtonian mechanics. Heisenberg realized that the rules of probability governing subatomic particles are born from an apparent paradox, whereby a particle's position and momentum, its rate of movement, cannot be simultaneously measured with perfect precision. Besides, according to this principle, on a quantum scale the mere act of observing changes what you are observing. As a result of this uncertainty, the world at small scales is chaotic and filled with fuzziness: there is much movement. Consequently, charges on atoms, the microscopic charges that exist in neutral objects, interact with each other through the laws of quantum mechanics.

A.P.:   Theoretically, entropy gets worse, never better; or do physicists want to turn things round and make entropy get better?

A.W.R.:   In this case, we are talking about a thermodynamic system that is in equilibrium; entropy by definition doesn't change. But yes, the idea is that we are trying to get hold of this disorder, this electromagnetic chaos, and do something useful with it. We are trying to understand it to see if we can develop some technology and get something innovative out of this chaos.

A.P.:   Alejandro, the people who read this will think we're talking about things that sound a bit alien.

A.W.R.:   About crazy things. ...

A.P.:   Crazy, yes. But anyone who has an iPhone knows that if it is rotated, the screen automatically changes from vertical to horizontal. This is due to micromachines, such as accelerometers, which are related to these Casimir forces. In other words, there are already applications being used by ordinary people. Don't you agree?

A.W.R.:   Yes, these forces have already been measured experimentally. Over the last four decades, dozens of experiments have been performed that have precisely measured Casimir forces. As technology continues to advance and the size of devices gets smaller, Casimir forces will play a much more prominent role. Right now there are devices with parts that function by taking advantage of these forces. But what produces the Casimir forces in the devices has nothing to do with gravity. Basically, these forces are the result of quantum mechanics and the electromagnetic field. Nevertheless, they are having a big impact on the macroscopic devices that we regularly employ.

A.P.: With Heisenberg's uncertainty principle, everything has become blurred, hasn't it?

A.W.R.: Exactly. The particle doesn't have a specific volume that is fixed and static in space. You can only state the probability of it being in a particular location, viewing the particle as a wave, not as a corpuscle. And by approaching it with quantum rules and wave dynamics, the strange behaviors we observe make more sense.

A.P.: To make sense for you in this way, maybe you were lucky to have been born after and not before Einstein because it would have been scientifically subversive to say these things eighty years ago, don't you think?

A.W.R.: Absolutely. For instance, there were physicists who believed that God didn't play dice, and Einstein was one of them.[2] Indeed, the consequences of quantum mechanics are very strange; it's very different from the macroscopic world that we experience on a daily basis.

A.P.: You don't talk about it as though it were anything strange. For you is it now?

A.W.R.: It seems natural to me now. For me now, the strange thing would be if quantum mechanics didn't exist. But this isn't trivial; it takes time to warm up to it. At this moment, I don't understand everything about quantum mechanics, but this ignorance is not ignorance of the theory of quantum mechanics but rather of how the world functions so strangely at such scales. The same ignorance concerning quantum mechanics is also still professed by some other physicists.

A.P.: There's something else I find very interesting. First there was a person, a theoretical physicist thinking about something; then there was somebody like Higgs. Many decades and billions of dollars or euros went into building an accelerator to find a particle that theory said existed. Now there are thousands and thousands of scientists in Europe working in a huge tunnel, where it seems to have finally been found, and Peter Higgs and François Englert were awarded the Nobel Prize in Physics in 2013. What's your opinion on this?

A.W.R.: I think this is part of science. The positron is an elementary particle, the antiparticle of the electron. Its existence was predicted by Paul Dirac in 1928. It was first discovered theoretically and later experimentally in 1932 by the North American physicist Carl David Anderson from photographs of the trail left by cosmic rays passing through a cloud chamber. In the scientific school that existed two hundred to three hundred years ago, experiments basically yielded to theory. Today we are in an age where theory is, in certain cases, much more advanced than experiments. And it is important to work on this kind of fundamental physics. Therefore, I do agree with people investing their money and time in studying fundamental physics. Nobody can tell you, "Look, this is what physicists should be working on; this is the branch of physics that matters," because history tells us something completely different. Sometimes the ideas that appear to be less important turn out to be the most revolutionary. A simple example is the laser.

### Part II: In the Department of Electrical Engineering, Princeton University

A.P.:   I met you at MIT. Now you are conducting research at another distinguished institution, Princeton University, on the campus where the Institute for Advanced Study is located. Here Albert Einstein worked and lived, as did many other physicists and mathematicians, such as Kurt Gödel, J. Robert Oppenheimer, and John von Neumann, etc. It's undoubtedly a great place for a physicist.

Although you haven't fully put the Casimir forces to one side, I know that you are researching questions related to power fluctuations of the electromagnetic field.

It also sounds exotic.

Can you briefly explain the topics you are currently working on?

A.W.R.:   My field of interest is optical fluctuations. Electromagnetic fluctuations include two types of processes: quantum processes and thermal processes. They are both important and are manifested in different ways. When devices are made smaller and smaller, these interactions become more and more important. We spoke about quantum fluctuations and how they lead to forces between objects. So, just as fluctuations lead to forces between objects, they can also lead to energy exchange between them. Planck, Kirchhoff, and other scientists in the late 1800s and the beginning of the 1900s developed the theory of thermal emission, related to fluctuations—the same kinds of fluctuations that lead to Casimir forces.

Just as in the case of Casimir forces, thermal fluctuations depend on the properties of materials and the shapes of the emitting objects. When you're looking at thermal radiation from an object into the far field—and this was well known in the 1900s—there is a maximum amount of thermal radiation that an object can emit, called a *black-body limit.*

A black body is a theoretical object that can emit with perfect efficiency into the far field. It doesn't exist in reality. It is also well known that if you have perfect emission, then you have to have perfect absorption. Therefore, a black body is an object that can perfectly absorb every ray of light coming into it at every frequency. We have this concept of a black body, which doesn't really exist in nature because no object can absorb light perfectly at every single frequency. This relationship between perfect absorption and perfect emission of thermal radiation was used for many decades and continues to play a key role in the design of solar absorbers and thermal emitters.

For example, the relationship is very important for solar photovoltaics and thermal panels for solar harvesting—in other words, for energy harvesting. It's basically prevalent in every single technology related to solar energy. The goal of solar absorbers and thermophotovoltaics is to create objects that can absorb radiation perfectly over some range of wavelengths so that this energy can be converted into electricity. This essentially uses the same ideas that physicists had developed over a hundred years ago, which we are using now to design surfaces and objects that can absorb light very efficiently and therefore emit light very efficiently, that is, thermal absorbers. However, our understanding of how to create a black body over a certain range of wavelengths is a work in progress. This applies to most objects. For example, if you take a piece of copper or gold, then that object is barely a perfect absorber and therefore doesn't have good emission properties. Instead, we design the surface of the object. We structure the surface of the gold or copper so that by changing the geometry, an object can be created that is as close as possible to a black body over some desired range of frequencies.

Until recently, it was believed that the maximum heat transfer between two objects would also be limited by the black body limit. However, in the 1950s Dirk Polder and Van Hove carried out a calculation in which they took two planar objects, one at a hot temperature, T1, and the other at a colder temperature, T2, and investigated their mutual heat transfer. Here, heat transfer refers to the heat radiated by the hot object and absorbed by the cold object, which varies as a function of the amount of separation between the two objects. They observed that as the objects got closer and closer to each other, below what we call the thermal wavelength—which is a wavelength associated with the higher of the two temperatures—you get hundreds to thousands of times more heat transfer from the hot object to the cool object, even going way beyond the black-body limit. Therefore, when you bring two small objects together, the heat transfer from one to another can exceed the black-body limit, which was established by Planck in the 1900s, by orders and orders of magnitude; it could be a billion times more. So you can actually extract a lot more energy from an object by putting it in the near field of another object than you would in the far field. By the near field I mean a very small separation exists between the two objects, in the micrometer range or smaller. Hence, just as Casimir forces increase significantly and are magnified at short length scales, so is energy transfer.

This was a revolutionary idea because now you're not limited. When you bring two small objects together at nanometric and micrometric separations, you can get heat transfers that are significantly larger than the black-body radiation limit.

A.P.:   Why is this field of quantum and thermal fluctuations important in practice, and what likely impact do you think it will have with respect to real technological applications in the future?

A.W.R.:   We know that over the past sixty years, optics has revolutionized many industries, including telecommunications and information technology. It is responsible for breakthroughs in medical science. The key is that we can manipulate light in very precise ways to control how it interacts with devices and how it behaves. But as you scale these systems and devices down to smaller and smaller sizes, it turns out that even the tiny fluctuations of matter—matter is constantly vibrating; there are charges, electrons and atoms—create optical fields and optical waves and electromagnetic fields whose interactions can no longer be neglected, whose interactions themselves can lead to interesting effects at those scales.

One of these effects is known as the Casimir effect, which involves electromagnetic fluctuations given off by matter as the matter vibrates, causing interactions or forces between objects, thus pushing objects away from or toward one another.[3] A vacuum is full of fluctuating electromagnetic fields. These are optical fields, released by the vibrations of matter, that persist even when the temperature of the materials is absolute zero, even when the only remaining source of fluctuations and vibrations is quantum mechanical and is due to the uncertainty principle. So what we are studying is just the familiar glow of matter, but we're studying the effects that result from the fact that these electromagnetic fields are everywhere.

I think there is great potential for the field. Just as we learned how to manipulate laser light to confine photons and to get information from one place to another, there is a lot to be done through harnessing and controlling quantum and thermal fluctuations by designing structures at the nanoscale.

A.P.: This revolutionary idea you've spoken about still hasn't reached the devices being used, but to make a projection, which areas of future applications are you thinking about in relation to the research you are conducting?

A.W.R.: One, for instance, is in the field of thermophotovoltaic energy generation.[4] A thermophotovoltaic device involves two objects. One is operating at a very high temperature, so it's heated either by heat, such as from the Sun or from a thermonuclear reactor, to a very high temperature, usually greater than 2000 Kelvin. The other object is an absorber, which is usually some kind of semiconductor with a band gap. The absorber absorbs the radiation coming from the hot object and converts it into electricity. That's the basic idea behind a thermophotovoltaic device. But all such devices currently operate with objects in the far field; that is, the emitters and absorber are very far apart, separated by hundreds of microns or more. Therefore, they are not taking advantage yet of the fact that as these objects are brought closer to one another, the heat transfer can be significantly larger, many times greater than the black-body limit.

The question, therefore, is whether or not our current theories and understanding of heat transfer will be able to make an impact on future thermovoltaic technology. I think it is very likely that it will because vast amounts of energy are embedded in material fluctuations that are not being exploited. Another application of heat transfer in the near field is cooling.

A very recent experiment, from 2015, demonstrated this idea. It showed that if you take a hot object and bring it close to a cold one, you get heat transfer from the hot to the cold object, and that heat transfer is essentially acting to cool the hot object because it is energy being released by the hot object. Every second that the Sun emits radiation to the Earth it is losing energy; thus the Sun is getting cooler with time, and at some point it will run out of energy as it will have released all of its energy in the form of thermal fluctuations. The same thing happens at the nano- or microscale. If you bring an object near a hot one, the cool object will act as a coolant; it will take heat away from the hot object. The more heat you can collect from the hot object, the more you cool it. So the idea that you can bring objects close together and absorb these thermal fluctuations more efficiently in the near field certainly has applications.

For example, in a computer or laptop, as transistors get smaller, their functionality is limited by effects associated with overheating, and a lot of energy is spent keeping the systems in your computer cool. If we could figure out a way to use this near-field heat transfer mechanism to absorb the heat and dump it into another system, a cooler system, this would have a big impact. Heat transfer and radiation have also been recently proposed as means to cool homes by taking advantage of the cosmos; it would work by dumping energy from hot objects here on Earth into the vacuum in space. This idea is not crazy at all and in fact was experimentally verified this year.

A.P.:   Today, on the International Space Station, no one would consider using any energy that wasn't solar. On Earth, however, we have long been tied to fossil fuels such as crude oil, and renewables make up only a small part of the energy we use.

Do you think that in the medium and long term, these types of revolutionary areas you've been researching have the potential to change the relationship between people and the energy we use?

A.W.R.:   I do think so, to a large degree. It's interesting that you mention space exploration and space technology taking advantage of solar energy to perform work. One of the systems used in space satellites is already taking advantage of the idea we're discussing, not necessarily of the enhanced near-field heat transfer that occurs between objects when you put them very close together but rather exploiting thermophotovoltaics, by heating up the emitter and using the energy that radiates from the emitter to the absorber to do work, to convert the energy to electricity. In fact, some of the thermophotovoltaic devices used in satellites employ thermonuclear reactors whose sole purpose is to heat up the thermophotovoltaic emitter. This is basically the way you convert thermal energy into work, into electricity. However, on Earth we have an abundance of solar energy, which means we would essentially be using the Sun to heat up these objects for us.

The problems we are working on right now do have the potential to completely change the way that we think about energy. Many of the systems that we use today are very inefficient in that a lot of energy is wasted as heat. I just mentioned the transistors and other devices inside computers, which typically release a lot of heat; much of this energy is unused and wasted. The problems we are discussing and the questions we are addressing revolve around how to use this otherwise wasted energy, which hasn't been tapped for centuries, at least not fully, and recycle it.

How do we use that thermal fluctuation, those thermal energy sources, to continue to do work, so as to make our systems more efficient?

However you want to think about it, whether in terms of cooling the system and increasing its efficiency or in terms of using the energy source and converting it into electricity, the basic idea we are working toward is that of using wasted energy (heat) and reusing it to do useful things.

Therefore, the idea of achieving higher energy efficiency, of using every bit of energy at our disposal, is definitely revolutionary, though the world is moving in that direction. Greater energy efficiency will revolutionize the way we view ourselves and our relationship to energy, technology, and the environment. Until very recently, efficiency wasn't one of the biggest metrics we used to gauge success; instead, we viewed technology through the lens of functionality. We wanted to create technology to function in a particular way, and sometimes it didn't matter how we got there. There is a growing movement to develop new technologies that not only expand functionality but also achieve higher energy efficiency. Every time we extract matter and energy from our planet, we want to make sure that all of it is being used for something that is useful.

Heat, in the form of thermal radiation, is a by-product of many of the technologies that we use today. My goal is to try to harness it and make it work for us.

A.P.:   With respect to this revolutionary change driven by discoveries, and the ideas you're explaining, do you think a change in thinking about energy expenditure will be necessary to take advantage of the scientific discoveries that are now emerging?

A.R.W.:   To a large degree there has already been a significant change in attitude about what we think of as a good technology.

There is actually a scientific movement right now that is moving closer and closer to the ideal I just articulated. More and more scientists and engineers are viewing themselves as agents of change and creators of technologies that operate on the basis of sustainability and higher standards of efficiency, which is essentially what I've been talking about. These metrics are being prioritized in many ways. Much work remains to be done in this area, and much remains to be proved. That is true, however, of any science that is still maturing. We do need more people to champion this cause and to recognize that by funding and supporting this kind of research, we as a society are making a statement about our priorities. I think to a large degree this is already the case.

It is very important that many people, both inside and outside the scientific community, understand why this type of research is crucial for our future, why it is "visionary" and decisive for our future relationship to energy and the environment, and why that relationship must be different and better than it has been so far.

A.P.:   Thank you very much, Alejandro!

A.R.W.:   You're welcome. It was a pleasure.

## Notes

1. Larry Hardesty, "Mysterious Quantum Forces Unraveled," *MIT News*, May 11, 2010, http://newsoffice.mit.edu/2010/casimir-0511.

2. The reference is to Ian Stewart, *Does God Play Dice? The New Mathematics of Chaos* (London: Blackwell, 1989), which suggests that simple systems, following precise rules, can nonetheless behave randomly.

3. In quantum field theory, the Casimir effect and the Casimir-Polder force are physical forces arising from a quantized field.

4. Thermophotovoltaic energy conversion is a direct conversion process from heat to electricity via photons. The process of generating thermophotovoltaic energy (TPV) is the direct result of converting heat into electricity using photons. A basic thermophotovoltaic system consists of a thermal emitter combined with a photovoltaic diode cell.

# 5   The Challenge of Climate Change

**Mario J. Molina and Adolfo Plasencia**

Mario J. Molina. Photograph courtesy of the Mario Molino Center.

*The ozone layer was the first example ever of a clearly global problem.*

*We had to be very patient because at first, our hypothesis [about the ozone layer] did not seem to matter to society.*

—*Mario J. Molina*

Mario J. Molina is Distinguished Professor of Chemistry and Biochemistry, the University of California, San Diego, and a 1995 corecipient of the Nobel Prize in Chemistry, recognizing the theory of ozone layer destruction, which was later proved experimentally. He also serves as President of Centro Mario Molina in Mexico City, a nonprofit organization dedicated to finding solutions to the challenges of environmental protection, energy use, and climate change.

He received a bachelor's degree in chemical engineering from the Universidad Autónoma de México, a postgraduate degree from the University of Freiburg, West Germany, and a doctorate in physical chemistry from the University of California, Berkeley. His work focuses on the effects of man-made components on the atmosphere.

Adolfo Plasencia:   Professor Molina, welcome to Valencia! You are one of the Nobel Prize winners among the juries of the Rey Jaime I Awards.

Mario J. Molina:   I am very happy to be here, thank you!

A.P.:   Professor Molina, yours is a very long scientific career. I would like to start our conversation by talking about your background. You were born in Mexico and you soon realized you wanted to be a chemist. I think you played with chemistry as a child. Then you thought that to be a good chemist and a good physicist, you had to learn German. So you decided to go to Germany. From Mexico you moved to Europe, and from Europe to the United States, where you toured the two coasts. That is quite a long journey!

What was that life travel for chemistry like? Can you sum it up?

M.J.M.:   I loved chemistry when I was a child. And I also liked mathematics, science, and physics, but then I started doing chemistry experiments, and fortunately, I studied chemistry before going to college and later at the Universidad Nacional Autónoma de México (UNAM), where I already knew I wanted to become a research scientist. That's why when I finished my degree in Mexico, I decided to go to Europe for a postgraduate degree.

A.P.:   You say that science is a great means of bringing the peoples of the world together. How can that be done?

M.J.M.:   The international scientific community can significantly contribute to bringing together people from all over the world. We can see this from the viewpoint of protecting the planet. When we talk about environmental issues, young people all over the world agree that our planet is fragile and that we must all collaborate to take care of the planet. And the same applies to knowledge. Science is universal. If interaction and communication among scientists, including young people, are encouraged, progress will be made toward knowledge. Besides, there is a major ethical factor for the scientific community in general respects: although science itself is neither good nor bad, it should be applied for the benefit of humankind.

A.P.:   Professor Molina, in 1974 you published a scientific paper with F. Sherwood Rowland that drew a lot of attention because your proposals concerning the ozone layer were surprising. They were not particularly welcome in some scientific circles, which thought that the calculations were a little far-fetched. What happened when your proposal turned out to be right?

M.J.M.:   I put forward a hypothesis together with my colleague, Sherwood Rowland. We predicted that something was going to happen to the ozone layer, which is very important for our planet, as it surrounds the whole planet and protects us from solar radiation.

Experts and scientists with information about these topics accepted the idea surprisingly quickly, but the part of the scientific community that was not very aware of such possibilities questioned whether we were doing something relevant to society and insinuated that perhaps we were just doing it as self-promotion. Of course, we applied all the rigor of the scientific method, writing papers and submitting them for review by the scientific community, not only proposing a hypothesis but also clarifying how it could be verified or refuted by means of experimental observations.

A.P.:   What was that first hypothesis about?

M.J.M.:   The first hypothesis was about some not naturally occurring industrial compounds called chlorofluorocarbons (CFCs,) which were being used at that time as coolants and as propellants in spray cans because of their great stability. They had replaced the toxic compounds previously used as refrigerants as they had no direct effects on people's health, but precisely because of their stability they were beginning to accumulate in the atmosphere; natural cleaning mechanisms such as rain and some chemical reactions did not have an effect on these compounds. That was part of the hypothesis.

We then concluded that they would be transported above the ozone layer, where there is high ultraviolet radiation, which breaks down molecules, and so that was also part of what we hypothesized, that ultraviolet radiation was going to destroy those compounds at those high altitudes.

The ozone layer is very important for that very reason. Without the ozone layer life would not have evolved the way we know it; ultraviolet radiation destroys molecules, like those at the very basis of life, such as DNA, for example. All these molecules are very sensitive to this radiation. So when the CFCs are above the ozone layer, these substances are likely to break apart. The by-products of that breakdown became the object of the second part of the hypothesis: they would have major consequences as a result of an amplification process; that is, relatively small amounts would affect very large quantities of the ozone layer at that height.

And that was indeed very worrying: the thinner the ozone layer, the stronger the intensity of the radiation reaching the planet's surface, with harmful consequences for ecological systems and even for humans, in the form of higher rates of skin cancer. Many biological systems are known to be sensitive to ultraviolet radiation.

A.P.:   A long time has gone by since the publication of your first article in 1974. It was not until 1995 that you and Sherwood were awarded the Nobel Prize, shared with Paul J. Crutzen, for your work in the same field. What happened during that time until the Swedish Academy finally certified that your predictions were right and that your calculations about the seriousness of the issue were real, giving you the Nobel Prize in Chemistry in 1995?

M.J.M.:   We had to be very patient because at first, our hypothesis did not seem to matter to society. On the one hand, the media played an important role, and also our communication with decision makers, first in the U.S. government and then with governments from the rest of the world via the United Nations and with the collaboration of the scientific community, trying to verify the hypothesis. It was gradually proved more and more clearly that what we were saying was actually happening, that the hypothesis was verifiable and that something was starting to happen to the ozone layer: the ozone hole was forming over Antarctica, where the stratosphere is more sensitive to the breakdown products of these compounds. There were observations that not only was the ozone layer changing but that the change was caused by the presence of these industrial compounds.

A.P.:   Science is usually associated with countries—American science, German science, Asian or Japanese science—but ozone depletion is a global issue.

Do you have the feeling that this was one of the first scientific issues with global relevance, as it affected opinions and societies all over the world?

M.J.M.:   Yes. The ozone layer was the first example ever of a clearly global problem. The compounds that affect ozone remain in the atmosphere for many decades, and since the mixing time is months or, at most, years between the hemispheres, the compounds disperse throughout the atmosphere regardless of their origin. But it was also the first time ever that society, at a truly international scale, reached an agreement to resolve a global environmental issue. And it was also the first time it was done successfully. This international agreement was ratified by almost every country, obliging them to stop the production of the compounds. And later on we also managed to verify, on the basis of scientific observations, that the amount of these compounds is already declining in the atmosphere, although they stay in it for many decades. So what remains today in the atmosphere are the compounds released in the last century, which are gradually disappearing.

But most important, this international collaboration did succeed, and therefore it became an important precedent underscoring that it is possible for society to agree and solve global problems that, given their very nature, demand a truly international consensus.

A.P.:   The culture we live in is that of the ephemeral. The preferred horizon for politicians is generally the time of the elections, the four years of their office. You are saying that we are now experiencing the effects of the pollution caused by the misuse of certain products in the last century, and that it is certainly a long-term problem.

How do you report something like that in that context? I know you have a very important role in communicating scientific issues to the political class, to the political powers. What can be done for such long-term issues to fit into the public culture of the ephemeral and the instantaneous in which we live today?

M.J.M.:   An important part in this process requires society to become aware of these problems; this should in turn be reflected in those who make decisions. Society demands that they do so. And in this case, the reasons are ethical: we are accountable to future generations; most people have children and grandchildren, and they want to leave them a legacy in which the environment is at least as favorable as it was to us. That is a big responsibility.

In addition to ethical implications, there are economic ones, which should also be long term. Considering that our natural resources are limited, the way things are at the moment, if we have a very short-term vision, we may destroy our resources, and in a few years' time that will also have a significant economic impact, in other words, it could even affect us in the course of a generation. So a very short-term vision is not acceptable, though sadly, it is all too common among many politicians who make decisions by taking into account only the immediate next years. Fortunately, there seems to be a tendency to counter this short-term view. It is called "sustainable development."[1] It is clearly necessary. And it should not take decades to implement. We are already noticing the consequences of actions taken decades ago as well as very recently and which have effects on the environment.

A.P.: How do you manage to communicate these inconvenient scientific truths to important politicians? I do know there is at least one politician you were successful with. I think former U.S. vice president Al Gore called you right away after you received the Nobel Prize to congratulate you. And then he became deeply involved and started his fight against climate change. Did you see his film, *An Inconvenient Truth,* which won two Oscars and was released at the same time his book, under the same title, was published? What do you think about the steps taken by Al Gore from the political arena and how, after some years, this issue is being tackled by world public opinion?

M.J.M.: Al Gore's film had a large impact, especially in the United States. For some years now Al Gore has been very actively publicizing the climate change issue. I knew him when he was a senator, before reaching the vice presidency. Subsequently I became a member of a scientific advisory group to the presidency, both to Bill Clinton and to Al Gore. Possibly, what really had an impact later, and especially recently, was that all this information was made accessible to people through more effective dissemination. And of course, what Al Gore did was based on assessments by the scientific community, to which he obviously had access as vice president. I think that was a very important step. A high-profile person, as vice president and later in other capacities, he took up the subject as an essential part of his own goals, trying to influence society so that it would take global warming seriously.

A.P.: Let's look at it from the perspective of the present. When you gave your first opinion and delivered the first serious warning about the danger to the ozone layer, even the scientific community said you were overdoing it a bit. I have the feeling that the documentary by Al Gore caused the American public, which is used to mass consumerism, to be less prone to say that the film's message was far-fetched. How has American society's opinion evolved in recent times in relation to the impact of what the film told us?

M.J.M.: I think the impact has finally been positive, from the point of view that many people simply did not have the correct information, owing to public relations campaigns intended to establish the idea that everything related to climate change was an exaggeration, the consequence of the activities of extremist or environmentalist groups ... that it was not a sensible opinion representative of the majority.

That is what Al Gore's film started to change. It was just a matter of communicating adequately the assessments carried out by scientific groups. The scientific community is not particularly good at communicating with the media, especially when they have to report on results that can be worrying. That job is frequently done by environmental organizations that are sometimes known for their exaggeration. They do not have a very good image with the public. Maybe that was the difference with the film; it put the public in contact with the scientific community, something that does not happen with environmental organizations all that much.

A.P.:   Scientists typically rely on science and are usually optimistic. I think you are optimistic. You say that climate change is not irreversible, how can this be conveyed to people? What should be done for it not to be irreversible?

M.J.M.:   We need to act as soon as possible. The risk of damages occurring to the planet, or the fact that the climate change might not be reversible, is a consequence of the fact that the phenomenon lasts for decades or even centuries. That risk increases as the average temperature of the planet's surface rises, which in turn is the result of the buildup of greenhouse gas emissions like carbon dioxide.

Again, much of this problem is the result of emissions during the last century. If we do not take actions, knowing the momentum in society, considering the importance and magnitude of fossil fuels—which are the source of most of these greenhouse gases affecting our climate—things will go wrong. Bearing in mind that these changes will take a long time, effective international consensus is absolutely urgent.

An initial agreement was already reached with the Kyoto Protocol, but the agreement was not ratified by the United States, the main source of these compounds.[2] In its first phase, the protocol stated only that developed countries were to put limits on their emissions. The second phase has to do with what we expect will happen with developing countries in the coming years, countries whose economic development is now very intense: China, India, and possibly Brazil and Mexico. They must commit to certain limitations too, not in their economic development but in the way they carry out that development because the environment cannot absorb or metabolize all of the resulting waste.

So, close collaboration between developed and developing countries is needed. All this can be achieved through an international agreement because we do have technologies that can boost economic growth without damaging the environment. We know that this can be done, but it requires a strong political will. It takes leadership, especially from developed countries. So yes, we can be optimistic because in principle, the solution is possible since we now have the means to do so.

A.P.:   Very good. Ever since the discovery of the hole in the ozone layer, you have devoted every year in your scientific life to working on climate. In keeping with Moore's law, the computing power available to us in digital technology has seen an exponential increase. Some of the largest supercomputers in the world today, with huge computing power, are engaged in climate simulation.[3]

What has changed in the current calculations compared with past ones and the amazing supercomputing tools we now have to simulate the planet's climate?

M.J.M.:   What has changed is precisely the potential to do more sophisticated calculations. That makes us more confident in models, in the predictions made by such models; and at the same time our understanding of the operation of the whole planet is more solid. It is not only the increased ability to perform complex calculations but also the accumulation of scientific results and observations. As happens when normal science makes progress, in this case the results are useful to many disciplines. This is another important step that will take time. To understand the operation of the planet, geologists, meteorologists, biologists, and others—a whole confluence of disciplines—are required. One way of synthesizing this is to condense it through these complex models that do require a huge and very sophisticated level of computing. Significant progress has already been made. As frequently happens in science, the veracity of these models can be checked by analyzing, for example, climates from the past, a discipline we call paleoclimatology. If we understand what happened in the glacial and interglacial periods, then we will be more confident in our models.

A.P.:   Another revolution that did not exist at a global level at the time is the revolution of the Internet. In your scientific experience, what do you think this global deployment of an instrument like the Internet means, a phenomenon that involves several billion people?

M.J.M.:   This has had important implications for science, as it was developed precisely to help the scientific community be efficient when scientists collaborated in research remotely.

The Internet is a very effective system of communication. It allows us to have scientific collaborations with researchers who are located far away. And it is much more efficient than before, because in many cases you can generate new knowledge in collaboration with experts from different disciplines, or even from the same discipline but with different views. Such collaboration no longer requires the physical presence of the groups, just very effective communication.

A.P.: Professor Molina, one of your roles now is to communicate the priorities of science to political leaders at an international level. Is this more or less work than what you had to do to win the Nobel Prize in Chemistry?

M.J.M.: It's a different type of work, a very big challenge, because it is not easy to reach such a level of communication. I think it is really the responsibility of scientists. I see it as a big responsibility, particularly after one has been recognized with a Nobel Prize; thanks to it, we probably have better access to important decision makers. And although we have to continue contributing to basic knowledge, as we have always done, through many years of experimentation, working closely with student groups, we have this other very important possibility. I think communication is something that the scientific community does not do efficiently enough. They do not do it well enough, considering the importance of many of the problems in society. So that's why I think it is a very important activity and it can also be very successful as far as specific results are concerned. In my case, I had the opportunity to work fairly closely with decision makers in Mexico. For example, in the metropolitan area of Mexico City, I have seen the results of improving air quality as a consequence of our exchanges with decision makers. So addressing top-level decision makers is definitely important. It can be rewarding when we contribute to the achievement of results that are beneficial to society.

A.P.: Nowadays, communication is omniscient. It is a very powerful tool, and new media have become an essential part of developed societies. How can the treatment given to science by the mass media and Internet-based media be improved at a state level?

M.J.M.: There is a lot of room for improvement. Science dissemination is a specialty in itself. It needs to be done not only with clarity but also in an entertaining way for society to become interested in these problems. In part, we have to change the widespread notion that science needs to be communicated to the world in a highly specialized, disciplinary language. Each discipline has its own vocabulary.

Like many of my colleagues, I think that if we understand a scientific problem very clearly, then it can be explained in basic terms, though an effort must always be made to synthesize the essence of new knowledge. In my experience, I can say that such communication is very interesting for a part of the population, but it is very boring if it is poorly or obscurely explained. People can fail to understand it because of the language used. Hence the need to communicate better with society, which, by the way, is a task that has a lot to do with education as well.

We have the same problem in education. There are examples of how well and how effectively we can work with children when they are interested and excited,  if they have fun instead of having to learn scientific facts by heart. There is a lot in common between these two very important aspects in our society.

A.P.:   Thank you very much this conversation, Professor Molina. Thank you.

M.J.M.:   Nice talking to you.

**Notes**

1. Sustainable development refers to meeting the development needs of humans while maintaining, not diminishing, the natural resources and environment on which the human economy depends.

2. The Kyoto Protocol is an international treaty that commits countries to reduce greenhouse gas emissions. It was adopted on December 11, 1997, in Kyoto, Japan, and went into effect on February 16, 2005. Though the United States signed the treaty, the U.S. Congress has not ratified the measure.

3. National Aeronautics and Space Administration, "NCCS Triples Supercomputer Performance for Earth Science Modeling," April 28, 2015, http://www.nccs.nasa.gov/images/discover_story_042815.pdf.

# 6   Graphene and Its "Family": The Finest Materials Ever to Exist

**Pablo Jarillo-Herrero and Adolfo Plasencia**

Pablo Jarillo-Herrero. Photograph by Adolfo Plasencia.

*To work in quantum physics, you have to have an open mind with regard to what is possible and what is not possible. … You have to be open to the fact that the impossible may indeed be possible.*

*Physicists always try to simplify things as much as possible in order to understand the essence, and then we add the complexities.*
—*Pablo Jarillo-Herrero*

Pablo Jarillo-Herrero is Mitsui Career Development Associate Professor of Physics, MIT, and Principal Investigator of the Jarillo-Herrero Group.

He received his master's degree in physics from the University of Valencia, Spain, a second master's degree from the University of California, San Diego, and a doctorate from the Delft University of Technology in the Netherlands. Subsequently he moved to Columbia University, New York, where he worked as a NanoResearch Initiative Fellow.

His research interests lie in the area of experimental condensed matter physics, in particular quantum electronic transport and optoelectronics in novel low-dimensional materials, such as graphene and topological insulators.

Among the awards he has received are the Spanish Royal Society Young Investigator Award (2006), an NSF Career Award (2008), and an Alfred P. Sloan Fellowship (2009).

Adolfo Plasencia:   Thank you, Pablo, for finding the time to speak to me again.

Pablo Jarillo-Herrero:   A pleasure.

A.P.:   Physics, I believe, is something of a vocation for you. You began studying physics in Spain, but a professor encouraged you to move on to discover new worlds. You took his advice and went on to study high-energy theory, the physics of particles, which is what you had originally been hoping to do. That's why you left Spain for the University of California, San Diego, and then went on to the University of Delft, where you came across nanoscience. What was that journey like, from high-energy theoretical physics, on the cosmological scale, to the experimental nanoscience of the physics of condensed matter, on the miniscule scale? There are many worlds within physics, aren't there?

P.J.-H.:   I wouldn't call them worlds, but yes, there are many disciplines within science. When I was in Valencia, groups conducting research in condensed matter physics were not as important as the theory groups. I arrived in San Diego and began to go to seminars on other subjects. I realized that other disciplines interested me more, so gradually I became convinced that I had to take the step toward condensed matter physics. Once I had taken that step, I took another giant step, toward the experimental side rather than the theoretical side. That's what took me to Delft. There I was carried away by experimental condensed matter physics and in particular nanoscience, because there was a research group, one of the best groups in the world, in that field. So yes, it was quite a roundabout route to arrive at the physics of nanoscience. But it was good for me, so I'm very happy.

A.P.: I know that you use mathematics a lot. Moreover, I believe you like mathematics a lot but are more excited by physics than by mathematics. What does physics have that mathematics does not, in your opinion?

P.J.-H.: Mathematics is a tool for understanding physics or to understand many other disciplines. Physics has a connection with reality and with the world. And sciences are sciences because of the scientific method, so that one can experiment and corroborate whether the world is as it is, or not. In mathematics, you can come up with a perfectly consistent theory but one that has nothing to do with reality, and what I like about science is that it is contrastable with reality. I mean, mathematics is fine and good, and without mathematics we could do almost nothing. It is the language that science uses to describe reality. In fact, many scientific theories, before making a discovery that later proved falsifiable, were perfectly coherent within their mathematical "apparatus."

I believe that for physicists, mathematics is a very useful tool. There have been many advances in physics that have been purely due to mathematics. In particle physics there is a famous example in which a series of particles had been discovered and physicists saw that this corresponded to a certain type of mathematics. It turned out that mathematics itself required other particles to exist, so they were then sought and found. This was a triumph for mathematics. Mathematics is a great tool, but what I personally like about science, and about physics in particular, is the connection with experiment and reality.

A.P.: One wonderful day for you—correct me if I am wrong—you came across quantum mechanics. I have heard you say that at the beginning of the twentieth century, a revolution broke out that was not only scientific but also conceptual and philosophical, the revolution of quantum mechanics. Can you explain why?

P.J.-H.: That's quite difficult to explain, but it is conceptual and philosophical. The most difficult thing about quantum mechanics was, at the beginning of the twentieth century, to convince people that the world and the universe were as we now believe they are, and that they also had these rare properties. Quantum mechanics tells you, for instance, that one thing can be in two places at the same time, and that notion conceptually is very difficult to assimilate, to understand.

A.P.: And to prove.

P.J.-H.: Conceptually it's difficult to prove; mathematically it's easy. It's also been shown experimentally, so there is no doubt—the world is like that. What happens is that our living experience does not correspond with quantum mechanics because, normally, its most pronounced phenomena occur in the microscopic world, or on a scale of energy very different from that which our natural and sensory perception responds to. It was also a philosophical revolution because it made clear that uncertainty is something very present.

A.P.: The physicist Carlo Rovelli claims: "We are deeply unpredictable beings, like most macroscopic systems. There is no incompatibility between free will and microscopic determinism."[1]

Quantum mechanics has introduced that uncertainty to us, and that is quite difficult for the rational mind to accept, is it not?

P.J.-H.: The consequences of the uncertainty of quantum mechanics for the behavior of the more complex systems, like the human being, are still being investigated. It is not known whether human uncertainty has an ultimate origin that is provided by the uncertainties of quantum mechanics or if it comes from another kind of more complex behavior with which quantum mechanics is only indirectly related. However, it is certain that in quantum mechanics, for example, one cannot know with precision where an electron is and at the same time what speed it's traveling at. We are accustomed to seeing a thing and saying "it is" in this place, and it is moving at this speed and in that "direction." In quantum mechanics, that cannot be done.

And if you aren't observing it, you cannot know.... In quantum mechanics the problem is that in order to know, one has to observe, but the moment you observe, you modify the behavior of the matter. All this is very difficult to understand and even difficult to imagine. But you have calculated it many times, you have tested it so many times that, in the end, you get used to it, and it no longer surprises you so much.

A.P.: As quantum physicists, you are laying down a good challenge to philosophers. Should the philosophers be saying something in return?

P.J.-H.: Philosophers can help provide a perspective on many things related to quantum mechanics, but fundamentally, quantum mechanics is a mathematical theory about mathematics and about what the universe and nature are like.

A.P.: The science philosopher Javier Echeverria states in his book *Entre cavernas* (Among Caves) that what quantum mechanics physicists say fits in perfectly, as far as he is concerned.[2]

P.J.-H.: Oh, very good!

A.P.: Another science philosopher, Thomas Kuhn, explains in his book *The Structure of Scientific Revolutions* that science usually scorns its contradictions. He says, "Until the scientist learns to see nature differently in something new, it will not be a full and truly scientific fact." Is it essential to work in quantum physics to see nature in a different way?

P.J.-H.: You have to be more open-minded because nearly all quantum behaviors are almost nonintuitive. You cannot let your intuition guide you much because it usually leads to the wrong conclusions. So you mum,st have an open mind with regard to what is possible and what is not. I think that's one of the fundamental principles; you have to be open to the fact that the impossible may be possible.

A.P.: To considering the impossible possible.

P.J.-H.: To what appears impossible being possible.

A.P.: Let's talk about a special case. The well-known physicist and Nobel Prize winner Richard Feynman said that nobody understands quantum physics because no matter how much people think about it, there's no way of relating it to anything similar in the normal world. According to your MIT Web page, your interest in research focuses on "quantum electronic transport and optoelectronics in novel low-dimensional materials, such as graphene and topological insulators." When I read that description of the specific field that you dedicate yourself to, I couldn't help thinking that Feynman is right.

P.J.-H.: Feynman was referring to something entirely different, but it isn't a question of nobody understanding it because it is so complex, because the words are complex, or even because the mathematics is complex. It's the concepts that are difficult to imagine. But what I actually do can be explained to novices.

A.P.: Here you have a novice. ...

P.J.-H.: Here's a simple example. When you take a battery and cable and connect it to a light bulb, the electrons circulate through the cable and lose energy in the bulb, which lights up for that reason. When transport by electrons is quantum, the transport takes place in a very different way.

A.P.: But what's all this about the quantum transport of electrons?

P.J.-H.: It all rests on the way in which electrons circulate through a material. When it takes place in a quantum way, they may not dissipate energy. This is a phenomenon that occurs in quantum mechanics, where electrons may travel through the inside of a conductor without losing energy. These are very anti-intuitive things and almost incomprehensible. Moreover, we are not used to them.

A.P.:   And they do so in accordance with different equations, don't they?

P.J.-H.:   Yes, they do so through the Schrödinger equation, which includes complex numbers in the mathematics, something that again people find difficulty in accepting, because almost everything to which they are accustomed is real number mathematics and the mathematics of quantum mechanics works with complex numbers, and those have their own specific reality, to put it one way.

A.P.:   But Pablo, why does this happen in the quantum transport of electrons? Is there any explanation that we can understand as to how the electrons comply with one equation in copper or silicon and another in graphene? Why do they do that with one and not the other?

P.J.-H.:   First of all, we have to accept that quantum physics also explains the traditional behavior of electrons; that is, it explains both quantum and classical behavior. To explain the classical behavior, a classical theory, let's say, is sufficient; it's easy. But in the quantum transport of electrons, the characteristic effects of quantum mechanics are the most unusual, for example, the fact that electrons can circulate through the inside of a material without colliding with the atoms inside that material, without crashing into anything.

A.P.:   But the effects of quantum mechanics have to do with a huge number of things. The technologies that they induce, that are created by those effects, we use in our daily lives, almost everyone, at home, all the time, and we don't even notice, do we?

P.J.-H.:   Exactly. People think that quantum mechanics is something esoteric, but mobile phones, laptop computers, lasers, GPS, and chips are all possible thanks to the technology of quantum physics. People use quantum physics every day but don't notice it.

A.P.:   Pablo, I know that you have a serious and personal relationship with carbon and its nanometric forms. Tell me what you do with carbon and its more exotic incarnations.

P.J.-H.:   Carbon is a very special element. It's not only for making diamonds, which are beautiful, it is also the main chemical element responsible for life, along with hydrogen and oxygen. And carbon, in terms of solid-state physics, is a very special element because it is the component that makes graphene, with which, in turn, carbon nanotubes, graphite, and fullerenes can be made. I did my doctoral research on carbon nanotubes, and when I finished, graphene was discovered. So, as I was already in love with carbon nanotubes, and graphene is their second cousin, I moved over to investigating graphene.

A.P.:   Graphene, which occupies most of your research time, is a material that has gone in a very few years from being completely unknown to being perhaps the best known of the new materials, given that in 2010, the Nobel Prize in Physics was awarded to the scientists Andre Geim and Konstantin Novoselov for their revolutionary discoveries about this material. As to its definition, what I like most is what I heard you say: "Graphene is the finest material that has ever existed, exists and will exist."

Why is graphene so important and why has it captured so much interest, including your interest as a researcher?

P.J.-H.:   Well, let me say that since our first meeting, other materials have been discovered that are just as thin as graphene. So, although graphene is the thinnest and finest that has ever existed, exists and will exist. ...

A.P.:   Because it only has one layer of atoms, right?

P.J.-H.:   Yes, but it does have "cousins" that are also as thin and therefore share that distinction—for example, hexagonal boron nitride.

A.P.:   So, that definition "perfect" that you used of a perfect material—we can keep it or not?

P.J.-H.:   We can keep it, but now it is no longer the only material that exists with that definition; there are others also. Graphene attracted a lot of physicists' attention at first and then engineers' attention because the electrons in it behave as ultrarelativistic particles, and that is something extremely unusual.

A.P.:   Ultrarelativistic?

P.J.-H.:   Ultrarelativistic, yes. You see, you have to join quantum physics with Einstein's theory of relativity in order to explain the electrons in graphene. This, before, was unnecessary for other materials, such as silicon, copper, iron, aluminum. There was no need to join the theory of quantum mechanics with the theory of special relativity in order to make a theory of relativistic quantum mechanics. It was necessary for other things—for the physics of particles—but not to study normal materials.

Electrons in graphene behave as if they were particles that have no mass and travel at a similar speed to the speed of light. It is not the speed of light, it's less, but they behave in a similar way to light particles, the photons. They "travel" like neutrinos, or, let's say, like particles without mass, and that is very weird. It's very unusual from the mathematical perspective and from the perspective of the consequences that it has for quantum transport and for devices or technologies that can be made, for example, using graphene.

A.P.:   And has that been shown experimentally?

P.J.-H.:   Yes, it has already been shown.

A.P.:   Because physicists would never say that a thing is as it is, if there were no physical experiment to prove it.

P.J.-H.:   If it is not shown experimentally, of course not. But this has been shown in many experiments, and let's say that the incredible properties that graphene has, from the electronic and optical perspective, are the direct result of those ultrarelativistic properties. That is why engineers, who normally do not think about ultrarelativistic physics, now have to think about it. Anyway, graphene is the best conductor that exists and is so because the electrons do not crash into obstacles within it. It is partly for that reason. And they don't crash in it because ultrarelativistic particles do not collide with obstacles. It's very strange behavior that is not shared by non-ultrarelativistic particles.

A.P.:   Information technology has so far consisted of hardware and electronics based on silicon and its miniaturization follows the rate of development pointed out in Moore's law more than fifty years ago. It's clear that soon it will be saturated, according to the logic of physics and geometry. Intel already manufactures chips with technology of 22 nanometers and has announced that by the end of this decade its chips will be 8 nanometers. It seems impossible that chip electronics can continue to produce smaller and smaller things for much longer.

Do you think that graphene chips would be a feasible alternative to the present electronics? Or should we not be thinking about making the same things that we make today with it?

P.J.-H.:   I think that graphene *chips*, as people currently think of graphene, will not be the alternative to the present silicon chips. Silicon is very good for what it is used for at the moment. It's being perfected and investment is going to remain there. If they arrive at 8 nanometers, I don't think that graphene will replace silicon. I believe the main applications for graphene are still to be discovered. It is a very unusual material that engineers are still fighting over because its characteristics are so different from those of all the other materials they have had before. They still do not clearly know how best to use it. And although now there are some very small applications for very specific things, let's say a "niche" market, the new properties are so extraordinary that I am sure they will be used in another way; a technology will be invented that makes better use of the extraordinary properties of graphene.

A.P.: Could one 'niche' be carbon nanotube batteries?

P.J.-H.: Carbon nanotubes are a material basically consisting of rolled up graphene.

A.P.: Graphene is now being manufactured. Everyone is boasting of it, is that not true?

P.J.-H.: Graphene can be used, for example, in batteries because in batteries it is essential to have a large surface in relation to the volume and, as graphene only is only one atom thick, a graphene surface has the largest surface with respect to the possible volume that you can imagine. This is very important in batteries and nanomechanical composites. But all these composites only use relatively basic properties of graphene. Graphene has much better advanced properties than those, and it will take time to discover and invent the applications that can make use of its best properties.

A.P.: That would be a far-ranging revolution, much further than people are imagining now, wouldn't it?

P.J.-H.: I believe so. Besides, graphene is not the only material. Over the last four or five years, we have learned that there are many other materials that are similar to graphene in that they are two-dimensional, ultrafine, flexible, with optical and electronic properties very different from those of three-dimensional materials. Graphene is just the flagship of a new generation of materials, but I believe that engineers are going to make much more use of these materials because they are easier to understand.

A.P.: And those materials were there just waiting to be discovered, weren't they?

P.J.-H.: Yes, they were there just waiting to be discovered. Someone just had to be brave enough to try.

A.P.: To take the risk of investigating them.

P.J.-H.: Exactly.

A.P.: That is to say, when something appears, a new challenge in research, it is usually because there are researchers who are less conservative and more willing to take risks, right?

P.J.-H.: That's it.

A.P.: They run the risk of their colleagues saying, "Where are you going?" or "You're crazy." They stick their necks out.

P.J.-H.:   That's true. In research as in other fields, the greater the risk, the greater the danger, but there is also a greater possibility of reward.

A.P.:   In another conversation in this book, I was speaking to the physicist Ignacio Cirac about the letter that Albert Einstein wrote to Max Born in 1926 in which Einstein said, "Quantum mechanics is very serious, but something inside me says that this is not the way." Ignacio replied that if Einstein were alive today and had been accustomed, as you say, he would no longer be so surprised that things of nature are as they are, that they are also quantum.[3]

Do you agree with Ignacio Cirac on this?

P.J.-H.:   Basically, yes. As I said earlier, I think we are nowadays brought up to study physics from the quantum physics principles and calculate quantum properties using its equations. Moreover, the development of computers and present visualization techniques offer many new ways of seeing quantum mechanics, of seeing the effects they have, ways that did not exist in Einstein's time. And perhaps, if he were living today, he would be one of the physicists occupied with the fundamentals of quantum mechanics.

A.P.:   But if Einstein were to appear through this door and you said to him that relativistic quantum physics now exists, he would probably laugh, wouldn't he?

P.J.-H.:   Relativistic quantum physics was already there in Einstein's time. It was discovered by the physicist Paul Dirac. But then it was only thought of as belonging to the field of high-energy particles, not to describing the electrons in a material that we all use, such as pencil lead.

A.P.:   But he would laugh, don't you think?

P.J.-H.:   Yes, I do. Einstein would have been delighted with the discovery of graphene. It would have appealed to him.

A.P.:   Returning to personal matters, Pablo, I think that for you physics—something so difficult to learn and master for many pupils and youngsters—has to do with emotion, passion. Normally a scientist contemplates things almost strictly from a rational plane. In your experience, do you think that one has to combine both halves of the brain, the more rational and the more emotional, in order to make a good nanophysicist? Is there a mystery? Is there emotion in the physics of condensed matter? How do you manage to devote that endless passion to something so pragmatic?

P.J.-H.:   Of course, there is mystery in the physics of condensed matter. That's where the fun lies, in not knowing what you are going to find when you undertake an experiment. There are many surprises, and clearly there is emotion in that. Because of the mathematics of quantum mechanics that we employ, we are obliged to make simplifications or approximations; one cannot completely study mathematically such a complicated system as a piece of silicon or graphene. So mathematics alone does not easily lead you to predict how an object will behave from the electronic, optical, or atomic perspective. We are in a discipline where there are continual surprises, where we discover things that we don't expect, and where mathematics only helps you to understand a posteriori. Surprises? Yes, every day.

A.P.:   But finally the rational half wins, doesn't it?

P.J.-H.:   Yes, in the final analysis. But don't forget that in the process of discovery, in research, intuition also plays a fundamental role. And not only intuition but also such simple things as the search for aesthetics or beauty. Normally we physicists like simple things.

A.P.:   Simple and elegant....

P.J.-H.:   Yes, simple and elegant. The search for simplicity and beauty is clearly a two-edged sword. Sometimes nature is not so simple and elegant and we have to accept that, but it is also true that many times the search for that simplicity or elegance, or even the beauty of behavior, leads us to discover something that really is so.

A.P.:   It can't always be minimal in quantum physics as in the minimalist painters, is that not right?

P.J.-H.:   Minimalists, no. There are many things that are complicated, and you try to understand them as best as you can. But we physicists tend to try to separate a little, to simplify things so that we know how to capture the essence of things. We remove noise and distractions from the essence. So the essence is usually something that can be explained with mathematics, that can be understood in a quantitative way, and later it turns out—

A.P.:   That it may be complicated.

P.J.-H.:   Yes, later it may turn out to be complicated. You can always add the complexities. Physicists always try to simplify things as much as possible in order to understand the essence, and then we add the complexities.

A.P.:   Many thanks, Pablo, for talking to us.

P.J.-H.:   My pleasure.

## Notes

1. Carlo Rovelli, "Free Will, Determinism, Quantum Theory and Statistical Fluctuations: A Physicist's Take," *Edge*, July 8, 2013, http://edge.org/conversation/free-will-determinism-quantum-theory-and-statistical-fluctuations-a-physicists-take.

2. Javier Echeverría had already used the terms *entre cavernas* (among caves), *nanocavernas* (nanocaves), and *nanocosmos* (nanocosmos) to refer to a new dimension of the universe, in which the laws of quantum mechanics rule. See Javier Echeverría, *Entre cavernas: De Platón al cerebro pasando por Internet* (*Among Caves: From Plato to the Brain via the Internet*) (Madrid: Triacastela, 2013).

3. Ignacio Cirac intervenes in dialogue 1.

# 7 The Laws of Thermodynamics Tell You What Is and What Is Not Possible

Avelino Corma and Adolfo Plasencia

Avelino Corma. Photograph courtesy of ITQ.

*Discovery is sometimes difficult because we tend to be biased. In trying to discover how nature works, we are not sufficiently open-minded to think freely.*

*The genome is an information code, obviously, but one that was built out of needs that arose through chemical reactions, which in turn corequired a series of processes necessary for life to unfold the way it is at the moment.*
—*Avelino Corma*

Avelino Corma is Professor at the Instituto de Tecnología Química, the Polytechnic University of Valencia, Spain. He earned his doctorate in chemistry in Madrid. His research focuses on heterogeneous catalysis, as basic research and for industrial application. He is an internationally recognized expert in solid acid and bifunctional catalysts for oil refining, petrochemistry, and chemical processes. He has published more than nine hundred research papers and is cited as inventor on more than one hundred patents.

Among the awards he has received are the Ciapetta and Houdry Awards of the North American Catalysis Society, the Gabor A. Somorjai Award for Creative Research in Catalysis from the American Chemical Society, the Royal Society of Chemistry Centenary Prize, the Solvay Pierre-Gilles de Gennes Prize for Science and Industry, the M. Boudart Award on Catalysis from the North American and European Catalysis Societies; the Gold Medal for Chemistry Research Career (2001–2010), the Prince of Asturias Award for Technical & Scientific Research (Spain, 2014), La Grande Médaille de l'Académie des Sciences de France (France, 2011), and the Spiers Memorial Award from the Royal Society of Chemistry (UK, 2016).

Adolfo Plasencia:   Avelino, thanks for having me. You have devoted your whole life to chemistry. Do you think it appropriate to refer to chemistry as being the creative, central science?

Avelino Corma:   Yes. For me, it is something I am passionate about. In a broad sense, chemistry is the discipline that allows me to delve into the knowledge of and the solution to problems in our society.

A.P.:   Albert Einstein said, "Most of the fundamental ideas of science are essentially simple." Do you think there are still ideas to discover that are at once fundamental and simple, in a chemistry as complex as the one you research in the twenty-first century?

A.C.:   These ideas often seem simple once we already know them, or once they are explained to us; in fact, discovery is sometimes difficult because we tend to be biased. In trying to discover how nature works, we are not sufficiently open-minded to think freely and to tackle new problems without preconceptions. If we were disciplined to do so, we would not take the wrong direction and we would reach newer, more original knowledge.

A.P.:   Avelino, in science, the past, time gone by, was *not* better. Do you agree with me?

A.C.:   Of course it was not. It was definitely more heroic in the past but not necessarily better. Whatever the case, we are the result of all those researchers and professors, and even anonymous heroes in science who struggled before us so that we could find what we now use as a starting point. And they must be given the credit for that.

A.P.:   You frequently talk about balance. There is a classic balance between basic or fundamental science—science seeking the truth in the long term, with no rush—and applied science, which leads directly to technology, to a specific implementation. How can this be translated to today's chemistry?

A.C.:   In our work at the Institute of Chemical Technology laboratory, we don't like making a sharp distinction between basic and applied research. To succeed, you need to make progress in both at the same time. Basic research is essential to increasing our knowledge. It is only through this knowledge that we can access ways and modes of application. But if we focus only on the application we will not be able to discover something that is truly new. Thinking about new hypotheses and testing them experimentally is what leads to creation. And considering these hypotheses and proving them is what satisfaction is all about.

A.P.:   What do you do at your laboratory?

A.C.:   We work in chemistry and, within chemistry, in catalysis. For those who are not familiar with it, catalysis is generated by catalysts. A catalyst is just a material capable of accelerating a reaction and increasing the reaction rate, directing it toward a result, that is, the specific compounds we seek. It is clear that the catalytic process develops at a molecular level, so if we want the process to fulfill its goal, we must work at the molecular level, which means we have to have a molecular design for our catalysts. The ultimate goal would be to attain "molecular recognition." That is to say, our catalyst should "recognize" the molecule it is to react with in such a way that it could activate only the relevant chemical bonds to transform it into the desired final substance.

A.P.:   Chemistry is also related to a lot more things than people think—to the brain, for instance. In neuroscience, the mind is said to have been generated by an emergence caused by a whole system, by the addition of chemical and electrical reactions.

Is that right? Is chemistry involved in the way our brain works?

A.C.:   Of course. Chemistry is the basis of life. And any process taking place in life has a very high chemical component. If we think about the way you and I are working at this very moment, there is a large system of a chain or cascade of chemical reactions taking place. My talking to you occurs because I activate a number of muscles; this in turn triggers a series of neurons, and all this happens through a series of chemical reactions and electrical impulses. Therefore, chemical reactions make up the very basis of life and make life as we know it today possible.

A.P.:   I'd like to continue on this theme. One of the basic revolutions in life sciences is the genome.

In your view, is the genome more physical or more chemical, more of an "information code"?

A.C.:   I would say it is probably all of these. It is an information code, obviously, but one that was built out of needs that arose through chemical reactions, which in turn corequired a series of processes necessary for life to unfold. If our individual genome is the way it is today, it is because certain chemical reactions took place over a long time, which led to the evolutionary construction of these molecules. Now, these molecules induce other chemical reactions, which allow life to develop at this moment in time.

A.P.:   Going back to your discipline, what approach or pathway is best suited for innovation in today's complex and very advanced chemistry?

A.C.:   Perhaps one possibility is to be positioned at the interface between disciplines, to be cross-fertilized by other fields such as material science, medicine, and physics. I believe that it is precisely these interfaces that provide new opportunities.

A.P.:   You are a cofounder of the Instituto de Tecnología Química (ITQ, Institute of Chemical Technology), an organization that you have headed and promoted for many years. Today it is a world-reference research center under the umbrella of both CSIC and the Polytechnic University of Valencia. It employs PhD students and researchers from the university where it is based, as well as from other national and international universities. Your main research area at ITQ is catalysis, both in basic and applied chemistry but also in chemical technology. And you also have strong links with companies on several continents. What is the best research approach for an organization like yours that manages to combine basic and applied chemistry and chemical technology, both today and in the near future?

A.C.:   That's correct. I indeed cofounded the Institute and I was its director for twenty-two years. The current director is Professor Fernando Rey García. As you rightly said, one of the most important activities at ITQ is catalysis, from both a fundamental and applied point of view. This said, the Institute is also very powerful in photochemistry and photocatalysis. In my opinion, the research approach that best combines basic and applied chemistry and chemical technology is one that starts by considering a problem from the point of view of providing new knowledge about nature itself. It seeks to answer fundamental questions about how a particular aspect of nature works. If we are going to deal with a problem of this kind, why not be ambitious and select a problem of fundamental importance? If we could choose, I would pick one that could also give us a solution to a very significant technological problem whose solution is needed or demanded by society. This is how basic and applied science—plus technology, in our case—are combined.

A.P.:   The first decade of the twenty-first century was marked by the powerful emergence onto the global scene of such countries as China, India, and Brazil, causing the demand for energy, fuel, and raw materials to grow dramatically.

How is your chemical science facing this huge and growing demand?

A.C.:   Chemical science, and chemistry in general, progress by trying to make the most of existing finite resources. To produce energy we continue to use fossil fuels. My view is that we will continue using them unless there is an absolutely sensational discovery in the coming twenty years. Therefore, it is our duty to use those energy producers in the most efficient way possible. To do this, we will have to develop fresh methods of extraction, refinery, and transformation. These should allow us, in the field of gas or oil, for example, to derive the maximum amount of fuel—in other words, energy—from the same amount of oil or hydrocarbon. We will therefore need catalysts that can improve the quality of the final products obtained while yielding more from the crude oil at our disposal. Or, in the case of coal, we will need to purify it so that emissions into the air and the environment include the smallest amount of emissions that are incompatible with the environment. That is, on the one hand, we must take advantage of what we have in a more rational way, and on the other, we must develop new energy sources in such a way that they also are economical to use.

In addition, and in relation to new sources of energy, we must make the most of solar power. Sunlight can be used through energy capture in photovoltaic or solar cells. It can also be used to dissociate water into hydrogen and oxygen, with the hydrogen then used as an energy carrier. So we need materials and catalysts. In this area too, we are trying to contribute to general knowledge and, if possible, to technology as well.

We need to get the message through to society that at the end of the day, we are all responsible for the use of energy, and it is an obligation for society to become aware of energy use and to realize that no improvement comes for free. Everything has a price. And we must start considering that price, together with its implications for our comfort. First, we should be able to save energy, to minimize its use. Are we ready to make the most logical decisions? The most logical thing is to start by saving. That is how we should start, in my opinion.

Then we must exploit existing resources in the best possible way and find new sources of energy. However, if we want to be realistic (and in the absence of some exceptional discovery), we have to think that our energy basket is and will be a combination of different power sources, including fossil hydrocarbons, nuclear power, hydraulic, wind, solar energy, and tidal or geothermal energy where possible. But only the first few tend to be within reach for many countries. The sum of them all will provide the total amount of energy that we can use. Hopefully, over time this energy basket will shift in the direction of renewables. That is what we are working on. But at the same time, we must guide, save, and reduce energy consumption.

A.P.: Regarding hydrogen as an energy carrier, I have heard you say that ultimately an acceptable, positive balance must be reached because the huge expectations about hydrogen a few years ago have not been fulfilled.

A.C.: Adolfo, the first law of thermodynamics says that to obtain energy, we first need to have energy. The energy balance must be fulfilled. It is always a prerequisite. For us to produce hydrogen, we first need to use energy. If we get it from water, we need to break down the water molecule—and that takes energy. If we want to break it down using electricity generated by means of fossil hydrocarbons, then it's back to square one. If we want to generate hydrogen from methane, we are still using fossil hydrocarbons. Essentially, what renewable energy source can produce this water dissociation and obtain hydrogen in a renewable way? Wind, solar, geothermal? Basically, we must be realistic; in the current climate it would have to be wind or solar. So, other than these two renewable sources, everything requires fossil hydrocarbons or nuclear power. That's our choice.

A.P.: And a price has to be paid for it.

A.C.: Of course, there is a price to pay. The equation is very simple: decreasing the energy we use by 10 or 15 percent, for example. A 15 percent energy reduction in developed countries with renewable energy, like Spain, would render nuclear power plants unnecessary. But we don't seem to be willing to reduce our energy consumption by such a figure.

A.P.:  Let's go back to science, to your passion: chemistry. For science, the study of immensely small things seems to be one of the roads to the future. Condensed matter physicists, such as Pablo Jarillo-Herrero, say that within the scale of a few nanometers, some rules of physics change completely with respect to those that apply to the world as seen by the naked eye.[1]

Does this apply to chemistry too? What is the difference between chemistry and, for example, physics in the range between 0 and 200 nanometers, an area that you are already investigating?

A.C.:  I believe that is precisely the meeting point of chemistry and physics. They meet in the realm of nanomaterials and smaller particles. They meet from a theoretical point of view: quantum mechanics has been developed by both physicists and chemists, and there is a discipline called chemical physics in which they converge in some ways. The same applies to the preparation of these nanomaterials. In such preparation, chemists use a certain number of techniques and physicists use others. More and more, physicists use techniques from chemistry, and chemists use those of physicists. Differences blur and disciplines come together. As I said, it is at that scale that chemistry and physics meet in such a way that the distinction between them no longer applies.

A.P.:  In one of your recent lectures, a researcher asked you a question about the possibilities of doing certain things with specific molecules. You told him, "Give it up. Matter is the way it is. It will do whatever it has to do and nothing else. You need to adapt to matter. There is no other way. You won't get it to do what it cannot do." It was something like that, I think.

A.C.:  Yes. There are a few basic principles that must be observed. I always say to my students, before studying your reaction from a kinetic standpoint, you must study the thermodynamics of the process. Thermodynamics tells us what is possible, what is not possible, and under what conditions it is possible. This needs to be taken into account before tackling any problem. Of course, you can always intervene, but if you need to use an enormous amount of energy to intervene in a process, and the performance you get is very low—thermodynamics is king—then the process may not be worth it.

A.P.:  When you imagine the scenario of "nanomatter" and think about physics, generally you think in "static mode." But if you regard it as chemistry, you're obliged to think about it as a dynamic scenario, a "film" instead of a "photograph"—something in motion, where things, reactions, happen. For example, I understand that you are coating gold particles by manipulating nanospheres in which the volume of a molecule can be synthesized, and covering them with other smaller molecules in a sort of nanoarchitecture.

Can you already handle matter in sizes that are almost unimaginable? How do you move things at that scale?

A.C.:   In the end, we have to go down to these levels in scale to move away from conventional, known scenarios, to find new possibilities for matter and molecular interactions. If we think about the properties of iron metal, or about an iron, cobalt, platinum, or gold oxide, its properties as metal are already known and they are very well defined. But a most interesting thing happens when one begins to reduce the size of these particles down to dimensions that are close to individual atoms. At that point, iron no longer has the features of a standard metal. Nor has it those of isolated molecules and even isolated atoms. Rather, it has a new behavior which corresponds to what we call nanoparticles and metal clusters.

A.P.:   But what you are describing is a quantum scenario, isn't it? In the quantum scenario, chemical behavior differs from that in our reality, doesn't it?

A.C.:   Quantum behavior is also part of the reality we live in. But in this case, that scale allows us to prepare reactive species and certain materials with unusual properties. We wouldn't have these if we worked with conventional materials, such as those derived from the metal elements that I mentioned, or bulk structures, or insulator atoms. This presents an opportunity to move forward. For this we need knowledge, techniques, and technology, to produce these nanoparticles and clusters in a stable way, because particles don't "want" to stay in this very small dimension if they have the possibility of joining others to form something bigger. It is a bit like what happens to us in society.

A.P.:   Then is there a "sociology of nanoparticles"?

A.C.:   Of course there is, because the minimum energy state these small particles look for is the one they find when they join together to form larger particles that are more stable, energywise. To study those unusual properties, we have to isolate them. That is what we are trying to do.

A.P.:   And how can you do it materially?

A.C.:   We are creating nanoreactors with zeolites to confine the nanoparticles and cause them to react selectively. Here we have to think about another type of nonconventional interaction. Take, for example weak interactions, which we call van der Waals interactions. They require little energy and they are extremely important because they stabilize reaction intermediates. Besides, since nanoparticles are confined to the pores inside nanoreactors, they are subject to very strong electric fields. As a result, they are "pre-activated," which would not happen in the medium if they were not confined. So interactions occur at the molecular scale, and with relatively long lifetimes, much longer than when normal particles are in much bigger places.

A.P.:   Described this way, these materials seem abstract, but they have an enormous impact in the real world, don't they?

A.C.:   Yes, they do. Without such porous nanomaterials, like zeolites, industries such as fine chemistry or oil refining would not exist as we know them today. Approximately 35 or 40 percent of the petrol and diesel used in the world are obtained through these nanomaterials. What is more, this kind of fuel has a better quality and octane rating—and better environmental properties too. It has less benzene and fewer, more inert olefins. We have also shifted from getting a 30 percent useful product out of each barrel of oil to 90 percent nowadays. Consequently, we are being at least three times more efficient in the use of natural energy resources. This is key from the point of view of preserving our raw materials and energy, which are finite.

A.P.:   Let's now move from the nanoscopic scale to the biggest things in our planetary scale. One of the concerns in today's world is understanding the planet as a finite whole, with limits, as something to protect. In this respect, green chemistry is providing solutions to many people's needs and at the same time respecting the planet's ecosystems. Given this, do you think people appreciate it enough? In everyday life, they enjoy thousands of solutions offered by chemistry but also criticize it?

A.C.:   Quite frequently, people criticize what they do not know, or their criticism is not sufficiently based on knowledge. In the case of chemistry, I think some things, such as pesticides, have had a more negative impact. It is generally believed that if these pesticides are released into the air, they will come into contact with people or with foodstuffs. People get worried about gases from car exhaust pipes and the like. However, when they criticize, they forget that it was chemistry and catalysis in particular that led to the discovery and allowed the synthesis of ammonia, without which agriculture would not have developed as it did, and as a result, the planet's population would not had grown as it did. Nor would we have had the yield per hectare that we can achieve nowadays, enabling us to feed so many more people. Chemistry made this possible. And whenever we have crops, we have pests.

Although we tend to say that any previous era is superior, that is not the case. For example, lead hydrogen arsenate was used for pest control for many years. It would be terrible to use it today. Chemistry has produced new molecules that are more and more specific to the insects that you want to control and less harmful to superior species. Besides this, chemistry has been capable of developing alternative pest control methods, such as pheromones, in which chemical ecology research, for example, has made interesting progress: chemistry can synthesize the sexual pheromones of insects and selectively achieve the elimination of these insects without harming other species, or human beings, of course.

We usually talk about gas emissions from either mobile sources or stationary sources. But we seem to forget that we all want to continue using our car. And that it was chemistry and catalysis that improved the quality of fuels and eliminated emissions from fixed physical sources in industries. Thanks to chemistry and catalysis, they have been minimized by three orders of magnitude and almost reach zero levels today. Therefore, when we talk about chemistry, first we must understand that today's standard of living or the fact that many diseases can be now cured is the result of chemistry. And in those cases where our activity generates products that can also be harmful to us, it is that same chemistry that is trying, successfully in many cases, to eliminate such products by transforming them into substances that are not harmful to the environment.

A.P.:   Cutting-edge chemistry also guarantees excitement, doesn't it?

A.C.:   The only thing I can tell you is that one of the biggest satisfactions is to make a hypothesis about how to solve a fundamental scientific problem and finally reach a solution that confirms the hypothesis. That is immensely satisfying. I don't know what I could compare it to—something big, for sure, particularly if you have a passion for creation and discovery.

I would say that science and technology are the answer to our current problems and those we will soon have to face and which often make us pessimistic about our future. I believe these problems have a solution from a basis of science and technology. Chemistry is a discipline that is fundamental specifically to the development of science and technology, and therefore also to the sustainable development of our world.

A.P.:　Thank you very much. It has been a pleasure. I hope you continue being so enthusiastic about chemistry.

A.C.:　I hope so too!

**Note**

1. Pablo Jarillo-Herrero intervenes in dialogue 6.

# 8   Wisdom Hewn in Ancient Stones

John Ochsendorf and Adolfo Plasencia

John Ochsendorf. Photograph by Adolfo Plasencia.

*When I enter an ancient building, such as the Agrippa's Pantheon in Rome, and I observe it from inside, it strikes me that the person who made this building was much cleverer than I and knew more than I do.*

*When we speak of the future of architecture, I believe it is essential to look back through history for inspiration.*
*—John Ochsendorf*

John Ochsendorf is Class of 1942 Professor of Architecture and Professor of Civil and Environmental Engineering, MIT. He received his master's degree from Princeton University and a doctorate from Cambridge University. He studied masonry vaulting in Spain in 1999–2000 on a Fulbright Scholarship and again in 2007–2008. After becoming the first engineer to win the Rome Prize from the American Academy in Rome, he spent a year studying vaulting in Italy. In 2008 he was named a MacArthur Fellow.

Ochsendorf conducts research on the mechanics and behavior of historical structures, especially masonry structures. His group is researching the dynamics of masonry buildings and the design of more sustainable infrastructure, drawing on examples ranging from the vaulted roofing of Gothic cathedrals to the hanging rope bridges of the Inca Empire. He is a world authority on structures and vaulting, combining his different knowledge in three fields of building construction: structural engineering, the historical evolution of architecture, and archaeology.

Ochsendorf founded the Guastavino Project at MIT, an initiative dedicated to documenting and preserving the tile vaulted works of the Guastavino Company, dating from the late nineteenth century and found in numerous buildings in the United States and across the world.[1] He has also taken part in sustainable architecture projects, such as the collaboration with Michael Ramage to construct the unique Pines Calyx building, a pod-shaped (calyx) event center in Dover, UK, that includes two Guastavino-style vaults. In another project he collaborated on building the Mapungubwe National Park Interpretive Center in South Africa, which received an award at Barcelona's 2008 World Architecture Festival.

Adolfo Plasencia:   John, thanks for receiving me here.

John Ochsendorf:   My pleasure! Welcome to MIT again.

A.P.:   John, you are a structural engineer who has also studied archaeology. You could, however, quite easily be an architect.

J.O.:   Although I teach in the MIT School of Architecture and Planning, I haven't actually studied architecture. I do, however, love the discipline and work a lot with architects.

A.P.:   In 2006, I published an essay on the emerging concepts of the MITUPV Exchange on the MIT OpenCourseWare site.[2] One of the concepts I identified there crossed my mind again when I was in Mallorca some years later, at the unforgettable "Stonemasonry in Context: The Artifex Workshop," with Miguel Ramis and Yung Ho Chang.[3] We were discussing how many "wisdoms" from the past have been "unlearned" today. In my essay, the concept discussed was unlearning. Basically the essay was about technology, but at the workshop I was reminded of it again as it occurred to me that we have also forgotten marvelous things in architecture and construction that would be well worth recovering. What do you think?

J.O.:   Yes, absolutely. For example, if we think about an arch and a dome, they are traditional forms with four thousand years of development. With the Industrial Revolution and its new materials of steel and reinforced concrete, we have in effect abandoned a four-thousand-year-old technology, and this has been a tremendous shock.

Today, a person with a PhD in engineering from a prestigious university has less idea of how an arch functions than a mason of today, or of a hundred or a thousand years ago. To take another example, there was a colonial era bridge in Mexico that had been there for three centuries and had survived earthquakes, modern traffic, and wars without any problem. Then, twenty to thirty years ago, an engineer with a master's degree in prestressed concrete bridges from MIT arrived, made some calculations, and said that this bridge, after three centuries, was no longer viable. He removed the bridge and made a new one of prestressed concrete because it was easier to calculate. My Mexican friends had the last laugh later on, saying that the original bridge had withstood wars, earthquakes, and modern traffic, but not the one built by a person with a master's degree from MIT.

If we overlook history, or how an arch really works, there is a danger in the economic, cultural, and training senses. We as engineers do not study history. Yet one cannot imagine a writer or a composer who does not know his or her field inside out. It's impossible. But in engineering, we know hardly anything about the history of our applied discipline, and it's an impressive history. We don't know the names or even the works of the great engineers of the past. I personally believe that if engineers and architects knew history, we could learn and understand many things, including both how to look after ancient buildings and how to construct new buildings.

A.P.:   John, you have spent a lot of time studying a building that is one of your favorites. The Pantheon in Rome was commissioned by Marcus Agrippa during the reign of Augustus and rebuilt by the emperor Hadrian. Could the Pantheon provide a lesson in the history that you are talking about? The building and its dome have been standing for two thousand years and continue to be used. Why do people who respect beauty, the past, or construction not learn a lesson from it? How is it possible that, as these examples exist, students at schools of architecture "unlearn" something like this?

J.O.:   If you study the history of a building like the Pantheon, its geometry and its symbolic elements, it turns out to be one of the most impressive buildings in the world. It has cracks within the dome, cracks of more or less

30 centimeters, quite large, that have been there more than two thousand years. They are not visible, but they are there. Even so, it remains perfectly stable. We do not, however, fully understand why it is so stable, nor do we fully know how safe it is. So one of the things we do here at MIT is to carry out tests and create new programs to better understand the structural safety of such ancient and impressive buildings. For example, if you consider the tools that we engineers have to make steel buildings, they don't tell us anything; they don't have the capacity to tell us enough about a building as old as the Pantheon. So what we have to do is rethink all of this and create new tools with software that takes the historical behavior of buildings into account.

A.P.:   Is it true there are cathedrals still standing that defy the programs of structural calculus in modern computers?

J.O.:   Yes, absolutely. We have no way of knowing at the moment just how stable they are. The Beauvais Cathedral in France, for example, or that of Palma de Mallorca are impressive buildings; they are mountains of stone laid in a stable form, but we do not really understand why they are so stable. The digital tools that we have don't work on them. We are, however, working a lot with computer programmers, writing new software so as to better understand how those buildings function. These are very difficult problems because the geometry has changed.

For example, the top of the Pantheon has dropped almost a meter over the last two thousand years, and that has deformed the dome a lot. We have been working with tools such as 3D Scan Laser to discover its precise geometry. It's a field of work that we are researching. Another field is the history of construction. I ask myself, How was it possible to build the Pantheon two thousand years ago? How was it possible to build a cathedral eight hundred years ago? The history of construction is wide open, and we think there are many ideas that can be taken from the past to create new things today.

A.P.:   That reminds me of the time I was on Easter Island with an archaeologist. He showed me three very important things. The first was a stone wall made from blocks that weighed several tons, with almost perfect joins. The second was that a civilization, completely cut off from the world, had developed that called itself "the umbilical cord of the world" (there is a place, Ahu Te Pito Kura, on the north coast of the island, with that name) because the people believed themselves to be the only inhabitants of the world. And the third thing he drew my attention to is an unanswered question: How did a lithic civilization that knew nothing about metal extract and hew sculptures of volcanic rock, some colossal, weighing almost eighty tons, and, from a quarry in a volcanic crater, move them fifteen miles over hilly land to their positions, and then stand them up? All the tests that have been undertaken to move similar grand structures have ended in failure. Perhaps it is because we have forgotten how these things were done twelve centuries ago.

Why do you think it is that that all this knowledge has been forgotten and that, as you said, a mason of ten centuries ago knew more than an engineer of today?

Why is accumulated knowledge not considered riches? Is it a question of fashion, of tools, or is it a cultural problem?

J.O.:  It's probably a cultural problem. I believe, in engineering, we have a misguided ideology of progress. According to the ideology of progress, for example, a brick arch belongs to the past and a modern building must be of titanium or steel. So, on the one hand, our thought is focused only on those new materials, and on the other hand, we are on top, in the most recent layer of history, and we think that we, the engineers of today, with all our computers and tools, are superior to people of two thousand years ago, who knew a lot but were basically ignorant. However, when I enter an ancient building, such as the Agrippa's Pantheon in Rome, and I observe it from inside, it strikes me that the person who made this building was much cleverer than I and knew more than I do.

A.P.:  He was cleverer and possessed the wisdom we have forgotten.

J.O.:  Exactly. I spent my PhD years at the University of Cambridge, England, studying domes and things like that. I began with the mathematicians' problems of the nineteenth century in France, studying buttresses and similar structures, which presented real problems at that time. But in the twentieth century, with steel and concrete, we have abandoned that field. When we refer to World Heritage sites, we have to understand how these ancient buildings functioned. Furthermore, if we are talking about sustainability, about passive technology, for example, in a climate where the weather is almost always good, there is a body of accumulated knowledge in the vernacular or traditional architecture about how to do these things, and that knowledge base has been growing for centuries and centuries.

However, when we enter our glass boxes in Houston, London, or Valencia, we see that they have nothing to do with place, with the site where they have been built. Knowledge of the site makes sense. When we speak of the future of architecture, I believe it is essential to look back through history for inspiration.

A.P.: Yung Ho Chang, an MIT professor, said at the Artifex workshop, "Stonemasonry in Context," in Mallorca, that we have to recover the values of the craftsman because he considers architecture to be a craft.[4] Do we have to put an end to the egos of architecture? Should we try to recover the humility that the stonemasons and the bricklayers had, those who knew both the materials and their trade?

J.O.: It's difficult to know why certain things happen with the training of architects today. In the United States, the architect is a god. The architects whose work I most like are those who work with people, who use their hands, who have technical ability in other fields, and who work as a team. Of course, a genius may appear at any given moment who can provide the total idea of a building, but I believe that making a building is not the job of a single person. One of the problems that we currently face is how to make buildings with lower carbon dioxide emissions. These difficult questions are beyond the ability of a single field to fully resolve. The only possibility for making buildings that people need nowadays is for there to be an ongoing dialogue among engineers, architects, biologists, information technology experts, historians, archaeologists. ... It's clear to me that we have a tremendous problem to overcome.

A.P.: Let's talk about a building that you and your team built with three conditions: it had to be built to last five hundred years, have zero energy consumption, and use construction techniques that took advantage of local and, if possible, craft materials. Tell me about this building that you were involved in and that, moreover, holds the record for sustainability.

J.O.: It's a building in England that we made several years ago. It's called the Pines Calyx building. You'll enjoy this story because you're from Valencia, and it has something to do with Valencia. We used the timbrel vault, which is a type of vault whose origins go back to fourteenth-century Valencia (the oldest documented example of this type of dome is found there, in the Capilla de los Jofre, in the Convento de Santo Domingo of 1382).[5] Later it was used in New Granada, Colombia, in the eighteenth century, and in the United States in the nineteenth and twentieth centuries. We faced a difficult problem: how to make a building with the materials available there? In the end, we made earthen walls, which are similar to concrete but with a much lower energy consumption, and we used hand-made clay bricks that were left over from a mine in England. This was an element true to the type of economic development of that same place. We contracted some Spanish masons from Extremadura to help us with the vaults and built two not very large vaults with a span of 13 meters. By doing so, we were able to considerably reduce their "embodied energy," that is, their level of carbon dioxide emissions emitted during construction. We reduced those emissions by 80 percent compared with the lowest levels of carbon dioxide emissions in building construction using local materials at that time in England.

It's obvious that if we don't have to go to China, Mongolia, Japan, or other distant sources to find materials, that has its advantages; it is much cheaper than other types of construction. Moreover, the money goes to the local people who build the building. This is sensible. In some way, this strategy might appear to be antiglobalization, but I don't think it is. It is trying to look for solutions that make sense in the area where one is going to build.

A.P.:   Who worked with you on this project?

J.O.:   An English architect and several English engineers, and a group of MIT students. We went there for two weeks, and a person from New Zealand did the vaulting. It was an experiment, but a very entertaining project that eventually won many awards as a sustainable, low-energy building. And thanks to that building of several years ago, we now have many more projects.

A.P.:   As you said, it may be seen as an antiglobalization building. Perhaps the great steel manufacturing and building systems industry have more power than the university schools of architecture and engineering that believe in sustainability? Do you think that the great global industries, for economic reasons, have imposed their power?

J.O.:   That's an interesting question. I hadn't given that much thought, but yes, I reckon so. For example, for people like us, as structural engineers, civil engineers, and also as architects, there would seem to be only two materials in the world, steel and reinforced concrete. Obviously, that is not the case. If someone invented wood tomorrow at MIT, it would be seen as the most amazing element in the history of the world. It's something that comes from trees, which consume carbon dioxide, that has all the desired properties of rigidity—wood is an incredible material, and so too is brick. They are simple materials that have been known for a long time and are materials on a human scale with huge possibilities. However, today we do not talk about those materials. That is partly due to our "ideology of progress," which doesn't allow us to think much about wood.

A.P.:   Let's turn now to a Valencia connection you have, the father-and-son duo with the surname Guastavino. They also had and expressed an "accumulated knowledge" that, unfortunately, we Valencians have almost forgotten. Tell us about your love for the work of these two Valencians.

J.O.:   The family Guastavino came from Valencia. The father, Rafael Guastavino Moreno, was born in 1842. He built several buildings in Catalonia, but at the age of forty-one he emigrated to the United States with his son, Rafael Guastavino Espósito. By the 1950s he and subsequently his son had built more than a thousand buildings in forty-one states, including more than two hundred in Manhattan and the most important buildings in the history of the United States. For example, Carnegie Hall has Guastavino vaulting. They built vaults just like the traditional ones of Valencia (here I have one of their bricks, which in the Valencia language is called a *rajola*; it has the brand name of the company that made it imprinted in relief with a picture of its two-layered brick vaults). Valencia provides the first documented example of this type of vault, built in 1382. And even as they worked with this traditional technology they continually sought to innovate, to make greater spans, new forms, using new materials; introducing, for example, Portland cement. They held twenty-six patents in the United States as they sought to make changes for the better, always innovative changes, such as in acoustics.

One of their important buildings is on Ellis Island, where all immigrants used to arrive in the United States, as my grandmother did, who came from Italy eighty years ago. Five capitol buildings in different states were built by Guastavino, and numerous universities, such as Harvard, Princeton, Yale, MIT, and the University of Chicago, have buildings by Guastavino. The main library at Harvard also has Guastavino vaulting, the original idea for which comes from Valencia. The Boston Public Library, one of the oldest and most important libraries in the United States, was vaulted by Rafael Guastavino. However, their history is not widely known here either: their buildings include churches, banks, and thousands of other such important constructions in the history of United States, but as architects, engineers, and builders, they have unfortunately been forgotten.

A.P.:   Moreover, these buildings were made at a time when it was difficult to move around the territory, in an age without planes, with very slow trains. The United States is very big, and we never truly appreciate the difficulties that people had to overcome.

J.O.:   Exactly. And they, one hundred years ago, had twenty offices in ten U.S. cities! A hundred years ago, at some point, they were making one hundred buildings at the same time, including very large train stations in Chicago, Buffalo, Detroit, and Boston. All the great U.S. cities built train stations using this incredible vaulting made by these Valencians. Today we are losing some of these old stations, such as Pennsylvania Station, which opened in 1909 with timbrel vaults built by the Guastavino Company. It is such a long and important history, and so little is known about it. I have been working for ten years with colleagues in Spain, architects and engineers who are interested in their history too, because they are also little known in Spain. I also wrote a book, *Guastavino Vaulting: The Art of Structural Tile,* which was published by Princeton Architectural Press. And I have a bet with my students at the MIT that if they find a Guastavino building or a vault that I don't know of, I'll invite them for a meal. We have found eighty buildings here in Boston—just a part of the important work they did. Almost every week we are finding new buildings because we still do not know many of their works. So it's an American, Valencian, and architectural story that is very important to world history but is still little known.

A.P.:   We'll try, through this conversation, to raise awareness of that Valencian and American story. Young people always want to be modern, and those at MIT even more so! How do you explain to students what the alchemy is between the knowledge of the past and the vision of the future, so as to continue to make progress without forgetting the best of previous generations?

J.O.:   That's a fascinating question, and it's true that students arriving at MIT to study engineering, for example with me, never imagine that I am going to work on stone vaulting. They never thought, prior to arriving at MIT, of building gothic vaults. However, with the training we have at present, it is almost impossible to make a vault like that today.

A.P.:   But people believe that is for economic reasons, but it isn't, is it?

J.O.:   No, absolutely not! It's because architects lack the knowledge of how to build them.

A.P.:   Miquel Ramis says he can prove that a stone arch can be cheaper than one made of steel or concrete.

J.O.:   Of course! And if we are talking about one or two hundred years ago, even more so, because they last much longer. For instance, the life of a reinforced concrete bridge or one of steel is fifty years, but a stone arch can last two thousand years!

A.P.:   And as to the alchemy of innovation and the wisdom of the past: how do you combine them?

J.O.:   It's difficult to explain, but the most important thing is to pose interesting problems. How does a Gothic vault behave in an earthquake, for example? It's a fascinating problem, and students are delighted with such questions. So if we pose really good questions, the alchemy begins to appear. Again, with the topic of sustainability, how do we make a building with low carbon dioxide emissions? That is a very difficult question.

A.P.:   Yes, because constructing a building throws waste into the atmosphere and worsens climate change. People are not aware of the energy required to make a building and the subsequent effects on climate, is that not so?

J.O.:   Yes. For example, here in the United States, buildings consume more electricity and energy than transport vehicles (cars, planes, trains, buses, motor boats). Of course, we are using buildings all the time but, in many cases, they are inefficient in terms of energy saving. There are, however, many ways of reducing their energy demands. We are moving in that direction, but it is not easy.

A.P.:   Are architects afraid of not being modern, perhaps?

J.O.:   Yes, and an interesting thing is today's problem, that architects want buildings with new shapes, with the "Guggenheim effect" of Bilbao. So for us the question is how to make a building with less material that consumes less energy and that also has an interesting shape.

As an example, the building we made in South Africa won an award for the best building in the world in 2009. It was made using vaulting.

The building is the Mapungubwe Museum, in South Africa, and is a World Heritage site. It has fifteen vaults made with African mud by local people and built by the same New Zealander who did the vaults in England, and it's considered a revolutionary building! Not long after we presented it as an idea for this way of thinking to the president of the World Bank. We said that instead of sending materials from China, such as cement or concrete, we look for materials and talent locally so as to make new things and build vaults with local materials in a passive building that does not use much electricity. That was a good idea. And little by little these buildings are gaining ground in architecture, but the problem remains of how to make buildings with a lower carbon footprint, but in an interesting and beautiful shape.

A.P.:   Thanks very much, John.

J.O.:   Thanks to you. It's my pleasure.

## Notes

1. In the late nineteenth century and first half of the twentieth century, Rafael Guastavino Moreno (1842–1908) and his son, Rafael Guastavino Espósito (1873–1950), were responsible for designing tile vaults in nearly a thousand buildings around the world, of which more than six hundred survive today. The remaining buildings are found in more than thirty U.S. states and include major landmarks such as the Ellis Island Registry Hall, the Oyster Bar in Grand Central Terminal, and the Boston Public Library. An ongoing database of Guastavino's works is being compiled on the website "Palaces for the People: Guastavino and America's Great Public Spaces," Harvard University, Harry Elkins Widener Memorial Library (http:// palacesforthepeople.com/project/harvard-university-harry-elkins-widener-memorial -library).

2. The MITUPV (Polytechnic University of Valencia) Exchange was a multimedia language-learning project initiated by the two institutions in 2000. Plasencia refers to his article titled "Transformar la formación humanística mediante la tecnología," which was published on MIT's OpenCourseWare site, "Brief Overview of the MITUPV Exchange / Additional Information about MITUPV Exchange" (in Spanish), http://ocw.mit.edu/courses/global-studies-and-languages/21g-703-spanish-iii-spring -2006/projects.

3. Yung Ho Chang, a professor of architecture at MIT, participates in dialogue 10, "Looking Forward in Architecture by Looking Back."

4. "Stonemasonry in Context: The Artifex Workshop," Inca, Mallorca, June 14–28, 2009 (http://www.artifexbalear.org/artifexworkshop09.htm).

5. See Mercedes Gómez-Ferrer, "The Origins of Tile Vaulting in Valencia," *Construction History* 24 (2009), https://www.academia.edu/6747084/the_origins_of _tile_vaulting; and Adolfo Plasencia, photo of the Capilla de los Jofre, in the Convento de Santo Domingo, https://www.flickr.com/photos/adolfoplasencia/ 29235948271.

# 9 Galileo Programme: Planning Uncertainty and Imagining the Possible and the Impossible

Javier Benedicto and Adolfo Plasencia

Javier Benedicto. Photograph by Adolfo Plasencia.

*There is a "Galileo system time notion" that we "manufacture" with our coordinated on-board and ground atomic clocks and constantly refer to as coordinated universal time, UTC. It is the "Galileo System Time" and is broadcast by the Galileo satellites worldwide.*

*Scientists who work in space ventures experience some frustration because those who imagine a mission rarely get to see the results, as they are unable to work with data that appears only twenty or thirty years later. The design time frame is very long, and this leads to a certain frustration, but it is extremely interesting.*

*It might be hard to believe, but we planned uncertainty in Galileo. In other words, uncertainty and possible issues that could arise.*

*—Javier Benedicto*

Javier Benedicto, Head of the Galileo Program Department, Directorate of the Galileo Program and Navigation Related Activities, is a Spanish engineer and an expert in the field of telecommunications. He graduated as an engineer from the School of Telecommunications Engineering of Barcelona, specializing in communication systems and microwaves.

His interest in satellite TV reception launched a broader interest in space satellites, and he joined the European Space Agency (ESA), which offered the challenge of becoming part of the European Union's first space program. Subsequently he moved on to the European Space Research and Technology Centre, where, after an interim period as director of EGNOS, the forerunner to the Galileo project, he was appointed director of the Galileo satellite navigation program at ESTEC and tasked with leading the system design.

The Galileo project provided its developers with many technological and diplomatic problems. In his position as Head of the Galileo Program Department, Benedicto has managed to get the fifteen EU countries to reach a consensus on a challenge of huge proportions, as the system's funding is shared by ESA and the EU. He now leads the deployment of the Galileo constellation of thirty satellites, which provides worldwide coverage through a new generation of GPS navigation system.

Adolfo Plasencia:   After the great amount of effort and time we have spent trying to meet up, we are finally here together. Thank you for receiving me Javier.

Javier Benedicto:   I'm delighted. Thank you!

A.P.:   Javier, after graduating as a telecommunications engineer in Barcelona with a specialty in communications and microwave systems, in 1985 you joined the European Space Agency (ESA), where you began managing the development of microwave equipment for television and cell phones, until in 1995 you were appointed director of the European Geostationary Navigation Overlay Service (EGNOS), a hint of what would later become the Galileo program. At that point you moved from Barcelona to Noordwijk, in the Netherlands, next to ESA's Research and Technology Centre (ESTEC); from there you went to the National Centre for Space Studies (CNES) in Toulouse, France. Finally, in 2000 you were named director of the Galileo satellite navigation program, and you returned to the Netherlands to run the system design.[1]

Your life has been linked to space for many years.

What was it that led you toward this?

J.B.:   You have to be given the opportunity to turn an idea into implementation, and then use this implementation to put a product on the market and close the loop between an idea, a draft, or formulas learned at university and a product, and then see this become marketable. This was how I got into the space industry. I wanted to go further, and that's why I tried out going to the ESA, which at that time was known as the European NASA. It has been a magnificent experience because in these environments, you are bound to come into contact with many disciplines of understanding and knowledge, which is very rewarding, and we are constantly working not only with engineers but also with scientists, doctors, and physiotherapists. All of this allows you to develop as a person and advance with your projects.

This has also steered me toward Europe, and I think the European scope is very important.

A.P.: Your current professional and private life have been marked in a good sense by the gigantic Galileo project, the most advanced satellite navigation system in the world today, which is promoted by the European Union.

Has the Galileo project changed your life in the long term?

J.B.: Yes, obviously: space projects last ten, twenty, thirty years or more, and it is very common that those who start them don't see them finish, or they see them end because they are finished by other people, and those who now use what they have done before are different people, from a different generation. Scientists who work in space ventures experience some frustration because those who imagine a mission rarely get to work on it, as they are unable to work with data that appears only twenty or thirty years later. But you have to understand that to achieve these big goals, we're talking about space discovery, designing unique objects; they are not repeats of others. These designs, therefore, carry a lot of risk; things fail, and we have to repeat. All this takes a long time; the design time frame is very long, and this leads to a certain frustration, but it is extremely interesting.

A.P.: Javier, the dimensions of the Galileo system are almost on the scale of an epic superhuman odyssey: An architecture covering the entire planet Earth with a constellation of thirty satellites at an altitude of 24,000 kilometers, each one going around the world every fourteen hours, in three 56-degree sloping planes with ten satellites in each plane. A global network of stations worldwide will receive the satellite signals, which will be sent to control centers in Europe, which in turn will compute the integrity data and synchronize the time signal (which travels at 300,000 kilometers per second) and triangulation of the satellites, thus retransmitting data to the satellites through a global network of fifteen up-link stations. It's a huge thing, almost difficult to grasp, with a precision an order of magnitude greater than that of the current GPS.

It seems almost impossible to achieve.

Is deploying the Galileo system like building something that seemed hardly possible, almost like science fiction becoming reality?

J.B.:   It is a system built on a scale never seen in Europe. Europe has contributed to the so-called constellation systems, that is, systems supported by a host of satellites. Before, Europe had never considered building its own satellite constellation. We want to provide services on a global scale for any user environment on Earth, so we also needed a ground system distributed around the world. This is why we have stations on the five continents and satellites that are permanently orbiting the Earth. It is a new dimension, and this dimension, in which industry participates, is going to open doors to other systems, which we are now starting to imagine.

A.P.:   For me, and I believe for anyone, there is a point that is quite astounding: the triangulation principle on which satellite navigation is based. I can calculate my exact position if I know my exact distance from three points whose exact location is known. If I want to know my distance from these three points, the most accurate way is to see how long it takes for the signal from these three points to arrive. However, in the Galileo system, these points are in three spinning satellites, in continuous movement around the Earth at an altitude of 24 kilometers, and the signal I'm talking about travels at the speed of light—300,000 kilometers per second.

Does it work like this? Because what I have described seems to be the embodiment of Arthur C. Clarke's famous phrase, "Any sufficiently advanced technology is indistinguishable from magic."[2]

Do you think that for many ordinary people, the science of an up-and-running Galileo system will be somewhat sublime, like magic?

J.B.:   It's true. When you look at the parameters from this perspective, it seems unbelievable. We are determining the position of satellites at an altitude of 23,300 kilometers with great accuracy, to within a few centimeters. We transmit signals that are synchronized in time to a precision of nanofractions of a second. In the end, all of this enables the system to operate and provide users an accuracy of a few centimeters, or a meter at most.

A.P.:   However, in a speech I've just heard you said that you create a time frame and you create a geometrical and space frame. You artificially create "space" and you create "time" for this project.

J.B.:   Exactly. What we are trying to do needs to be understood. We are trying to help users wherever they may be to calculate their position. But their
position: where and when? The where and when is information that we
must provide.

Nevertheless, the Earth is a planet that changes its constitution. Tectonic
plates and continents are constantly moving. There are certain drifts that
aren't trivial. Land masses are moving around a centimeter per year and,
depending on this, they move more to the north or to the south, to the east
or to the west. If we want to determine the positions of elements on Earth
to a high precision and stability, we have to continuously supply information such as the geodetic reference, where the person is with a time reference, and how this evolves over time.

We also build this information and supply it to the users so they can
trace it on their maps and then determine their position using this map.

A.P.:   However, it doesn't take very long to reach a satellite at a height of
24,000 kilometers at 315,000 kilometers per second.

Can you measure the lag between one distance and another?

At this speed, you must achieve an amazing precision in time measurements. Is that right?

J.B.:   We achieve extremely high precision in measuring time and its accuracy, and in measuring the satellite's position in orbit as well. You have to
understand that the Galileo system is designed for the orbits to be very
stable; in other words, they require very little correction from the ground
for satellites to be where they should be. There is a set of factors that can
achieve this kind of precision. We also rely on support from external
means with laser reflectors using telescopes on the ground. All this combines with techniques which, from a control center, allow us to restore the
orbits of the satellites and synchronize the atomic clocks that travel in the
satellites; and we can achieve this with the precisions we talked about
before.

A.P.:   But you don't measure in minutes or hours.

J.B.:   No, we don't. We measure in nanoseconds. This is where the complexity lies. What is also very complex is having to think that the Galileo
ground segment is like a huge computer, but on a world scale. We are constantly receiving information from the satellites spinning around in their
three planes. All this information has to be synchronized and connected to
a control center. All of this happens in real time; it's like a huge computer
that recovers information and synchronizes it so that a single product
comes out at the end, which is your position.

A.P.:   In the end, you not only have very good engineers. You also have very good watchmakers, don't you?

J.B.:   Absolutely, and it isn't a joke because the atomic clocks in the Galileo satellites are made in Switzerland. It might seem surprising, but they are the masters of watchmaking, including that with atomic technology.

A.P.:   The economic scales of investment in the European Galileo system match those of its science and engineering. I'll mention some: the definition and validation phase (years 2001 and 2002) cost the EU and ESA €80 million; development, starting in 2003, around €2.4 billion. The deployment phase/FOC phase (Full Operational Capability) cost a total of €3.405 billion. And the system that operates on the Galileo architecture, EGNOS, cost around €1.1 billion.

These investments involved thousands of engineers, technicians, and workers, facilities, companies and universities.

Was it hard to set up something on the scale of the Galileo project? Because the countries that led the project, France, Italy and Spain, encountered a range of problems, from technical to diplomatic, as creators and promoters sought to convince and bring the fifteen EU countries to consensus. Is that right?

What was the most complicated issue?

J.B.:   It has been very difficult to convince and align the political viewpoints in order for this will of European sovereignty to come about. It is not enough to simply say that we have to have our own GPS.

If there was a conflict tomorrow, the White House or other world powers would be likely to modify the characteristics of the GPS system in such a way that GPS navigation would not be precise in the area of conflict, and maybe also Europe. This is, obviously, something that we cannot accept.

People should be aware that this is a project particularly about infrastructures. It's like a freeway or a power station. In the end, the freeway has to be built so that vehicles and trucks can circulate, so that trading can be carried out and society can use it.

This is a basic infrastructure. After, with satellite navigation, you can use it in a cell phone or in your TomTom or a GPS navigation device, to land an aircraft at an airport, or in a hospital, or in a telecommunications network. One service of the Galileo system that isn't often mentioned is the time reference, which provides a universal time for everyone who has the same receiver. This is very important for synchronizing.

A.P.: The "universal timeframe" that you manufacture. …

J.B.: Exactly, there is a "Galileo system time notion" that we "manufacture" with our coordinated on-board and ground atomic clocks and constantly refer to as coordinated universal time, UTC. It is the "Galileo System Time" and is broadcast by the Galileo satellites worldwide. So, this time is used, for example, in financial transactions. The moment in which a company or an individual makes a financial transaction is very important to win a market someone else is competing for.

It is a basic information infrastructure that, later on, can be applied to all facets of social and economic life.

It has been necessary to convince people that it is worth investing in. It's strategic for Europe; in other words, it is something you have to know how to do yourself; it's something you don't need to buy elsewhere. You have to know how to do it yourself and, once you've carried it out, you have to operate and guarantee the service on your own; you can't rely on anyone else for this.

A.P.: And this really moves the leading edge of industry in technology and engineering.

J.B.: It moves industry and an endless number of applications because we shouldn't consider only the people who work in space. In total, if you counted the people who work in European space, we wouldn't fill half a sports stadium. What's important is all the business and activity that is generated around this and the new applications it opens up.

A global navigation system is of course used by all Europe, but furthermore, it allows European technology to be sold all over the world and products to be developed that work all over the world.

A.P.: And to turn Europe into a leading player on a world scale.

J.B.: That's right. You have to be a player to get respect and have a place at the table in order to discuss, to establish, to convince other countries which policies are good. Not in the sense of the wheeling-and-dealing politics but rather social, economic, and educational policies, or policies about how we should evolve as humans, as a civilization, to be something in the world.

That's why it's necessary to have the means and to demonstrate that you know how to do it by doing it.

A.P.: And managing to get space to be considered more important than the freeways—was that difficult too?

J.B.: Everything is important; in the end, everything is complementary—that is, for a transport company to operate, a freeway is necessary. For air traffic to operate better, satellite navigation is necessary, as this will allow you to go from point A to point B using less fuel, using less time, doing it more safely, and finally, reducing costs and the use of energy, which increases efficiency. All of these structures complement each other.

A.P.: And to find a person who has an accident on a mountainside, right?

J.B.: Indeed, there are now applications that are not measured in terms of economic benefit but rather in terms of social benefit for the population, for persons. With Galileo, we provide a significant improvement to what currently exists for finding and rescuing a person in difficulty, who would at present have to wait for hours to realize that he or she had a problem. With Galileo, this person can activate a beacon. The signal is immediately detected and the rescue center will inform the person that it is aware of the situation and has set in motion the appropriate operations to implement the rescue. All of these are applications we consider to be important.

A.P.: What you're saying is fascinating, but we also have to talk about the difficulties.

Let's talk about the difficulties that were made for the Galileo project by non-Europeans. From what I've read, the main pressure came from the United States, which felt that its world monopoly on navigation using its military GPS system was under threat, following the virtual disappearance of the old Russian system, GLONASS. It turns out that the GPS was going to become obsolete when compared to Galileo; Galileo allowed a precision in civil terms of five meters for locating the position of a vehicle, building, or person, whereas the American system allowed for 50 meters. Its basic services were also conceived as free, like those of the GPS, but not in the case of commercial or professional use, in exchange, naturally, for enhanced performance

Was it like that? Can you tell us about the hurdles you had to overcome?

J.B.: Indeed, this was one of the main elements. The systems and technologies are bargaining tools and vectors of influence for exerting force. The U.S. administration is currently in discussions with China and negotiating with Beijing, and it is obvious that behind the scenes, high-level agreements are being negotiated with industry. In other words, politics aren't carried out without weapons—"weapons" or means in the economic and industrial sense, which allow you to negotiate and offer something in exchange for something else.

A.P.:   Politics isn't naïve.

J.B.:   It isn't naïve, and it feeds off this type of thing. Without this, politics has no power.

It's true that when we set Galileo in motion, we did it with great determination, and we met with a lot of resistance from the United States.

It's also true that in 2004 we signed a cooperation agreement with the United States to guarantee, for the good of the users, the systems would be compatible, that is, that they wouldn't be harmful to each other and would be interoperable. In other words, a user would be able to combine the signals from two Galileo satellites and two GPS satellites and join them up as if they were four satellites from the same system.

And that's the way it is. We've achieved it, technically and politically. We are receiving huge support from the international satellite navigation community, manufacturers of receivers and users of the technology, for Galileo to succeed, as it has seen that we've done it in an open way. It is not a project generated by military motives. GPS *was* initially motivated by defense strategies, and in fact, this was what most worried the United States. They said, Let's see; what is Europe up to? It's like when a country is building nuclear weapons. What are they going to do with this? Are they doing it to attack? And whom are they going to sell this technology to?

Let's not forget that satellite navigation is also used to guide missiles nowadays, and obviously, these are very delicate issues.

We also had many problems with the Chinese, but eventually we achieved a cooperation agreement with them concerning these points. We will also be intraoperative with their systems. Now they have finished building their own system, BeiDou. And it's normal that they should do this. Nobody can claim that what they do is going to be unique and used all over the world. I think the open policy of the Galileo project will show its benefits in the future, because we are seeing that all satellite navigation products are now including it—many smart phones today come with Galileo installed—because they know we're going to do it and that we combine with everybody else, that we join in. It's not a question of this or GPS: Galileo and GPS, strength lies in the unity.

A.P.:   Let's talk about other difficulties that came up.

Last year two satellites were launched, completing the fourth validation phase of Galileo. Because of problems with the Russian launch rockets, the two satellites didn't reach orbit. In cases like this, the uncertainty will be of a size to match the scope of the project.

How do you manage the uncertainty in these cases? How do you live with it?

Do you have guidelines in ESA and Galileo for cases in which you need to troubleshoot the impossible?

J.B.: We do. It might be hard to believe, but we planned uncertainty in Galileo. In other words, uncertainty and possible issues that could arise.

Possible problems are identified a priori, and we are ready to cope with them. If a launch goes wrong at any given moment, a crisis plan is activated in minutes. The crisis plan is deployed both in communication and technical areas. We know what has to be done, we have cases which have been "broken down" and we have to go to them, "role our sleeves up," "take the bull by the horns," and make the thing work. Obviously, there are always gray areas that you may not have studied but we try, as far as possible, to imagine them. In these programs, there are risk analyses; a lot of resources are required for failure mode risk analysis. And risk analyses envisage the possible and the impossible. We have many scenarios of things that can go wrong.

There is a manual of everything that could go wrong. And though it's very hard to get people involved in this, experience shows that statistically, one day or another, you'll be faced by one of these failures, and it is necessary to try and see what set of problems you might come up against and how to use this manual to deal with them.

A.P.: In your speech, you said something very interesting concerning an incident in which, once you've found out what happened, you say, How silly! How come we never realized? In other words, no matter how many manuals there are, reality can always surprise you with some idea that wouldn't be found in even the thickest manual. Is that right?

J.B.: Yes. There are things you can't plan for, you haven't envisaged, and maybe you cannot control. There are "statistical effects." I always say this: for a launch to go well, millions of things have got to go well, everything has to go well. If one of them goes wrong, a diode, a capacitor … or a button not pressed properly, the whole program can go down the drain.

A.P.: What happened in the specific case you mentioned?

J.B.: We know what happened. What we're trying to ensure is that nothing similar happens in other design areas; that is the difficulty.

What happened. ... The rocket has a series of engines. The ascent path is very long. We have a long mission in which, between lift-off and the "final burn," as we call it, almost four hours go by. We have a rocket up there that is pushing and pushing toward a very complex orbit, with a very complicated trajectory and set of dynamics. A series of engines perform in the different stages. One of these motors is fed by hydrazine fuel, traveling along a small pipe. This pipe was too close to a liquid helium pipe that acts as a refrigerant for a propulsion system. The liquid helium froze the hydrazine that was traveling along the pipeline. When the hydrazine froze, the engine didn't have enough thrust and changed the upward trajectory, meaning the satellite ended up being fired at a different, lower point, far from what we expected. The good thing was that we had ground stations that began to investigate using programs they had already designed for satellite searching. We found out what happened and this has been rectified now. It will not happen again, but other things might. I'm sure that in the history of Galileo there will be other flaws on a different scale that we will be able to cope with.

A.P.:   Now that we've spoken about the worst things, let's speak about the best, about the benefits.

From a strategic point of view, Galileo will endow Europe with independence in satellite navigation, a sector that has become very important for its economy (around 7 percent of the EU's GDP in 2009), and it will improve many aspects of well-being for its citizens. Moreover, according to studies, Galileo will provide the EU economy with around €90 billion during the first twenty years of operation. (Galileo will be operating in a world market of satellite navigation products and services, which is currently valued at €124 billion.)

Are these benefits in the region of the forecasts?

What kinds of applications and services do you think the Galileo system will provide for Europeans and non-Europeans, and what is an approximate timescale for these benefits to start being felt?

What specific things could it change in people's lives?

J.B.:   The benefits you've listed are indeed corroborated in our market studies by analyses made in other countries such as the United States, Japan, Russia, Australia. Obviously, it's justification for the investments that have been made. The investment return is calculated on several scales, and therefore we consider it and make the decision to carry it out, to do it, to go ahead.

It is interesting to reflect on the extent to which this technology is going to change some of the things we're doing. We are already seeing the growth in systems called "location-based services" (LBS), services that are based on a location. There are many of these things we already use in our smart phones as well as in other professional applications that rely on knowing the position of an individual or an object. Galileo will contribute to this in two main aspects: on the one hand, the positioning precision is going to be much greater, and on the other, the Galileo signals are going to allow a position to be located even when there is no direct satellite visibility, such as inside buildings. This is important.

Today, when you go inside a building, or garage, you have no cell phone or GPS coverage, and the same thing happens when you enter a department store or a business office. This will be achieved with Galileo.

The signals from Galileo are designed with much more important bandwidths, which work in different band frequencies.

The combination of all these technical means lets you "navigate" inside buildings. In fact, you're navigating using the signals that bounce off the walls of the room you are in.

At the present, this cannot be done with GPS. The future GPS system, which will start functioning in 2025, is already moving in this direction, but Galileo is implementing this technology now. The Galileo satellites we have in space already transmit this sort of signal.

Automatic vehicles will be a reality soon, yet in order for there to be automatic vehicles, we must be able to determine the vehicle's position in absolute terms and relative to other vehicles much more precisely than is done today and with much more reliability. Great care is needed because inside a vehicle there are people, and we have to be sure the vehicle knows where it is going and doesn't make mistakes.

Another important social application is in preventing train crashes. I get shocked so often these days when I read that two trains have crashed head on. Just how is this possible? It should be known before it happens that two trains are on the same line. And, of course, we shouldn't forget that two sets of train tracks are very often separated by one or two meters, and if you want to know whether the train is on one set of tracks or the one next to it, you cannot use a system that is precise to one or two meters; you have to use a measuring system with a higher order of magnitude relative to that being measured.

Therefore, a system of satellite navigation with decimetric precision is essential to be able to access these types of applications: automated guided vehicles, automated guided trains, automated aircraft landing, all with a high level of reliability and integrity.

A.P.: In other words, Galileo is no longer precise to meters but to centimeters.

J.B.: Galileo is indeed heading toward an accuracy at the decimetric level to be able to accommodate all the types of applications we have today. When you're in your vehicle and you tell your GPS navigation device your home address, you are indeed taken there; but if you notice, there's an error in how you do it with respect to the exact address you wanted to go to, of a few meters. And this is not sufficient to manage the types of applications I've mentioned.

A.P.: You need surgical precision, almost.

J.B.: We need an astounding precision of a few centimeters with satellites that are thousands of miles away.

A.P.: Would I exaggerate if I said that this global leadership, which is missing in European political leaders, is only being provided, for the time being, by European science and technology in very symbolic examples, such as at CERN, where the Internet was invented and the Higgs boson was discovered, and at ESA, with its recent feat and, of course, the ambition, scope, and deployment of its Galileo system?

J.B.: Europe is the continent of Galileo, of Newton, of Kepler. We have led science in the past, and I believe that today we have everything needed to continue doing so. We have great intellectual capacities and research capacities. We have great ideas and we've shown through projects such as Rosetta that we can go very far. The Galileo project is also an example of infrastructures on a global scale that we are setting up for the benefit of all humanity.

We spoke earlier about time frames of ten, twenty, or thirty years. But here we're talking about the future of planet Earth with a view to thousands of years. It is important to make long-distance projections. And to do this we have to know how to go backward and maybe find out where we came from, how the planets in the Solar System were created. And this is what these types of missions are used for, nothing else.

So we have the capacity to set ourselves some very, very ambitious goals and really make them successful, such as that demonstrated by the Rosetta program.[3] I think this is the direction that Europe should move in.

A.P.: Thank you very much, Javier. It has been a pleasure.

J.B.: The same for me. Many thanks; it's been a pleasure for me too.

## Notes

1. Galileo is Europe's global satellite navigation system, providing a GPS service that is under civilian control. It is interoperable with the U.S. and Russian global satellite navigation systems (http://www.esa.int/Our_Activities/Navigation/Galileo/What_is _Galileo).

2. The science fiction writer Sir Arthur C. Clarke promulgated three rules to help people think about the claims of developments in science. His third and most famous law, any sufficiently advanced technology is indistinguishable from magic, he is said to have added because Newton had three laws.

3. The European Space Agency's International Rosetta Mission has sent a spacecraft on a ten-year traverse of the Solar System, to rendezvous with Comet 67P/ Churyumov-Gerasimenko and then remain in close proximity to the icy nucleus as it plunges toward the Sun (http://www.esa.int/Our_Activities/Space_Science/ Rosetta).

# 10 Looking Forward in Architecture by Looking Back

Yung Ho Chang and Adolfo Plasencia

Yung Ho Chang. Photograph by Adolfo Plasencia.

*The rational and irrational processes are not two separate notions but rather just one. One might push the rational process to the extreme, to the point that it becomes its own opposition.*

*Digital space is no substitute for architectural space. To me that seems pretty obvious.*
*—Yung Ho Chang*

Yung Ho Chang is an architect and Professor of Architecture and former Head, MIT Department of Architecture; Professor of the 1000-Scholar Plan, Tongji University, Shanghai; and former Head and Professor of the Graduate Center of Architecture of Peking University. His degrees in environmental design and architecture are from U.S. institutions, including the University of California, Berkeley. He established China's first independent architectural firm, Atelier FCJZ, which undertakes a variety of projects, ranging from private residences to museums and government building, as well as experimentation in furniture, product, stage, and graphic design.

Chang is the recipient of numerous awards, including the UNESCO Prize for the Promotion of the Arts (2000), a China Architectural Arts Award (for the Hebei Education Publishing House, 2004), and a *Business Week/Architectural Record* China Award (for Villa Shizilin, 2006).

*Note:* This dialogue took place on the beautiful Mediterranean island of Mallorca, off the eastern coast of Spain, within the framework of the inspiring "Stonemasonry in Context: The Artifex Workshop," sponsored by Artifexbalear and its director, Miguel Ramis, in June 2009.[1] This text was revised by Yung Ho Chang in June 2015.

Adolfo Plasencia:   Professor Chang, welcome to Spain. And welcome to the wonderful island of Mallorca. You are here to participate in the "Stonemasonry in Context" workshop, organized by a courageous group that promotes the traditional use of stone as an element of the Mediterranean heritage. This essential material is part of the collective memory of every civilization that has emerged around the Mare Nostrum, as the inhabitants of both its shores have called this sea of ours for thousands of years.

In its online statement, the workshop is defined as "an in-depth discussion that explores the intersection between nature and creativeness."

Do you think that creativeness and all that it achieves is the result of architecture's inherent creative processes, determined by a rational human mind? Or are emotions more decisive in shaping the outcome of that creative process? Some people say that at the end of the day, commitment and hard work are the determining forces. Which factor is for you most crucial for maximum success in an architectural creative process? Hard work and self-discipline? Technical knowledge? Emotion and sensitivity? Or the constant effort required for the creative process?

Yung Ho Chang:   That's a whole set of questions. I think I can answer maybe three elements of the first issue. First, as far as hard work and self-discipline are concerned, they are actually the bedrock for any work; you can't do anything without a readiness to work hard and without exercising self-discipline. But as for the notion of rational and perhaps irrational approaches to architecture, I do believe that you need both; that's why architecture is so attractive. However, personally, I think such rational and irrational processes are not two separate notions but rather just one. One might push the rational process to the extreme, to the point that it becomes its own opposition, so what is a rational way of thinking in the end becomes irrational—and the outcome is at one and the same time both rational and irrational.

A.P.: Aristotle, one of the greatest minds to emerge from the Mediterranean world, said, "When you are compelled to build something well, you become a good architect."

Do you think Aristotle was right? That building well—in other words, becoming involved in the practical activity of constructing something—is the best way to become a good architect?"

Y.H.C.: That's actually a harder question than it seems because, for society in general, good buildings are very important—maybe more important than good architecture, if we can make such a distinction between buildings and architecture. However, for students, I think that it's vital to acquire a knowledge of what good buildings are as part of the design foundation. We take a good look at examples of successful buildings in the world; we examine buildings that were well built, then we consider buildings that work very well—and we assume that these buildings may also be capable of inspiring people. If a building can move people, that's truly something that goes beyond the basic function of architecture, which is also something that's very necessary for us. To put it another way, for basic learning and a clear understanding of the quality of a building, and for society, again, well built and functional are crucial. But for the discipline we have to go further than that. You might say it's a *game* for architects, but a very serious one. In a way, it's the Olympic Games for architects to go beyond that point.

A.P.: Mies van der Rohe is universally considered to be a great master of architecture, not only for his work but also for his thoughts.

He once said, "Each material possesses its own unique characteristics, and you must know them in order to work with it properly." He also remarked, "We conclude that nothing is achieved from the material alone, but only from the correct use of that material. New materials do not guarantee better results, as the worth of each material is tied to what we create with it. Just as with materials, we also must learn about the nature of our goals."[2]

Do you think that, with the abundance of new materials and technology, we sometimes forget to learn the nature of our goals, as argued by Mies van der Rohe?

Y.H.C.: There are two very important points here. First, in general, I very much agree with Mies van der Rohe; I'm very much influenced by his way of thinking about architecture. Architecture is the art of building, I too think this way. If you agree with this statement, it follows that architecture is really about *materiality*; materiality is the basis of architecture. Second, as for new materials, Mies himself embraced the new materials of his time—steel, glass, and so on—in such a way that he not only used them to create architecture in the traditional sense, he perceived a new architecture in such materials. And of course he became very influential, a super-significant milestone for modern architecture. So today, when we look at new materials, we have to look beyond their novelty. At the same time, we have to examine old materials in the light of the new materials available today and do this in a very democratic way and on equal terms. Because a material is new, it's not necessarily better. Similarly, because a material is old, that doesn't necessarily make it more interesting. Old and new materials have to be considered in the same light and examined on the same terms, and then that examination needs to go further so that we explore their deepest, most fundamental quality.

A.P.:     There is an architectural tradition that has been recently revived in a contemporary context. The architect Steven Holl, who received the BBVA Foundation Frontiers of Knowledge Award, said that he does not want to build here and there on the basis of his own style. He'd rather have no personal style, as each architecture should emerge uniquely at each specific place, in each location.[3]

The German philosopher Martin Heidegger, in his essay "Build, Inhabit, Think," argued: "The location is not already there before the bridge is. Before the bridge stands, there are of course many spots along the stream that can be occupied by something. One of them proves to be a location, and does so because of the bridge....a location comes into existence only by virtue of the bridge."[4]

Do you think that architecture and place should be linked, as Steven Holl says? Furthermore, with the appearance of each work of architecture, does a place emerge that did not exist before, as Heidegger maintains?

Y.H.C.:     I have two comments. The first is about the notion of style. It's not that architects don't have styles but rather that styles have to be based on something: an attitude, an understanding of architecture, strategies, and so on. I know Steven Holl personally as a friend quite well, so I understand his argument. It's also very important to understand that it's not someone who changes the look of a building who qualifies as a more progressive architect but rather one who may have a more consistent use of certain forms and materials yet is trying to really focus on certain issues. Such architects are architects without styles because they are really working on *issues* rather than questions of style.

The second comment relates to this notion of how architecture might change a place or not, or make a place. I've had that experience myself. We did a project called "Commune" by the Great Wall outside Beijing. Once the project was completed, a lot of people asked me and the client how we picked a special valley outside Beijing in which to undertake it. The truth is, it was not a special valley; it was a typical valley outside Beijing. There were hundreds of valleys just like that. That said, a work of architecture, when carefully placed into the landscape and working with it, can give the locale and landscape a further legibility. That's actually what architecture is all about.

A.P.:   Professor Chang, in your opinion, is architecture a discipline that can be simultaneously eternal and in constant evolution? What do you think?

Y.H.C.:   I agree with your statement. Let me explain why. Right now, we're talking about architecture in a strictly academic context, so we are not talking about the market, the economy, society, and such. However, architecture is totally rooted in society, so it's part of the economy, culture, and so on, and it evolves with changes in society, technology, and the like. So much for the change part. And then it's not so much that the physical buildings are eternal; they aren't in any way, and indeed may actually have rather short lives. But the need for a settlement, a city, a building is constant. Not too long ago there was a tendency to talk about how digital space, cyberspace, would take over physical space. It was a seductive argument but it didn't make sense. Why shouldn't we have both? Indeed we do have both, we enjoy having both, because in the end, digital space is no substitute for architectural space. To me that seems pretty obvious.

A.P.:   I have seen the website of your Architecture Atelier.[5] Although I am not able to understand some of your comments because I can't read Chinese, I think I can make two observations. The first is that you agree with the statement by Mies van der Rohe that "less is more." Your website is very elegant and synthetic, and the architecture featured there exemplifies this very principle. My second observation is that the chronology of the website, the order of access to your projects and works, is not the usual one; rather, it seems to be chronologically reversed or randomly arranged.

Are these impressions of mine accurate?

Y.H.C.:   You have two questions there. The first question is about "less is more." That statement has been frequently challenged during the post-modernist era. People argued, "Why not 'more is more,' 'less is less,' or 'less is more'?" As a purely aesthetic statement, I think there is room for such questions. But in reality, two things dictate the practice of architecture. First, architecture, the actual building, is always complex. It is not about a single idea, however pure that idea may appear. In reality, architecture is all about the synthesis of space and usage, structural engineering and mechanical engineering. And it will always be that way. So no matter how simple the original conception is, in architecture it ends up being complex, if not complicated. Even so, having a simpler, clearer idea to begin with is very important. I really believe that's what Mies van der Rohe meant and that he used it in an aesthetic, "minimalist" sense. So I agree with that, although more in terms of the design process rather than as a fixed aesthetic. As for your second point, I have to say if I redo my website I probably would think about reordering it, but at the time it was set up, we didn't have much work to show. Then again, that's what I understand as interaction: letting people look for something rather than it being presented straightforwardly, so that what you might call "our office on the Internet" achieves a labyrinth-like quality. I'm actually open-minded about it. The website does have some kind of inherent architectural quality that is probably very true to the way I understand architecture. In a way it has to do with the way that we put it together. It's all about a very simple structure to present a rather complex experience. I hope that I can keep that on my website—and do it even better in my buildings.

A.P.:   The architect Rem Koolhaas, winner of the Pritzker Architecture Prize, said: "The outcome of contemporary architecture is 'junk space': an infinite, undistinguishable interior space defined by innovations such as air conditioning, escalators, acoustic ceilings ... which have damaged architecture's credibility forever."[6]

Do you agree with him?

Do you think that such excesses of modernization are affecting the credibility of architecture?

Y.H.C.:   In a way that's true because architecture initially—or rather buildings initially—were there to create a shelter for people. It was all about comfort, and then somehow the task of providing comfort was taken over by technology—by things like air conditioning, insulation, almost everything except architecture. It was a problem then and still is one. Some elements of this technology have been around for rather a long time but we architects have not yet fully digested their possible impact on architectural design, so we have been reacting rather passively. We all like to imagine our pure space, but in the end we get all this extraneous stuff, and therein lies a problem. If we can't solve it, maybe the next generation of architects can, so that we actually manage to fully integrate the mechanical systems into architecture within contemporary issues of importance such as energy and sustainability. In this way architecture remains as powerful as it was before while working effectively together with all the engineering disciplines. Maybe I'm overly optimistic but I think it's possible.

A.P.:   Finally, what do you think about the friendly atmosphere and, you might say, the "heroism" of the "Stonemasonry in Context" workshop, devised by Miquel Ramis, Luis Berríos-Negrón, and Michael Ramage, here in Mallorca?

Y.H.C.:   I'd like to use a couple of words that Luis Berríos used when translating for me: one is *heroic* and the other is *materialize*. Heroic or not, for me the whole idea of heroism today is very different from that of ancient times, in both the West and China. But I think that whoever takes on a larger than life challenge is pretty heroic. I think there is a level of idealism in this project that challenges the status quo in architectural education. It's important and it's clearly heroic, especially today, when a lot of our students, because of the way we teach—I can't blame others because I'm part of the system now—may actually forget that architecture, however conceptual it might be, is deeply rooted in materiality. So that's why this term "materialize" is very important. It's not so much a matter of thinking about abstract ideas or big ideas but rather whether such an idea is firmly rooted in something tangible.

On the surface, there appears to be something almost innocent about this workshop: to get an idea of the traditional, age-old craft of stonework. But in fact this poses several questions. Number one: Is craft still relevant and important for architecture today? And then we can ask how the past and present are connected. This is really interesting since ultimately it leads you to the question, what really is architecture? Is it something physical, tangible, so that it's always about crafts—although crafts can, of course, evolve? Or is it more an intellectual exercise, in a way that's attached only to the metaphysical world and has nothing to do with the physical world? These are important questions. I think workshops like this for students are very important experiences for them to consider in depth such questions, which they may or may not have had opportunity to reflect on. In a sense it's a process of materialization, meaning that by the end, some issues will have been clarified in their minds, so it's a very, very rewarding experience for them. Finally, I'd like to add that I've never been to Mallorca before. And I've never been to a workshop like this, so I'll tell my own students about my own experiences here, where I'm learning quite a lot.

A.P.:   Professor Chang, thank you very much for your time. Thanks for this conversation.

Y.H.C.:   Thanks to you.

**Notes**

1. "Stonemasonry in Context: The Artifex Workshop," Inca, Mallorca, June 14–28, 2009, http://www.artifexbalear.org/artifexworkshop09.htm.

2. Debbie Mills and Margo Warminski with the Greenhills Historical Society, *Greenhills* (Charleston: Arcadia publishing, 2013), 43.

3. Antón García-Abril, "Steven Holl," *El Cultural*, June 12, 2009, http://www.elcultural.com/revista/arte/Steven-Holl/25466.

4. Martin Heidegger, *Poetry, Language, Thought* (New York: Harper Collins), 151–152.

5. Atelier FCJZ FCJZ / 非常建筑 (http://www.fcjz.com).

6. Rem Koolhaas, "Junkspace: Modernization's Fall-out," Arquitectura Viva 74 (October 9, 2000), 23–31. http://www.arquitecturaviva.com/en/Shop/Issue/Details/74.

# 11   The Seamless Coupling of Bits and Atoms

**Hiroshi Ishii and Adolfo Plasencia**

Hiroshi Ishii. Photograph by Adolfo Plasencia.

*I was just captivated by this simple artifact, the abacus, because it's so playful and tangible. It symbolizes my vision of Tangible Bits. Tangible Bits implies making information tangible—you can touch and manipulate it.[1]*

*I would say the MIT Media Lab is a place to invent a future of your own.*
*—Hiroshi Ishii*

Hiroshi Ishii is the Jerome B. Wiesner Professor of Media Arts and Sciences and Associate Director of the MIT Media Lab, founder of the Tangible Media Group, and Codirector of the Things That Think consortium.

His research focuses on the design of seamless interfaces between humans, digital information, and the physical environment. His team seeks to change the "painted bits" of graphical user interfaces (GUIs) to "tangible bits" by giving physical form to digital information. In 2012 he presented "Radical Atoms," a new vision that takes a leap beyond Tangible Bits by assuming a hypothetical generation of materials that can change form and appearance dynamically, becoming as reconfigurable as pixels on a screen.

Adolfo Plasencia:   Professor Ishii, you were born in Tokyo; you are an electrical engineer and also a computer engineer. You have conducted research in Bonn, Germany, in Toronto, Canada, and at MIT. You specialize in the relationship between humans and computers.

Which culture "understands" computers better, the Asian culture, the European, or the American? Is this relationship more a question of the universal human condition or of a particular culture and each person's individual background?

Hiroshi Ishii:   The question of how a culture may influence the design of human-computer interactions is indeed a deep one. As I see it, in the early stage of human-computer interactions, which mainly involved the use of a keyboard and characters, there were definite advantages for cultures that utilized alphabets like the roman one with only twenty-six characters. However, with the advancement of human-computer interaction techniques and also the development of multimedia technology, computers today can not only interpret text type from the keyboard but also understand what is happening in both video and audio. Therefore, for example, in the theater, a computer can now capture the motion of dancers or actors, then project the video or sound accordingly. So I do believe the rich modalities of these interactions have become the most important element from now on. In this way, those who appreciate the use of the body or the voice can also enjoy the advantages of human-computer interaction.

A.P.:   Professor Ishii, you started working at the renowned MIT Media Lab in 1995, where you founded the Tangible Media Group. This offered a fresh view on human-computer interaction, in a project you named "Tangible Bits."

That expression is an oxymoron, isn't it? Because conceptually and in essence, bits are intangible. What is the meaning of Tangible Bits?

Are you just being provocative?

H.I.:   I intentionally chose this term, which at first sight appears to be an oxymoron or contradiction. You are 100 percent right that bits are intangible; you cannot touch, smell, feel, or taste them, which is why I try to make them tangible and confer added value.

A.P.:   For quite some time now, ever since 2002, you have been the coordinator of the Things That Think consortium. I think it is a great name, but to some it might sound like a provocation.

Do you think artifacts will ever be able to think?

H.I.:   Yes. Most of artifacts, including current computers, have no awareness about what's happening around them; for example, as we speak right now, we are in Valencia, and we are talking and interacting. But this pen on the table has no idea where it is or what is happening around the table. I think that computers are now becoming more aware, context-aware, and more sensitive to what's going on around them, so that they can provide more appropriate, smarter services. That's one important direction.

An important idea that came out of the Things That Think consortium is what we termed "Things to Think With." Things to Think With argues that instead of making things smart or intelligent, why don't we make people think better or more smartly using "augmented objects" or artifacts? So the key idea at the end of the day is that since human beings are more intelligent and smart, why don't we design things that help people become even smarter or more intelligent or more productive? That is the key idea.

A.P.:   I had the opportunity to visit the Media Lab in the company of James Patten, one of your more outstanding students. He showed me around and was my guide for what proved to be a magical visit. At first glance—and bear in mind that I was an outsider—the artifacts that I was able to see on my visit to the Tangible Media Group and the Media Lab seemed to fall into four basic categories: software, electronics, image-light-information, and inert physical materials.

What kind of alchemy keeps them in equilibrium?

H.I.:   Another very good question. Perhaps the most important thing is having a very clear concept and a vision that helps people connect all the components, such as sensors, actuators, computations, and physical artifacts. For example, in the context of urban planning, Dr. John Underkoffler and I came up with the concept of the digital shadow, having a model of the buildings that cast a shadow. With this concept it became very clear what objects have to be sensed and tracked, what computation needs to take place, then what kind of shadow should be projected from the video projector. Key concepts and accurate architecture were both of fundamental importance.

A.P.:   If I may, I would like to ask you a more personal question. On your website there is a picture of you when you were two years old. There is an abacus in that photograph, your PDA, so to speak, at the time. I did not know you were going to bring an abacus to our meeting.

Please tell us the story behind the abacus.

Were you already interested in arithmetic at such an early age?

H.I.:   I wish I could say yes, I understood all of arithmetic and mathematics at the age of two. But the truth is I was just captivated by this simple artifact, the abacus, because it's so playful. Today it helps me in two ways: one is the key idea of Tangible Bits. Tangible Bits implies making information tangible—you can touch and manipulate it. This abacus is a good example of how information bits are tangible so that you can directly compute using your fingers and hands. The second concept is "affordance," a term borrowed from psychology.[2] Simply by taking hold of this abacus, it becomes very clear what you can do with it, without the need to understand a complex theory. For example, it can become a musical instrument or a toy train, or you can use it as a back scratcher, all things that I could do at the age of two. So one of the important things that we are studying at MIT is how to replicate the clarity of affordance inherent in the abacus in human-computer interactions.

A.P.:   Professor Ishii, I have a feeling that you, like me, love movies. Do you remember Kubrick's film *2001: A Space Odyssey*? The machine in the film, HAL 9000, uses natural language to communicate with humans. The movie took for granted that computers and humans would interact, but in it there are no personal computers or PDAs or smartphones, and astronauts on the *Discovery* wrote in pencil on a piece of paper.

How is it that today's reality is so very different from predictions in a film that, in its time, looked so plausible and realistic?

H.I.:   I savor divergence or difference of viewpoints about the future. For me, it's entirely natural. I also think it's healthy that there should be variety in the different versions of the future that people predict. You can think about different roads in which only one version of the future exists, and which could have been defined by Microsoft or Intel. But such a road would be boring, so I much prefer that people like Steve Jobs, Arthur C. Clarke, Stanley Kubrick, or myself have a variety of different versions of the future, from which people can select.

A.P.:   Is a natural language for all languages an irresolvable interface? Is it just a matter of time to find a solution? Will it take more research with sufficient computing capacity? Or does a critical level of complexity need to be overcome?

Do you think a machine will ever pass the Turing test as far as language is concerned?

H.I.:   That's also a very profound question. Perhaps machines may one day pass the Turing test. You also mentioned languages, and language is becoming very important because information retrievers are using all the vast number of information resources available on the Web these days. The question is, if natural language is really the ultimate interface, maybe the answer depends on what application you are going to design. But fundamentally, natural language requires huge commonsense knowledge in order to interpret the language appropriately and according to context; this is a very big challenge, and many researchers are still struggling to find the answer.

A.P.:   I know that this is a huge challenge for researchers. In your opinion, what is the next step in the attempt to advance the human-computer interface through spoken language?

H.I.:   I think it depends on which metaphor you choose to use when describing human-computer interaction. At this moment, you and I are interacting using natural language because we are both human beings who use language to communicate. If you want to make a computer become like a secretary or a friend, then maybe natural language might be a good choice. However, since computers lack common sense or experience of the real world, it's very difficult for them to understand or interpret anything that depends on subtle nuances of language. Perhaps we should start thinking about the computer not as a human-like creature but as a simple tool, like a pen or a pencil or this abacus. That is why my group takes the view that computers are more like a simple tool or a medium to help people express ideas or communicate. For this reason, it rejects the anthropomorphic notion of making the computer resemble a human being or intelligent body.

A.P.:   In your laboratory, to create some artifacts, I believe that you, like Professor Douglas Engelbart, use the "augmented reality" concept, which is something that appeals to me too. For example, in a Things That Think project, there are some objects that are "digitally augmented."

Can you tell me what these digitally augmented objects look like?

H.I.:   One important design principle is to utilize people's existing under-standing and knowledge about objects or the immediate environment. Without any digital augmentation, everybody already knows what a pen is and what paper is for, so using this knowledge, then adding augmented capabilities, seems an important way of approaching this. For example, this pen is a tool to write my ideas on a piece of paper. But imagine for a moment that this pen has a huge amount of memory and can recall every idea that I've written down throughout my lifetime—and that I can retrieve all of them. In this way, the pen can become an interesting medium for memoriz-ing important ideas.

A.P.:   Among the huge number of projects undertaken by the Tangible Media Group, I have long been fascinated by Topobo.[3] I saw the first proto-types personally at the Media Lab. Topobo is like a modular volumetric puzzle, with a simple but powerful idea: it has embedded kinetic memory, a kind of memory capable of copying and pasting physical motion.

Conceptually speaking, what type of memory is it, and how might it be used?

H.I.:   I'm very glad that you liked Topobo, which was designed by Hayes Raffle and Amanda Parkes, both PhD graduates in the Tangible Media Group. You mentioned memory, and specifically kinetic memory—in other words, memory that captures motion and movement for replay later. This notion of "record and play" came from another project called curlybot, undertaken by Phil Frei, who graduated from the Tangible Media Group a few years ago.[4] By providing people with a simple mechanism of record and play, they can manipulate physical objects; then the objects remember the motions that they made. In this way, people can learn a variety of complex ideas like differential geometry or how to design motion (as you also can by using Topobo). We are very excited about this simple function of memory, which captures motion and explores dynamic behavior in a variety of toys and robots.

A.P.:   At the Tangible Media Group, you do not only invent artifacts, whether tangible or intangible. When you invent an entirely new thing, you must also invent a word that expresses the essence of such an artifact.

Is it as difficult to come up with these new words as inventing the artifacts themselves?

H.I.:   I believe that a new concept or artefact and its name are inseparable; they are as one because we have to communicate the new concept, idea, or philosophy to others, hoping that people will remember that idea through its name and also use it in their own work.

I chose the term tangible as a key word in my group because at that time all the human-computer interfaces used pixels that are intangible. So making information tangible has quite a strong message. Also, it's very simple to create a word that is easy to remember, so today in the world of human-computer interactions many people are using the term tangibles.

A.P.:   Arthur C. Clarke said "Any sufficiently advanced technology is indistinguishable from magic in your research group."[5]

Do you perform magic at your lab?

H.I.:   Maybe, yes! I strongly agree with this famous phrase of Arthur C. Clarke. Additionally, magic brings something of a sense of wonder and surprise to people, and this in turn makes them more engaged in the interaction. So I think that this magic quality is very important and helps us a lot when we are designing interaction.

A.P.:   My last question is about the vision of the Media Lab. Referring to the MIT Media Lab, Nicholas Negroponte said that what characterized people there was the fact that they were guided by passion. Michael Hawley, a former professor at the MIT Media Lab, where he was cofounder and director of the Things That Think project, told us that MIT is a place where "people have their head and their heart in the right place."

In your opinion, what are the main characteristics of those who work with you in the Media Lab's Tangible Media Group?

H.I.:   I would say the MIT Media Lab is a place to invent a future of your own.

A.P.:   Thank you for sharing your ideas.

H.I.:   My pleasure.

**Notes**

1. The raison d'être of Tangible Bits is articulated on its website, to wit:

Tangible Bits is our vision of Human Computer Interaction (HCI) which guides our research in the Tangible Media Group. People have developed sophisticated skills for sensing and manipulating our physical environments. However, most of these skills are not employed by traditional GUI (Graphical User Interface). Tangible Bits seeks to build upon these skills by giving physical form to digital information, seamlessly coupling the dual worlds of bits and atoms." (http://tangible.media.mit.edu/project/tangible-bits)

See also "Toward Seamless Interfaces between People, Bits and Atoms," http://web.media.mit.edu/~anjchang/ti01/ishii-chi97-tangbits.pdf.

2. *Affordance* refers to all actions that are physically possible and those that we are aware of (https://en.wikipedia.org/wiki/Affordance).

3. Topobo is described on the Tangible Media Group's website as "a 3D constructive assembly system with kinetic memory, the ability to record and playback physical motion" (http://tangible.media.mit.edu/project/topobo).

4. curlybot is described on the Tangible Media Group's website as "a toy that can record and playback physical motion. As one plays with it, it remembers how it has been moved and can replay that movement with all the intricacies of the original gesture; every pause, acceleration, and even the shaking in the user's hand, is recorded. curlybot then repeats that gesture indefinitely creating beautiful and expressive patterns" (http://tangible.media.mit.edu/project/curlybot).

5. The science fiction writer Sir Arthur C. Clarke promulgated three rules to help people think about the claims of developments in science. His third and most famous law, any sufficiently advanced technology is indistinguishable from magic, he is said to have added because Newton had three laws.

# II Information

Information scientists describe a hierarchy of the known and the knowable, with data at the base of the pyramid and information at a higher level. At a third level of refinement is knowledge, and finally, at the top, is wisdom. Whether or not one agrees that this pyramid accurately describes how we understand our complex world, one thing is clear: we are drowning in data. Humanity now produces, gathers, and stores incredible amounts of data, but extracting information or knowledge from it is harder. We may be more informed, but not for that reason any wiser.

In 1999, Manuel Castells published his book *The Information Age*, which described what he called the "network society."[1] The personal computer had begun to permeate all latitudes. That same year, scientists at the SETI project (Search for Extra-Terrestrial Intelligence) came up with a brilliant and somewhat utopian idea. It entailed undertaking, on a grand scale, the search for some form of intelligence beyond our Solar System based on tracking data.[2] The immense parabolic radio telescope at Arecibo, Puerto Rico, could gather data at a superhuman scale. But the computational capacity to process the data posed a problem, to which the SETI scientists came up with an unprecedented solution.

Their brilliant idea was to launch the SETI@home project, which made a worldwide appeal to volunteers to form a distributed community of PC users who would allow the project to make use of the "free time" of their domestic computers. The resulting distributed global network would process the profusion of data from that grand telescope. The application that allowed this was a simple screensaver that processed the radio signals from the telescope, once broken down into a multitude of two-minute data packages. These were sent to the voluntary collaborators throughout the world who made their computers available. The results of that space signal processing were then collected at the University of California, Berkeley. The hope was to discover whether, in any of the data sequences captured in space, there emerged something that could be recognized as the product of intelligent beings.

SETI offers a good metaphor for how science tackles the superhuman quantity of data that our available technology is capable of gathering. To give you an idea of the project's social scope, more than five million users in more than 233 different countries have participated in this project by providing more than 308,000 million computer hours on their machines so far. According to Wikipedia, on June 23, 2013, and with more than 278,832 computers active in the system (1.4 million in total), SETI@home is the world's fastest supercomputer, capable of calculating 33.86 pet-aflops.[3] To date, the most promising signal analyzed by SETI has been the SHGb02+14a, which originated in the Pisces and Aries constellations 1,000 light-years from Earth.[4] That means that it originated at least one thousand years ago when, here on Earth, we were undergoing the pre-Medieval period.

The SETI project offers a romantic vision of what is possible in the search for extraterrestrial life. The project was largely conceived by Frank Drake, who created the Drake equation to determine the number of technological civilizations that could exist in our galaxy.[5] In 1974, Drake and Carl Sagan broadcast the "Arecibo message" from the Arecibo radio telescope, a two-minute-long communication that was sent in the direction of the astronomical object M13, an accumulation of some 500,000 very old stars. The communication, based on Leibniz's paradigm that "everything can be expressed by a one or its absence," and having a length of 1,679 bits, contained information on the Solar System, the Earth, and the human species, and was written in binary language (zeros and ones). The assumption was that binary code might be sufficiently universal, a language the supposed extraterrestrial intelligence would be able to translate. The experiment was largely symbolic, though, since M13 is situated 25,000 light-years from Earth. Any immediate response, traveling at the speed of light, would not reach Earth for another 50,000 years. Who knows what will have become of humanity in 50,000 years!

Human beings currently live in a world in which we are confronting amounts of information of a previously unthinkable magnitude. What else can we understand about this information age?

Currently, most of us are connected to millions of others, with disconnected areas of the planet becoming rare. We function as de facto creators, distributors, and consumers of digital data.

Finding signals in the noise may become more difficult as our data proliferate. I hope that the following conversations can offer some knowledge, and even some wisdom. They reflect on topics as diverse as the culture of convergence; the logic of computer science as opposed to that of physics, open knowledge, software algorithms that make the implicit explicit, the emergence of nonbiological intelligences, or "remembering" the future, the challenge of openly spreading knowledge, the potential of technology to make the world a better place, and encryption as a human right.

Having learned from many skeptical scientists, I am cautious about technology and the claims of its promoters. As with any tool, technology may be used—and is being used—for good as well as for ill. But I also bear in mind opinions like that of Dieter Bohn, who says we should not see technology as a tool but as an instrument.[6] Technology must be used like a violin, not like a hammer, as something with which to create new things. Umberto Eco has argued that one idea immediately summons another, and one random book makes you want to read another.[7] In a similar way, we can consider the generative nature of technology without losing sight of its unintended consequences. In many respects, my interlocutors in the forthcoming pages were much more optimistic than I was before speaking with them. I believe this is because they know much more than I do. Knowing more does not have to lead to pessimism, quite the contrary. I hope the following dialogues convey some of that optimism.

## Notes

1. M. Castells, *The Rise of the Network Society,* Vol. 1, *The Information Age: Economy, Society, and Culture* (Oxford: Blackwell Publishing, 2010).

2. SETI Institute, "Who We Are / Our Mission," last modified January 6, 2016, http://www.seti.org/about-us.

3. A petaflop is the computation of one quadrillion floating point operations per second (FLOPS) (http://www.webopedia.com/TERM/P/petaflop.html).

4. "Radio Source SHGb02+14ª," last modified August 19, 2015, https://en.wikipedia.org/wiki/Radio_source_SHGb02%2B14a.

5. The Drake Equation, invented by Frank Drake while he was working as an astronomer at the National Radio Astronomy Observatory in Green Bank, West Virginia, offers a way to bound the number of possible technological civilizations in the universe by "identifying specific factors thought to play a role in the development of such civilizations." SETI Institute, "The Drake equation," last modified January 6, 2016, http://www.seti.org/drakeequation.

6. Dieter Bohn, "Technology Isn't a Tool, It's an Instrument," *The Verge*, October 11, 2015, http://www.theverge.com/2015/10/11/9493425/technology-isnt-a-tool-its -an-instrument.

7. Lila Azam Zanganeh, "Umberto Eco: The Art of Fiction No. 197," *The Paris Review* 185 (Sumer 2008), http://www.theparisreview.org/interviews/5856/the-art-of-fiction -no-197-umberto-eco.

# 12   Convergence Culture: Where Old and New Media Collide

Henry Jenkins and Adolfo Plasencia

Henry Jenkins. Photograph by Douglas Morgenstern.

*How to widen access to the new social skills and cultural competences undoubtedly constitutes the new "hidden curriculum."*

*It's not a question of replacing traditional cultural literacy with media literacy. It's a question of expanding how we learn and how we process that information.*
*—Henry Jenkins*

Henry Jenkins is the Provost Professor of Communication, Journalism, Cinematic Arts and Education at the University of Southern California. He was formerly the Peter de Florez Professor of Humanities at MIT and Co-Director of the program in Comparative Media Studies. He holds a master's degree in communication studies from the University of Iowa and a doctorate in communication arts from the University of Wisconsin, Madison.

Jenkins served as Principal Investigator for Project New Media Literacies, a group that originated as part of the MacArthur Digital Media and Learning Initiative, and has also led the Convergence Culture Consortium, which built bridges between academic researchers and the media industry to help inform the rethinking of consumer relations in an age of participatory culture. Jenkins wrote a white paper on learning in a participatory culture that has become the springboard for discussions about new media literacies around the world, and his group has developed and tested instructional materials and professional development programs to put these concepts into action.

Among his publications are *Textual Poachers: Television Fans and Participatory Culture* (Routledge, 1992), *Hop on Pop: The Politics and Pleasures of Popular Culture* (Duke University Press, 2003), and *Convergence Culture: Where Old and New Media Collide* (NYU Press, 2006), *Fans, Bloggers and Gamers: Exploring Participatory Culture* (NYU Press, 2006), and *Spreadable Media: Creating Meaning and Value in a Networked Culture* (NYU Press, 2006). The last titles return to the question of media audiences and participatory cultures at a moment when fans and fan activities are central to the way the culture industries operate.

*Note:* The original dialogue took place May 16, 2007, in the office of Henry Jenkins, Comparative Media Studies, MIT. This text was revised by Henry Jenkins in June 2015.

Adolfo Plasencia:   Here we are with Professor Henry Jenkins. Thanks for agreeing to this dialogue.

Henry Jenkins:   It's a pleasure.

A.P.:   Henry, I know you're interested in the effect caused by the combination of various media on the education of children and adolescents, and that you've identified eleven skills that young people need in order to move within a participatory culture on a global scale.

What are the opportunities and the troublesome aspects, if any, that should be taken into account in all this?

H.J.:   The world has been immersed in a prolonged period of transition as far as digital and communications media are concerned. Changes in media technologies, in fact, are rooted in how we process knowledge, how we construct information, how we participate in the community and in society, and, of course, in how we learn. Young people all over the world have been at the center of all these changes.

There are many young people who adopt new technologies, trying out and testing things spontaneously, such as game fans and bloggers. They are forging new relationships with information and the communities that have grown up around information media. The challenges we face are threefold. First, new skills are unequally distributed in all cultures of the world; not even within North American culture are they distributed evenly, and on the world scale, they are distributed even less so. This is what we call the "participation gap."

At the end of the twentieth century, we were talking about a digital divide. In the American context, the digital divide has mostly been overcome, if by digital divide we understand access to digital technology, as most young Americans now have access to computers in schools and public libraries, if not necessarily at home. However, as we close that divide, it is increasingly clear that, socially and culturally, another breach is opening up. Most of us have twenty-four-hour access to new technologies, which constitute a massive and powerful learning tool, but the same learning opportunities are not available to the children who have only ten-minute access a day in a public library, with limited broadband and compulsory filters and with no possibility of publishing things. This poses an obstacle to full participation in the technologies that are now available.

So the challenge is how to widen access to these new social skills and cultural competences, which undoubtedly constitutes the new "hidden curriculum." In the same way as young people in the past who had access to opera recordings and visits to museums and dinner-table chats on politics did better in school and were seen by their teachers as better students, young people today who have access to virtual universes, online metaverses, or blogging technology, fan communities, and all those things that require new skills are better prepared for the future, and for the same reason, children who do not have access to these experiences are left behind in the classroom. And those who are acquiring those skills are often not allowed to tap their best ways of learning when they enter the classroom.

Second, there are questions about digital ethics. A huge number of young people are producing media content and share what they create with a public beyond their friends and family. Many of them do not have adults behind them to guide them or to give them ideas on the ethical decisions they have to make as participants in online communities and as media producers and distributors. That, I believe, is the fundamental challenge.

Young people don't need adults looking over their shoulders all the time but do need adults to watch their backs. They need adult mentors, someone who understands the risk they are taking and the ethical decisions they make, someone who can direct them to resources that can help them grow. These are the areas that the Media Literacy Project at MIT pioneered. Our intention was to encourage thought on this new media literacies both in and out of school, to help young people see the opportunities represented by the new digital landscape, but also to reflect on the unequal distribution of skills, the so-called participatory gap, with the aim of encouraging young people to think about the ethical decisions they are taking.

A.P.:   For eleven years in the MITUPV Exchange between MIT and the Polytechnic University of Valencia (UPV), Spain, we participated in an initiative that you cofounded and were the driving force of during your period at MIT, called Digital Humanities: Transforming Humanities Education.[1] That expression was a very provocative concept for some traditionalists in the humanities faculties in Spain.

In what ways is technology transforming traditional humanities?

H.J.:   There are many things in teaching humanities that we wouldn't like to change: there is deep respect for the past, a profound immersion in culture, and growing sensitivity to diversity. These are fundamental things in the teaching of humanities, and we must respect and continue to transmit them in the twenty-first century. It's not a question of doing away with Shakespeare or Cervantes, or of rejecting the history of our societies, or of replacing traditional cultural literacy with media literacy. It's a question of expanding how we learn and how we process that information.

Take, for example, a project such as that of Peter Donaldson's Shakespeare Electronic Archives. This brings together the original texts of Shakespeare's plays, the original quartos and folios, with hundreds, if not thousands, of images from the archives of the Folger Shakespeare library and videos of performances of his both on-stage and on-screen. Students can then access that information on the Internet, extract the information that is relevant to them, comment on it, study it in depth, and share that information with other students. Digital humanities can also connect people from different fields through programs such as MIT's foreign languages projects, which connect students who study in America with students who speak Spanish or French in their native countries and allow them to share materials, comment on each other's materials, and enter into conversations with each other. This helps the humanities enter the networking era of culture.

It helps us make use of the global dimensions of the new media landscape in such a way that we may learn from each other. So it's not a question of changing humanities but of using the new tools that allow us to tackle the new pedagogical opportunities that the media provide, the prospects they offer for studying the same things that we used to study but in a much deeper way. That has nothing to do with throwing away the past; on the contrary, it is committing ourselves to the past in new ways and making use of new media so as to make our understanding of traditional humanities deeper and richer.

A.P.:   Henry, let's talk now about your book, *Convergence Culture: Where Old and New Media Collide*, which has become a classic.[2] You refer to convergence not in the customary senses as convergence of media but in terms of production and cultural interaction.

Do you think that the technology industry has imposed its hardware and software paradigms (artifacts and tools replacing the role of thought), with the result that there is neglect of what is most relevant, namely, the intellectual processes and cultural interactions of users who share electronic communication spaces?

H.J.:   I think that when many in industry speak of convergence they are usually talking about the imagined black box through which the content of all the communication media flows. That's the wrong way of looking at it, reducing it to a technological question instead of talking about it as a cultural issue. In my living room I have more black boxes than ever, heaped one on top of another with remote controls incompatible with each other that get lost down the side of the sofa and with cables all intertwined like spaghetti. My students carry around more devices in their backpacks than the soldiers in Vietnam did! Today's smart phone is the technological equivalent of a Swiss army knife: it has all the functions imaginable and does everything badly. (*Note*: I made this comment before the introduction of the iPhone and other smart phone technologies, which significantly improved the functioning and usability of these devices.)

We are in a period of profound and prolonged change in which technological convergence is being developed, but at the same time technological change is incapable of keeping up with what I call convergence culture. In "convergence culture" there are ways of thinking of, about, and through digital media that open up new opportunities for us. It's about a

world where every story, every sound, every image, every brand, every relationship is spread across the maximum number of media channels. Convergence culture is taking shape in teenagers' bedrooms and simultaneously in
the boardrooms of major communications companies; that is, it's taking
shape both through business practices that wish to exploit synergies across
diverse media channels and through consumers who want to consume the
media they want when they want it, where they want it, and in, the form
in which they want it.

As we learn to live with a new convergence culture, we need new ways of
thinking, new ways of connecting with each other. One of them is what the
theoretician Pierre Lévy calls "collective intelligence." As he says, collective
intelligence is what arises in a networked society where nobody knows
everything, everyone knows something, and what is known by any one of
the members is available to the group as needed on an ad hoc basis. So
online communities are pooling knowledge and skills, tapping diverse
experiences and expertise, and working through complex problems together
in ways that would be impossible in isolation. This way of treating information as a shared community resource serves the needs of the convergence
culture I am describing.

I'm very interested in how we change cognitively, how we produce
culture, how we tell stories in new ways, how we respond to changes in the
global cultural landscape. I think that what we are seeing, as we take advantage of all the new things that digital media provide us with, is that we are
changing how we process knowledge and how we create culture. And that's
what most interests me, this concept of convergence culture that I am
trying to describe in the book.

A.P.:   Another question related to the book: do you think that the collision
between old and new media can produce cross-fertilization?

H.J.:   I believe what we are witnessing is the birth of a new cultural ecosystem. However, the question of what is called collision or intersection is not
defined. My original title was "Where New Media Intersect with the Old."
It was my editor who thought it would be better to introduce the word "collide." But there's a lot in play in these two verbs. On the one hand, of
course, there's the antagonism between the established commercial media
and the new forms of communication, which are more grassroots and participatory. Everyone seems to imagine a more participatory culture but
nobody agrees on the conditions and forms of participation.

I think what we are currently seeing is that consumers are going to take
the contents of communications media as the raw material for their own

cultural production and social participation, and yes, it's true, there is cross-fertilization between the old and new media, but there are also tensions as the new and traditional media compete for consumer attention. We are seeing expanded expectations on the part of consumers about the active role they can and should be playing in shaping the social and political agenda.

A.P.:   Referring to another book of yours titled *Fans, Bloggers and Gamers*: in Spanish there is no word for the last two terms, so we use the English words.[3] Although they appear to be recent terms, the ancient Greeks had fan clubs for their mythological heroes and gods, the remains of which I was able to see during my visit to the Valley of the Temples in Greece and to the best conserved theater, at Epidaurus.

Do you think that the current phenomena of fans and bloggers are generational?

Are they ephemeral, or do they transcend specific epochs and local cultures?

H.J.:   I'm sure all these phenomena operate in a much wider history. As for fans, people over thousands and thousands of years have told the stories that interested them. They took the sagas of great cultural heroes and elaborated them in their own language for their own purposes in order to transmit their own messages. That's how culture has grown. So we can trace the route, let's say, of King Arthur from when he was an obscure figure in the footnotes of some chronicle of the kings of England right up to the fully developed mythology that now surrounds King Arthur.

This comes about through that elaboration of the basic story. This is what the fan culture means today; they are people who are passionately interested in stories with certain characters, and they rewrite the stories and tell them in their own terms. In the history of cultural production, this period of large companies regulating author's rights will turn out to be a bubble, a blip. It is a parenthesis in the middle of a greater evolution of our present culture.

Blogging is the same. It comes from that hunger that human beings feel for commenting on politics, life, and culture among themselves so as to tell their own stories.

Gaming is a much more complicated question. Of course, games are one of the oldest forms of human culture. We define play in a wide sense and suggest that humans are not the only species that play, as we know animals play too, but games are rule-governed activities that seem uniquely human. There are different things that computer games have provided us with that show the fundamental nature of play, new technological possibilities, new links and affiliations that are changing what we understand by play. I think that the gap between a board game and a computer game is quite large, and in part this has to do with the capacity to create animation by computer, stories that emerge in real time, complex artificial intelligence, and so forth. In many aspects digital games on the Web go much further in terms of what we call cultural expression than the games we had in the past.

A.P.:   With regard to authorship, before the Renaissance, cultural and artistic works were not signed by their authors. European cathedrals didn't have the architect's signature at the bottom of the façade. Today, in Spain, and in Europe in general, the author is marketed almost as a sacred entity, and the law supports that paradigm of intellectual property. Web 2.0 culture and Wikipedia are counterexamples of a different cultural construction. What do you think of open and collective cultural production along the lines of Lessig's model and his warning about the power of large media forces to enclose and limit culture and creativity?

H.J.:   This is probably one of the most important cultural fights of our time. This tension between a legal culture that emphasizes exclusive rights controlling intellectual property goes against the new forms of culture and new forms of technology that stimulate collaboration, shared authorship, and grassroots circulation of media content. I think that's where we are as a society, and I believe you are absolutely right in saying that that focus on collaborative authorship taps some very old values and practices. The idea of culture being a collaborative creation, of artists in dialogue with their culture, dates back to the beginning of humanity. The idea of an author's rights is an aberration. In short, I think the legal structures will buckle under the challenge because it's going to be very difficult for media industries to hold the reins or regulate that culture of "remixing" and fight against this emerging commitment to shared participation and collaboration in producing culture. It'll be fascinating—over the next decade or two—to observe how this conflict gets resolved.

A.P.:   Lawrence Lessig gave a talk in Barcelona on the Apple iTunes advertisement, "Rip. Mix. Burn." Today it might be "Rip. Mix. Post." How do you see this paradigm of sharing playing out among lay users and among the scientific, academic, and literary community?

H.J.:   *Time* magazine annually chooses a personality of the year. In 2006 the cover read: "You. Yes, you. You control the Information Age. Welcome to your world." According to that cover, *Time* makes you the "person of the year." The choice of that pronoun has always intrigued me because "you" in English is both singular and plural, and that is significant. "You" means you as individual and you as a group. And when we think of "you" in You-Tube, the first question is, perhaps, does it stand for individualized personal expression or shared cultural production? What interests me and what is radical about YouTube is that idea of a space where everyone shares media with everyone else, where the media produced by the fans, the media produced by educational and governmental bodies, and the media produced by companies coexist.

It's what Yochai Benkler would call "mixed media ecology," where the walls separating different types of media producers fall down and the role of the public is to curate and circulate content.[4] This is bottom-up creativity. The problem is that legal culture tries to stifle the public's active role in shaping our culture. That is going to be one of our main challenges in the future.

Business groups increasingly tell me they are worried because they fear losing control of media flows. The first message I try to convey to them is that they have already lost control. The contents are at the disposition of whoever wants them. Anyone can take the content and do what they like with it. The best thing for them to do is to return to the conversation, become part of the dialogue that surrounds the media. Then I tell those business groups: "In the modern age, if it doesn't spread, it's dead. If the contents do not circulate, they are trapped." It's no longer simply about capturing consumers and maintaining their loyalty; it's about enabling consumers to circulate your messages as resources for their own conversations and in the process, invest them with new meaning and value.

That causes an interesting tension. Is it collaboration or is it a collision between the mass media and the new digital media? Or is a new relationship emerging? What happens on YouTube is not intrinsically a fight between consumers and producers. Perhaps it is simply a new configuration of the logic by which our culture operates, a logic that makes the interest of the consumer coincide with the economic interests of the companies in a new way.

A.P.:   You state that YouTube is not just a distribution channel, it's a source of cultural creation and a prime example of what you call "convergence culture." It is, of course, a huge form of collective audiovisual narration where the spectators have taken over the role of editors. If this is the case, what is the relevance of Marshall McLuhan's "The medium is the message"?

H.J.:   I think that McLuhan was only partly right when he said that. He was a great enthusiast of a kind of deterministic technological model, according to which the introduction of a new technology radically alters the cultural setting. Although for fifty years now we have been studying his work and basing our media studies on him, we have discovered that the same technology acts in very different ways in different cultural contexts, which we cannot predict. In absolute terms, when a technology is developed, we do not know how it is going to be used. What pioneering users are doing—the pioneers who adopt and adapt a new technology—paves the way for the second stage of technological development.

In convergence culture, I distinguish between interactivity and participation, which is a good way of approaching that tension established between technological change and cultural change. Participation is a property of cultures. A culture can be more or less participatory, and as a new medium gets inserted into a culture, it may be used in ways that are more passive or active, depending on the tendencies of that community. Interactivity is a property of technology. It has to do with how technologies are designed to be used. The introduction of a new technology opens up new affordances, new capabilities for its use, but cultures determine what users it attracts and how it is used.

A.P.:   Henry, I'd like to ask many more questions, but because of time constraints, this is the last.

While reading your most recent interviews and writings, I have the feeling that the current explosion of new media, social media, and the Web 2.0 has enthralled you with what the ancient Greeks called "ecstasies." or "enthusiasms."

You seem to be enthusiastic and happy about what is happening and what you are experiencing in this field. Is that correct?

Can the rest of us become as enthusiastic, or are there warning signs?

H.J.:   I describe my own points of view as those of a critical utopian. The concept of utopia implies inventing a better society, imagining a world that would be an improvement on the one we live in. Many people think that utopians are naïve and unrealistic, but that's not what I say. That's not the essential part of the critical utopian. If we can formulate a model of a future world in which we would like to live, and if we can use it as a landmark for measuring our current problems, identifying the faults in our present system, then we can use it as a roadmap for knowing what changes have to be made in order to arrive there.

A naïve utopian, by contrast, is an eternal optimist who imagines a better world and sits down to wait to see what happens.

What I am suggesting is that we recognize the progressive trends in our culture. We should be identifying the potential for cultural and social transformation. Our intellectuals should become intermediaries who help facilitate meaningful change by inserting themselves into the important conversations of our era. We should imagine a better world and work to achieve it.

I now see a strong potential for the public in terms of cultural production, new ways of relating with each other, new forms of politics, new types of economic life and of legal culture that could arise from this type of collision between new and old media, as described by *Convergence Culture*. I want to show people where that potential lies and document that potential by documenting various sites where change is taking place.

So we can say: Well, what do we have to do in the educational sector to ensure that all children have access to the new social skills and to cultural competencies necessary to become full participants in the world today?

We can ask ourselves what the legal structure would have to be like to allow us to have the freedom of expression that seems to be the promise of this new media landscape. And we can ask what the new type of political organization would have to be like for implementing and returning democracy to the people so as to allow them to have greater control over the government, and to make their leaders more responsible. That is the challenge. It's not a question of living in a perfect society but of using what we observe about the present to reinvent the world, and I think that the challenge is to keep our ideals open and return to the real world and say, "Look how far we are from that ideal. What are the steps that we have to take so as to be able to arrive at a better society?"

A.P.:   Thanks, Henry! I hope to see you in Spain soon.

H.J.:   Thanks to you.

## Notes

1. The MITUPV Exchange was an online cultural and language-learning exchange between students at the Polytechnic University of Valencia, Spain, and MIT students. The Transforming Humanities Education initiative of the MIT Comparative Media Studies program focuses on emergent practices to help teachers exploit the potentials of new media to engage students in learning, and also so that teachers can make the subject contents and skill sets of the humanities more engaging for their students. (http://web.mit.edu/21fms/Research/transform.html).

2. Henry Jenkins, *Convergence Culture; Where Old and New Media Collide* (NYU Press, 2006). See also Jenkins, "Welcome to Convergence Culture," blog post, *Confessions of an Aca-Fan: The Official Weblog of Henry Jenkins,* June 19, 2006, http://henryjenkins .org/2006/06/welcome_to_convergence_culture.html.

3. Henry Jenkins, *Fans, Bloggers, and Gamers: Media Consumers in a Digital Age* (NYU Press, 2006).

4. Yochai Benkler is Berkman Professor of Entrepreneurial Legal Studies at Harvard Law School and faculty codirector of the Berkman Center for Internet and Society at Harvard University. In a blog post, Jenkins wrote, "Yochai Benkler and others articulate in this context a hybrid or new mixed media ecology, typified by a global digital culture," "The Ethics of the Sociable Web and the Shifting Roles of Media Theorists," blog post, *Confessions of an Aca-Fan: The Official Weblog of Henry Jenkins,* July 10, 2007, http://henryjenkins.org/2007/07/the_ethics_of_the_sociable _web.html.

# 13    The Logic of Physics versus the Logic of Computer Science

**Bebo White and Adolfo Plasencia**

Bebo White. Photograph courtesy of B.W.

*Yes ... [computing] is an approximative science. In fact, traditional scientists, such as physicists, often argue whether computer science is a science at all.*

*At the very beginning, we thought people would not remember how to spell HTTP, it was so strange at the time. Today it is so universal nobody seems to have any problems.*
*—Bebo White*

Bebo White is a Departmental Associate (Emeritus) at the SLAC National Accelerator Laboratory, Stanford University, and Visiting Professor of

Computer Science at the University of Hong Kong. Working as a computational physicist, he first became involved with emerging Web technology while on sabbatical at CERN in 1989. Upon his return to the United States he became part of the team that established the first non-European website, at SLAC.

His academic research interests have evolved in parallel with Web technology and include computational physics, high-energy physics, Web science, Linked Open Data, "The Internet/Web of Things," online security, human-computer interactions, Web-based teaching and learning, cybercurrencies, and Web engineering.

Adolfo Plasencia:   Bebo, thanks for meeting me for this conversation.

Bebo White:   My pleasure.

A.P.:   You are a computational physicist and an information systems analyst at SLAC,[1] but your background is actually in physics; you trained as a physicist, but you are also interested in computing. Am I right?

You come from the world of physics, don't you?

B.W.:   Yes, I am originally a physicist with an interest in instrumentation and in data analysis produced by instruments. That is how I arrived in the IT world. And based as I am in the area of San Francisco, you can imagine how hard it is to avoid computers.

A.P.:   Bebo, was it by chance that you became a physicist and that the Web was invented at CERN, the particle center? Or was it your sabbatical as a physicist that led you to CERN and to get involved with the Web project?

B.W.:   It was physics, definitely. I remember the first time I signed up for a computer course. I thought it was so simplistic, so silly. The Web was invented at CERN, which is primarily a physics laboratory.[2] I was working at SLAC at the time, and the need for SLAC to collaborate with other international laboratories was the reason we got involved with the Web at such an early stage. Our website at SLAC was the first site not only in the United States but in the whole Western Hemisphere, and the fifth in the world.

A.P.:   Yes, we are going to talk about that. What I meant is, was it the logic of things, of reality, or was it chance that took you to the Web?

B.W.:   Let me tell you a little story and you will understand. A major contribution of SLAC to the high-energy physics (HEP) community is a preprints database, used by physicists from all over the world. However, prior to the Web, the access to that database was extremely awkward. When

SLAC physicists visited CERN, the physicists at CERN would complain about how difficult it was to use SLAC's preprints database.

Tim Berners-Lee showed us during a visit to CERN (before 1990) that with a Web browser it was possible to access the database independent of the user's operating system. This was important since the HEP labs were inconsistent in the computer operating systems they used. By using the Web interface, issues with access to the preprints database miraculously disappeared.

A.P.:   Bebo, is the logic of physics the same as the logic of computing? Some argue that computing is an approximative science. What do you think?

B.W.:   Yes, I agree, it is an approximative science. In fact, traditional scientists, such as physicists, often argue whether computer science is a science at all. You actually want to know whether computer science adheres to the scientific method, don't you? However, I suspect that the algorithmic rigor (or logic) of computer science could have made me a better physicist. I am very interested in Web science, which tries to understand the evolution and possible future of the Web in scientific terms. There is also no doubt that Web technology—and all computing—has had a major impact on how scientific research is done. That's why I am fascinated by computational science.

A.P.:   The International Academy of Digital Arts and Sciences gave you the Webby Award in San Francisco, the "Oscar" of digital arts and sciences. Other awardees have included the music star David Bowie, the great filmmaker Francis Ford Coppola, the actress Susan Sarandon, and many more from their field. How do you feel being among such a group?

B.W.:   Well, it is not exactly an award; it is more like being a member of a committee. The goal is to see how the Web—not only the Web but digital technology in general—can help artists such as David Bowie or Francis Ford Coppola advance their art. I am very happy to be included in this group of people because the digital world and the Web are relevant not just to science and technology; they also have an impact on the art forms these artists use to communicate with people, for example in music or films.

A.P.:   Bebo, the twenty-fifth anniversary of the Web was recently celebrated. You were at CERN in the late 1980s, collaborating in the development of the hypertext transfer protocol, or HTTP, with a very small group of people, including Tim Berners-Lee and Robert Cailliau, the "fathers of the Web." According to the site WorldWideWebSize there are now more than 4.71 billion websites, and according to Google's total index there are more than 46 billion sites with URLs that begin with those letters, http.

How was your experience at CERN, twenty-five years ago? And what do you think about its amazing effects? Almost a third of the planet uses HTTP.

What are your memories and your feelings now?

B.W.:   It was a unique experience to work with people like Tim and Robert. Tim had a dream, and Robert took that dream and helped promote it. There is a book by Ted Nelson—who invented the terms "hypertext" and "hypermedia"—in which Tim is depicted as Don Quixote and Robert as Sancho Panza. Tim had all the great ideas and was like on another planet, and Robert, like Sancho, helped him keep his feet on the ground to make them happen.[3]

I am still amazed by the size of the Web nowadays. At the very beginning, we thought people would not remember how to spell HTTP, something so strange at the time. Today it is so universal that nobody seems to have any problems whatsoever.

A.P.:   Bebo, you took the Web technology with you, back to California, when you returned from CERN, and you collaborated in the launch of the first website in the United States. You are considered to be the "first American Webmaster" and one of the founders of the discipline of Web engineering. I think people did not realize at the beginning how important it was. They thought it was just a nice computer toy, as you once said in an interview.

Was it really like that?

B.W.:   Yes, at the beginning, physicists at SLAC/Stanford saw it as an exercise dreamed up by some computer geeks, but when they realized it made their job a lot easier and that they had more time for doing physics instead of computer work, they were anxious to adopt its use.

A.P.:   You said that you think it is important for the Web—developed by a small group of pioneers at CERN, you included—to keep its democratic nature. In an interview you said: you go to a website and at first you can't tell if it is the website of a large corporation or if a thirteen-year-old made it at home in the garage. Both the corporation and the child can use the same medium, the Web.

Is it important for you to have a free and open Internet?

B.W.:   There are two parts to that question. As for the first part, an open Internet is essential. In fact, some previous systems did not work for that reason, so it is definitely essential.

Then there is the issue of the validity or provenance of the data on it; that is another important aspect. To know if such data are valid, you need to be sure where they come from. The validity of data is essential, especially now with the increasing use of the Web for science. A student must be confident that references are valid, that they come from a qualified authority. So, for example, if you are sick and you want to check the opinion that someone gives you about your illness, you should be able to tell whether that person really knows what he or she is are talking about, or if it is someone who simply has an opinion.

A.P.:   But in today's world, on Web 2.0, Wikipedia is a high-level way of learning and knowing things, finding knowledge and information, I think. And yet the academic world and the big publishers criticized Wikipedia at the beginning. Some still do: they say it is not a proper encyclopedia, that the information is not reliable enough. What do you think?

B.W.:   The case of Wikipedia is very interesting because no one could have predicted that something like it would eventually exist and effectively make traditional encyclopedias obsolete. It is a huge collaborative environment and effort. Even the greatest encyclopedias have editors, and those encyclopedias reflect the views of their editors. With Wikipedia there is not a single editor, there are potentially thousands of editors, and the intention or the usual effect is that the valid information is finally condensed. Can I give you another example?

A.P.:   Of course.

B.W.:   You mentioned Web 2.0; a good example is Amazon. Amazon's main business is to sell books, but it allows its users to post reviews on the books Amazon sells. So in the end, all these reviews become part of the corpus of a book. And then a user or a potential buyer of the book finds and reads the reviews, and so he or she also becomes part of this whole network of knowledge about the book.

A.P.:   Bebo, let me go back to the first period of the Internet, and then we will talk about the present. You said that being so close to the explosive growth of the Internet at the time, right in the place where it all started, was exciting and overwhelming. You said it was like being in the eye of a hurricane, of a great storm. The explosion of Web 2.0, the Social Web, something which is now happening all over the world—do you think it is equally amazing?

B.W.:  Yes. At the beginning, the Web was fundamentally a method for making information readily and easily available; that was it. However, few people realize that Tim B-L's original proposal included a social component. With Web 2.0, we have seen that component realized. The so-called Social Web is more a space where people can meet up and interact with each other. It is a fundamental part of the Web's evolution and is here to stay.

A.P.:  Let's talk about the Social Web, Web 2.0. Social media are booming, it is the "hype cycle." Billions of people interact on social media. If they were countries, Facebook and its 1.71 billion users would be the first-largest country in the world; Whatsapp, with a user base of more than one billion would be second; Facebook Messenger with 1 billion active users per month would be third; QQ, the instant messaging service of China's social media, with 899 billion active users monthly , would rank fourth; WeChat, a chinese instant messaging cross-platform with 806 million users, would be fifth; QZone, the Chinese social network, with 652 million, would be sixth.[4] All of them have more "inhabitants" than the United States. What do you think about this? Does this change people's relationships in today's world? Why do you think international public spheres do not seem to care about this?

B.W.:  Well, relationships change in line with uses on the Web. That is, some people participate as a way of advertising themselves; some people want to be stars and have lots of contacts. But for many other people it is the way they communicate and cooperate with one another. A case that illustrates this very well is Barack Obama's early campaign and its use of social media; it did help him win the presidency of the United States. I think that the Web has significantly impacted all of human communication, not just what happens online. It has opened the door for significant discussion to a global audience on such issues as privacy, censorship, human rights, and much more.

A.P.:  Lately, in many European technology forums I attended, everybody seems to talk about the economic crisis. Usually, in IW3C2 Conferences[5]— the World Wide Web meetings (and you go to most of them)—you don't hear a word about the crisis, not even in European ones.

Is it because of the optimism that characterizes people like Tim Berners-Lee, Vinton Cerf, and yourself?

B.W.:  The WWW Conference Series is primarily an academic research conference dominated by computer scientists. Discussions of economic issues have appeared (largely in keynotes, workshops, panels, and the like) but are not core to the conference mission (except when they impact research funding). The economic issue does appear in the Web science community where the audience includes social scientists and economists. It also appears in discussions about e-government (or open government) and the Web and society.

A.P.:   I'd like to turn now to one or two topics that are seen as quite critical these days. One of these concerns the HTTP protocol. Not so long ago, Dean Barker posted an article titled "We Suck at HTTP" in which he criticized those who develop apps for having abandoned the standards that underlie HTTP and URL. Baker wrote, "Narcissism runs rampant in this industry, and our willingness to throw away and ignore some of the core philosophies of HTTP is just one manifestation of this."[6] Then again, the *New York Times* writer Conor Dougherty expressed much the same opinion in a recent article in which he said, "Unlike web pages, mobile apps do not have links. They do not have web addresses. They live in worlds by themselves ("walled gardens"), largely cut off from one another and the broader Internet. And so it is much harder to share the information found on them."[7]

Bebo, do you think that Barker is right to be indignant at the way those who develop apps disregard and reject the usual standards? Do you feel that there is indeed a problem of narcissism in the industry, or is it more a question of lack of knowledge and training on the part of the app developers of today that makes them disregard an important aspect of software development?

B.W.:   I so disagree with Conor Dougherty! Just because apps don't have links does not mean they do not communicate with or access information from the Web. The Web is simply hidden from the user's view (or recognition). The global database or library that is the Web is fundamentally accessed by URLs and URIs, so that means HTTP. It will be very counterproductive if app developers in their haste to get to the market decided to overlook or ignore the standards process.

A.P.:   Another topic of debate today is the complaint by one or two major Internet companies that have argued that the two full years required by W3C to establish standards for HTML5 is far too long for a fast-developing technological industry based on the Internet. They have called into question the scientific methods used by the W3C scientific community to determine standards. They argue that they're simply unwilling to wait so long and instead, at Google's initiative, intend to create a group to establish standards within the time limit that "the industry requires."

Do you think that these companies are being pressurized to some extent by Wall Street or by a desire for greater speed that comes from shareholders? Or is it really more that these interest groups don't like such standards being determined by W3C and the scientific community, which have greater regard for what Internet users really need than for corporate requirements? Could it be that industry is keen to assume the power that W3C currently holds to determine Internet standards in order to shape them to its own liking? What do you think of such an argument?

B.W.:   One of the greatest threats to the future of the Web and to bringing it to its fullest potential is to ignore or bypass open standards. The success of the Web is in part due to the fact that it is not a corporate product—that was always Tim B-L's intention. If its features and functionalities are driven by business models, special interest groups, or corporate competition, we would likely lose control of one of humanity's greatest resources because of a loss of interoperability and a development process based on self-interest. We all need to work together to keep this from happening.

A.P.:   Bebo, thank you ever so much for this conversation.

B.W.:   Thank you very much.

## Notes

1. Information on Stanford University's SLAC National Accelerator Laboratory can be found on its website (https://www6.slac.stanford.edu/about/slac-overview.aspx).

2. CERN, the European Organization for Nuclear Research, located near Geneva, describes itself and its research programs this way:

At CERN ... physicists and engineers are probing the fundamental structure of the universe. They use the world's largest and most complex scientific instruments to study the basic constituents of matter—the fundamental particles. The particles are made to collide together at close to the speed of light. The process gives the physicists clues about how the particles interact, and provides insights into the fundamental laws of nature. (http://home.cern/about)

3. Ted Nelson, *Geeks Bearing Gifts: How the Computer World Got This Way* (Sausalito, CA: Mindful Press, 2008).

4. Statista, "Leading Social Networks Worldwide as of September 2016," http://www.statista.com/statistics/272014/global-social-networks-ranked-by-number-of-users.

5. IW3C2 refers to the International World Wide Web Conferences Steering Committee (http://www.iw3c2.org/conferences).

6. Deane Barker, "We Suck at HTTP," *Gadgetopia*, January 7, 2015, http://gadgetopia.com/post/9236.

7. Conor Dougherty, "Apps Everywhere, but No Unifying Link," *New York Times,* January 5, 2015, http://nyti.ms/1xNyadV.

## 14 The Pillars of MIT: Innovation, Radical Meritocracy, and Open Knowledge

Hal Abelson and Adolfo Plasencia

Hal Abelson. Photograph by Joi Ito. (CC BY 2.0)

*My core fundamental values are academic freedom and being deeply intellectually honest.*

*I think we're only at the very, very beginning of what's happening with the digital revolution. It's still special that someone's connected to the Internet.*
—*Hal Abelson*

Hal Abelson is the Class of 1922 Professor of Computer Science and Engineering in the Department of Electrical Engineering and Computer Science at MIT. He holds an AB degree from Princeton University and a PhD in mathematics from MIT.

Abelson has been active since the 1970s in using computation as a conceptual framework in teaching. He directed the first implementation of the children's computer language Logo for the Apple computer, which made the language widely available on personal computers beginning in 1981. Together with MIT colleague Gerald Sussman, Abelson developed the computer science subject, Structure and Interpretation of Computer Programs, which is organized around the notion that a computer language is primarily a formal medium for expressing ideas about methodology rather than just a way to get a computer to perform operations. This work served as MIT's own introductory computer science subject from 1980 until 2007.

Abelson is cochair of the MIT Council on Educational Technology, which oversees MIT's strategic educational technology activities and investments. In this capacity he played key roles in fostering MIT institutional educational technology initiatives such MIT OpenCourseWare and DSpace. He co-authored the 2008 book *Blown to Bits* (Addison-Wesley, 2008; now available through the Creative Commons), which describes the cultural and political disruptions caused by the information explosion. Abelson also teaches the MIT course Ethics and Law on the Electronic Frontier. He is a founding director of Creative Commons, Public Knowledge, and the Free Software Foundation and a former director of the Center for Democracy and Technology, organizations devoted to strengthening the global intellectual commons.

His awards include the Bose Award (MIT's School of Engineering teaching award, 1992), the Taylor L. Booth Education Award of the IEEE Computer Society (1995), the ACM Special Interest Group on Computer Science Education Award for Outstanding Contribution to Computer Science Education (2012), and the ACM Karl Karlstrom Outstanding Educator Award (2011).

*Note:* This dialogue took place in May 2006 in the office of Hal Abelson in the Ray and Maria Stata Center, the headquarters of the MIT Computer Science and Artificial Intelligence Laboratory (CSAIL). This text was revised by Hal Abelson in June 2015.

Adolfo Plasencia:   Thanks for agreeing to this conversation. I'm extremely pleased to be here with you. At first I had thought about preparing the questions, including those in your biography. However, it's so impressive that we could do a hundred interviews alone concerning your background. Therefore, it is probably best to start by discussing your career.

What is it you most like and have most enjoyed from the experiences gained and achievements during your career?

Hal Abelson:    I came to MIT as a graduate student when I was twenty-two years old and I studied mathematics. I got a PhD in the Mathematics Department and I became very interested in computing and also in educational computing. I worked on a project called Logo, contributed to the development of the Logo Turtle, and then got involved in research at MIT in educational computing. After that, I started working in the Computer Science Department at MIT. I worked on developing what has become MIT's major introductory computer science course, the major software course, for the last twenty-five years now. Since then I've become interested in the way MIT organizes its educational computing activities all across MIT.

I'm the chairman of the committee at MIT that thinks about our educational technology strategy. I run a very large joint project that MIT has with Microsoft, and I'm also involved in a project that MIT has with Hewlett-Packard Company.

A.P.:    I'm curious about something: You are a great expert in computational sciences, as well as in mathematics and many other sciences.

Do you think that the computation and digitalization of the world which is being brought about by computing in our century will help us understand the world better?

H.A.:    We can always hope that the digital revolution will make the world better, but I'm reminded that when they laid the first transatlantic telegraph cable, there were lots of predictions about how the telegraph cable was going to bring world peace immediately. So we can hope, but human beings are very, very complicated. One of the potentials of the digital revolution, however, is for people to share information much more, and I hope that what happens with the Internet and digital technology is that people become involved in open sharing and begin to really understand what people in other countries have to contribute to each other.

A.P.:    There is something interesting in what you're saying. I know that you have run courses relating technology and engineering with ethics.[1] In your career and your writings and projects, there is a relationship according to which knowledge should be shared and not privatized. Moreover, you are one of the founders of the Creative Commons.

Could you give us your opinion about this issue? Is it necessary to fight for shared and open knowledge?

And is this one of the main ways forward for future technology, ethics and open knowledge?

H.A.:   The great potential of the Internet is for people to share and for people to build on each other's work. That is true of artists building on each other's work to create a culture and that's true of scientists building on each other's work to promote knowledge.

One of the things that motivated us in starting Creative Commons was the understanding that the technology now permits sharing at the same time that the legal structures are becoming much more strict in preventing sharing. So the whole idea of Creative Commons is to create a legal structure that makes it very, very easy for people to share, and then to integrate that with the Internet.

Because of international copyright law these days, if you see something on the Internet that is not labelled, that is not indicated, that is not marked, you have to assume that you cannot use it. So this enormous resource whereby people could work together suddenly is boxed in by a copyright regime that's very much controlled by the metaphors that are given to us by the entertainment company, and Creative Commons is trying to create an alternative for that. We believe that if you make it easy for people to share, then they will.

A.P.:   There is a big debate in Spain over issues such as intellectual property rights in music. For example, in music distribution, the middlemen have managed to get legislation in favor of paying a fee by claiming they are defending the musicians, which has become a huge controversial topic. Things are viewed completely differently here.

What would you say to Spanish creators who are starting to become aware of Creative Commons, to make them understand it is a way to help defend their creations, and not the opposite?

H.A.:   Any artist relates to intellectual property in two ways: One is that an artist produces something and wants to be compensated for that, wants to be able to earn a living doing that, and that's perfectly good and perfectly reasonable. But artists also are consumers of intellectual property. If I want to create a song that's based on someone else's song, or that incorporates parts of another song, that is also a need for an artist. In the United States there's a whole genre of music—I don't know about Spain—called *sampling*, whereby people create pieces that are made by gluing together other songs. There's a whole music form that is called sampling, which in the early 1990s was almost destroyed by a very rigid interpretation of copyright.

What we need is a balance. What we hear in the debate about art and compensation, we hear only the first thing, we hear only about the artist as someone wanting to make a living and we never hear about the artist as someone who also needs to be able to draw on other people's work. Therefore, a balance must be achieved, and Creative Commons is trying to make that balance, and is trying to do so in a particular way, by giving the choice to the artist. If an artist wants, in order to get publicity, to make it easy for someone to copy her song and for free, she can do that. If you want to let people distribute it if they don't resell it, you can do that. If you want people not to be able to distribute it but you want to allow other people to take little pieces of it and use it, you can do that.

The whole idea of Creative Commons is to give the choice to the artist because what happens right now, even though everyone talks about protecting the rights of the artist, is that the artists do not have a choice. In the United States they are imprisoned by the record label. I don't know what it is like in Spain. I think they're imprisoned by collection societies. But they don't have choice. So much of the rhetoric you hear about protecting the artist is not really about the interests of the artist; it's about the interests of the middlemen who control the artist. The people who find Creative Commons a little bit threatening are not the artists themselves but the people who would be moved out of their position of power over the artists.

A.P.: I think we are currently living through another historic moment for the Internet. We are moving away from the age of the Internet, or the Web invented by Tim Berners-Lee, to the age of Internet 2.0, or Web 2.0 (formulated by Tim O'Reilly), or the Social Web, which is conquering civil society through its meaningful contents and ability to share everything.

Do you think the existence of a part of Internet that is open and free for everyone depends on us as individuals, or do you think that it will come automatically as the Internet evolves?

H.A.: Again, the issue of the shared vision of the Internet is that as the technology makes it easier to share, the legal structure in reaction to that makes it more and more difficult to share. For example twenty years ago no one ever thought about copyright. It was not in the public discussion. A few professional publishers thought about it, but people didn't care. It wasn't part of our vision. And now, suddenly, because of the ability of everyone to share, the legal regime comes down and tries to reeducate the people, tries to tell them that there's this wonderful value called copyright. That didn't exist twenty years ago. So, as much as we can be optimistic about the good effects the network can have, we have to constantly be careful that the legal regime—in order to support old, obsolete methods of doing busines—is going to compensate for the technology by getting us into a worse legal regime, and that's happening right now.

It's important for people to present an alternative view to show how sharing can work. More than that, how sharing used to work, how the culture used to be. You didn't use to worry if you made a new song that was based on folk material. In the United States, for example, it's impossible to create a new folk song because what people would have done in 1900 was to take materials out of the culture, and creating new things is almost impossible today because everyone is saying that the modern equivalent of the popular culture is owned, it's not something that people can build on. Therefore, that's what we have to worry about.

I'd love to be very optimistic about the technology and if it's purely a technology story then you could say everything is wonderful. But of course, it's not purely a technology story, and some very powerful interests are threatened by technology. We have to get into a dialogue with them and to seek a balance.

A.P.:   Your name is associated with breakthroughs in teaching technology. Your work as a professor is very important for me and is at the same level as your work as a researcher.

Could you speak to us a bit about your role in teaching with regard to these advances in ways of teaching technology and engineering?

H.A.:   MIT is a very special place, and the thing that is most special about it is that there is not a big divide between teaching and research. The most innovative subjects at MIT start in research, and more than that, when you think very hard about how to explain something to a student, that can often lead to new research ideas. So at MIT we tend not to distinguish. Most students, about 60 percent, are involved in research projects. Many of the faculty view teaching their courses, even the elementary courses, as a source of new research ideas. I would say the key to MIT is that you don't make a distinction between teaching and research. Every student is a potential colleague and every colleague is a potential student before whom you work out your ideas. If there's one magic thing to MIT, I would say it is that.

Another thing that's special about MIT is that it is ruthlessly democratic and based on meritocracy. Everyone who comes in is constantly having to prove themselves by showing that they can master the material. There are no people who come in with special privileges in that sense. I think it's those two things that are very much at the core of what makes MIT unique.

That same meritocracy is part of the culture that leads to open sharing. Therefore, MIT has a commitment to take all of its course material and place it on the Internet for free. And there are two aspects to that. One aspect is again part of the meritocracy and openness. As you say, it should be available to everyone. And the other part of it is to say that the material that we put on the Internet is not really an MIT education. Real MIT education is coming to MIT and becoming part of the community.

If we put all these things together in terms of predictions about where things are going, I certainly hope that universities will become more and more open about their material. And I certainly hope that universities will share more, and also learn to collaborate more. One of the things that people are starting to look at, when we think about the Internet and communication, people talk about multinational corporations and how it's possible that a multinational company will suddenly be able to take up projects that span many continents and manage them—how you could have a department that is really spread across many different countries and many different cultures. We haven't thought about that very much with respect to universities, and that's an interesting place to look.

What's the analogue of making a multinational product? What's the analogue of doing product design that has to respect many cultures? What's the analogue of that when you're not doing manufacturing or doing education?

That is something that universities will have to learn, and I think they will learn.

A.P.:   I'd like to ask you something. In Spain, the way engineering subjects are named is different from here. For example, I have experience, thanks to a project with Douglas Morgenstern, with the nature of courses over here. Course 6, which is connected to you, is about computational engineering and electrical engineering, although it would appear to be much more than this. At first sight, the naming of subjects might simply be a part of engineering, but I believe that it hides many more things.

What is the alchemy of the components of this potent course of which you are a professor?

H.A.:   One of the special things about Course 6, which makes MIT different from most universities, and most universities in the United States also, is that we are one department that is both electrical engineering and computer science.[2] And the thing that's special about that is that so much of what's happening in the current ground of technology and the world through semiconductors and everything is that the ideas profoundly involve both the physical electrical engineering-electronics side and the logical computer science side. It's very special in the department that we do not make that split. That's a choice that's certainly different from how things are done in most universities in the United States.

We are seeing that again in the revolution in biology. Biology is now transforming computer science both in terms of the algorithms that people are thinking about designing and in terms of the actual way you implement things, if you think about DNA computing, for example.

So Course 6, as a department, is able to easily let those interactions happen and have our students understand how those happen. Therefore, that's one thing that is very special. The other thing about Course 6 really goes back to the late 1950s and the early 1960s, when some very far-sighted people in the department decided to change the emphasis of the department from essentially power engineering to communications engineering. That's what allowed MIT electrical engineering to become a leader at the same time that the electronics communication revolution was happening. And for that reason, I think that Course 6 is an outstanding department, and much of that traces to the choice made in the late 1950s and early 1960s.

A.P.:  In Spain there is always controversy between the researchers and the administrators of science. You know both fields. You are a professor, but you also have to be a member of committees that obtain funding for researchers and for science.

Is it difficult to combine being a researcher and a scientific thinker, as well as being involved in administration, which is possibly less fun?

H.A.:  MIT has always been a university that's very, very engaged in the real world and very engaged in industry. That goes back to the founding of MIT in the 1860s, and it's consciously that.

There are very positive parts to it and then there are challenges. The positive parts are that you get a good understanding of the issues of the real world; you don't feel divorced from it, you don't feel as though you've been put aside on some theoretical plane where you're not understanding the realities of what's going on.

The challenges are that you also have to divorce yourself from the particular interests of the companies you're engaged with because, again, they're very powerful and very compelling. What you need is a constant sense of academic integrity that says even though I'm involved in this world, even though I understand what they're doing, my core fundamental values are academic freedom and being deeply intellectually honest.

Often, when you're working with a company, you find the stream of what is really true and what you're discovering is going against the immediate interests of the company, and despite that, you have to maintain your core integrity.

I think that's the real challenge.

A.P.:   I'd like to look at the issue of scientific thinking. I'm going to tell a little story. In a scene from Plato's *Phaedrus*, Hermes, the alleged inventor of writing is going to present his invention to the pharaoh Thamus, thinking that the pharaoh would lavish riches on him and be filled with joy. According to Umberto Eco, in this scene, the pharaoh expresses his concern: "'My skillful Theut,' he said, 'memory is a great gift that ought to be kept alive by training it continuously. With your invention people will not be obliged any longer to train memory. They will remember things not because of an internal effort, but by mere virtue of an external device.'" That is, the pharaoh thinks that with the invention of writing, people will have an external memory and will therefore be unable to remember what it had previously been able to recall.[3]

Let's imagine that we have the scope to see the revolution in the Internet, the digital revolution.

Do you think this is one of the revolutions in the Internet and that the digital revolution is on a similar historic scale to what the invention of the written word would be, for example?

H.A.:   I don't think that the Internet and the digital revolution rates at the level of the invention of writing, which is an incredibly fundamental thing. But I do think it's comparable to the invention of the printing press.

It has the same potential for democratizing and spreading knowledge, and taking things that were very restricted, to a small group and suddenly spreading them.

But the printing press—we don't have to go to writing—is a pretty good comparison historically, and we can aspire to that. I think we're going to see it; I think we're only at the very, very beginning of what's happening with the digital revolution. It's still special that someone's connected to the Internet.

We could be heading for a time where that just happens normally, where you always have access to all of the world's knowledge. Normally, you can get in touch with anyone in the world. You don't even think about it; it's seamless. And that's just starting to happen. I couldn't even predict what it will be like in ten years, when the next thing happens. Right now, we're talking about continuous connectivity; we're getting to the point where we think it's normal that someone could be sending video all of the time, and I don't even know what comes after that. But it's a very big change, and we're only at the beginning of it.

A.P.:   I would like to get some of your personal impressions. You have been at the forefront of technology sciences for years, especially in the computing field, which is one of the pillars of the huge change we are undergoing.

Do you still have the capacity for surprise with the passing of time?

Does the speed and evolution of technology still keep surprising you, or are you no longer as shocked or excited as you were at the beginning?

H.A.:   If anything, the pace of change is accelerating and things are becoming even more interesting. For example, fifteen years ago you thought about communication over the Internet as something like email, and it was a very exciting change that you could send a message to someone quickly. Now I can be doing telephone conversations all over the world, twenty-four hours a day, very cheaply. So you don't even think about it. That's an enormous change. You may have seen these really fancy videoconferencing rooms where you walk into a room and on the other side of the room is the other side of the world. And you take it as something for granted. That doesn't feel to me like a slowing down of change. It feels like an acceleration of change.

Going back to education, we now have students in Singapore and Kenya who are using the laboratory equipment at MIT as a natural thing.[4] Again, that doesn't feel to me like a slowing down of change. I think that if anything, it's a speeding up of change.

A.P.:   The movement or change concerning ethics and aesthetics established by the computer industry was strongly challenged by Richard Stallman's articulation of the four freedoms of free software. It is now a movement made up of millions of people linking up on the Web all over the world. It is very exciting for me to be here because it's like saying "everything began here."

Do you get the same feeling over here as we do in Europe that this is almost like a pinnacle, the Kilimanjaro of computing, and it's right here in this building, in this university?

H.A.:   From my perspective, it feels like an enormous responsibility. Maybe it's arrogant to say that people pay a lot of attention to MIT, but I think they do.

I think that it makes a very important difference how we at MIT, how I personally, behave. When we came up with OpenCourseWare, we consciously thought about that as a kind of leadership, and it made a difference how we described it. It made a difference what type of policies we do.

For example, we could have said that open courseware is just about MIT's material, just about MIT's curriculum. But MIT is the kind of place where you think about yourself as being as a leader. So a very important part of OpenCourseWare now is encouraging other universities to do the same kind of things we are doing.

It's just a tiny example of how you think about stuff at MIT. You don't think about it as something that you alone are doing, that's going to have an effect only on you. But you're constantly saying, How can this be an example? If you'd like other people to follow, how can you encourage them to follow? And that's a very kind of special thing at MIT.

It's a responsibility, but then again, it's a responsibility that comes from earning a position of leadership, which is something you have to constantly redo every day.

A.P.:   I'm from Valencia in Spain, and Valencia participates in a project with MIT called MITUPV Exchange. Thanks to you, among other people, two thousand Valencian people have been able to interact, collaborate, and share things with students from MIT. I would like to say thanks to you on behalf of these students from Valencia

H.A.:   You're very welcome. For the MITUPV website program, my favorite thing about it is that it was really made by students and was not something made by the administrators; it really was the students taking the lead and putting things together. And that's the kind of partnership we like to do, so thank you very much!

A.P.:   Thank you very much. I hope that you will include Valencia among your trips to Europe. Thanks.

H.A.:   Thank you very much!

## Notes

1. The MIT course, Ethics and the Law on the Electronic Frontier, can be found at http://ocw.mit.edu/courses/electrical-engineering-and-computer-science/6-805 -ethics-and-the-law-on-the-electronic-frontier-fall-2005.

2. MIT's department of Electrical Engineering and Computer Science, the famous Course 6, is the largest MIT department, and computer science makes up more than half (http://www.eecs.mit.edu).

3. Umberto Eco, "From Internet to Gutenberg 1996," lecture presented at Columbia University, the Italian Academy for Advanced Studies in America, November 12, 1996, http://www.umbertoeco.com/en/from-internet-to-gutenberg-1996.html.

4. The MIT Fab Lab is teaching people in Pakistan, Afghanistan, and Kenya how to make their own DIY wireless Internet networks (http://inhabitat.com/mit-fab-lab -helping-villagers-build-diy-wireless-internet-networks-with-found-materials/ fabfi32).

## 15  We Need Algorithms That Can Make Explicit What Is Implicit

Bernardo Cuenca Grau and Adolfo Plasencia

Bernardo Cuenca Grau. Photograph by Adolfo Plasencia.

*In technology, decisions cannot be overly bureaucratized. … Bureaucratization confuses process with progress.*

*To a great extent, whether a scientific idea in technology thrives or not depends much on chance. Some very good technologies have not made any progress while worse ones seem to have flourished.*
*—Bernardo Cuenca Grau*

Bernardo Cuenca Grau is professor at the Department of Computer Science of University of Oxford and currently holds a University Research Fellowship awarded by the British Royal Society. He is a supernumerary fellow of Oriel College and a faculty member of the Information Systems group at that institution.

His research interests are in knowledge representation, ontologies and ontology languages, description logics, automated reasoning, and applications in areas such as biomedical information systems and the Semantic Web.

Adolfo Plasencia:   Bernardo, I would like to have a conversation with you about computing, the Semantic Web, and ontologies.[1]

Bernardo Cuenca Grau:   Thanks! It's a pleasure for me to talk about those things.

A.P.:   Your research focuses on knowledge representation and ontology-based technologies and their application to the Semantic Web.[2] Let's talk about ontologies.[3]

In philosophy, ontology is the study of the nature of being. However, in computer science, which is your field, ontology has a different meaning. It refers to the formulation of a comprehensive and rigorous conceptual schema within a given context or domain.

Bernardo, how do you understand ontology? And what does it mean in your research? Is it also associated with metaphysics or just with computing?

B.C.G.:   To me, ontology is just a document containing a description which—in a formal language—describes a domain. Let me give you an example that everybody can understand. Imagine you want to write in a language that a computer can understand … let's say about biomedicine. If, for example, I want to say that psychosis is a type of mental disorder, I couldn't do it in natural language; I would have to find a way to represent it so that a computer could process the information.

A.P.:   And what a computer can process can be processed by another computer. So the idea is for any computer to be able to process it.

B.C.G.:   For a machine to be able to process it. But what does "process" mean? Process, in our case, means automated reasoning. For example, let's say that schizophrenia is a type of psychosis; psychosis is a type of mental disorder, and if someone has been diagnosed with psychosis correctly, then an algorithm can infer that that person has a mental condition.

Normally, such automatic inferences—and that was a very simple example—can be very complicated. At a large scale, it is something that would be impossible to do manually. That is why we need algorithms that can make "explicit" what is "implicit," and that is valuable because you are then able to know exactly what the implications of what you said are.

A.P.:   Make explicit what is implicit. ...

B.C.G.:   Yes, and this has a huge impact not only on the Semantic Web but also on bioinformatics. There is a huge ontology with over half a million concepts called SNOMED.[4] The NHS, the UK's health system, is digitizing the medical records of all of its patients using SNOMED. In this system, when a doctor assigns a diagnosis, only standardized terms can be used; all doctors use the same terms so that it is easier to share information between departments, for example. You can also make more sophisticated inquiries that would be impossible with a traditional database. This application, which cannot really be called a Semantic Web, is very important in bioinformatics and also in ontologies.

A.P.:   Speaking of ontologies. ... In technology, ontologies are also associated with artificial intelligence (AI) and knowledge representation, which is another area of expertise of yours. Has the digital revolution changed knowledge representation disciplines? What is knowledge representation for you, and how does it fit in this field?

B.C.G.:   Obviously, ontology-based technologies are part of knowledge representation. Knowledge representation has been studied for quite a few years now.[5] It is halfway between mathematics, philosophy, and computer science.

A.P.:   Not a bad meeting point. ...

B.C.G.:   Yes, a nice intersection. In knowledge representation conferences, some people have a philosophy background and some are pure mathematicians. You see people interested in computing and people who don't care about computers. It is a very interdisciplinary field.

As far as I am concerned, my interest is in the application; that is, we want what we do to serve a specific purpose. We really are—not just me but several people in my group—in a most unconventional situation. We are neither theorists nor pure mathematicians. When we attend very formal conferences, we are treated as hackers in a way. We also are not "application guys." When we go to conferences such as those on the Semantic Web, which are more application-focused, then we are the theorists. ... So we are definitely at a midpoint. We try to understand enough about theory to be able to develop it and design algorithms that work, and we then use that theory and program prototypes, and that's it. We fill the gap between theory and pure practice.

We do not want to do pure theory without a direct application. We are not interested in creating a product from a proof-of-concept prototype, either. As researchers, that's not our job.

A.P.: Wolfram MathWorld is an outstanding collaborative environment on the Internet. They say that the universe could be modeled with the logic of computing software much better than using conventional mathematical terms or models, which is how it has been mostly done so far. Bernardo, isn't this a bold statement? Is computational logic different from that of mathematics?

B.C.G.: Computational logic and mathematical logic largely overlap. In our case, at a theoretical level, we deal with questions such as, is it is possible to design an algorithm for a problem, which is purely computational theory; such questions are part of mathematics. But what we are ultimately interested in is designing computer programs, and it is there that our work differs from mathematics. What we do, as far as logic goes, is more meta-mathematics. We explore the boundaries of mathematics.

That is also done by pure mathematicians and philosophers. But we want to see which logic languages are suitable for practical applications.

A.P.: So in some mathematics and pure science conferences, you are treated as hackers. Is the Internet the only place where Wolfram can claim that modeling the universe with algorithms might be more effective? You couldn't make that type of statement in a pure science congress, could you? You would be expelled.

B.C.G.: Let me tell you something about the hacker thing. I remember talking to a British professor once—a theorist, in my opinion. He was telling me a story. He had attended a conference on category theory, and another scientist said to him: "Ah, but what you do is 'fixing' the language, right?" Meaning, do you really study specific languages? Because we study the properties of all types of languages.

In other words, the researcher who was a theorist to me was a hacker to the scientist in his story. There are so many layers. …

A.P.: Bernardo, let's talk about the digital revolution. Hal Abelson, one of the most important scientists of the MIT CSAIL group, told me in his conversation for this book that the digital revolution, combined with the Internet—with billions of people already connected—is comparable to the revolution of the printing press, but that it does not compare to the invention of writing.[6] The science philosopher Javier Echeverria says it does form part of the invention of writing.[7] What do you think?

B.C.G.: The invention of writing started it all, but the digital revolution— our access to information—it's amazing. Here's a very specific example. I use Wikipedia a lot. It takes only a click to know something, to obtain knowledge, or to read about anything. In the past you had to go to the library, find things in an encyclopedia, investigate. … And there were many things you couldn't find. Besides, Wikipedia includes articles in different languages where you find different perspectives on the same topic. It has really changed everything, and so has Internet searching. As for social media, on Facebook you can find people you haven't seen for ages.

It's hard to say if it compares or not to the invention of writing or the printing press, but it is indeed one of the greatest revolutions of the modern world.

A.P.: Now that you have mentioned that example, It reminds me of another conversation I had with Jimmy Wales and what he did when founding Wikipedia.

We are here speaking about orthodoxy in science, the layers between scientists and the relationship between a hacker and other scientists. Jimmy Wales took a gamble with Wikipedia; he went against the flow and stood up for utter trust and the radical decentralization of knowledge. I think Wikipedia is an example of how surprising this revolution can be.

Why are things still being published in the media saying that the contents of Wikipedia are not secure or not validated? Don't you think this type of information comes from some commercial powers who want that utter trust and support from millions of people who use Wikipedia every day not to be so strong?

B.C.G.: Perhaps. I don't know. I always use Wikipedia and I find quality information, much more reliable than what you find on some websites because in Wikipedia, if something is not reliable, you can always discuss it. There are public discussions, and if something does not seem right, you can discuss about it and see previous discussions. You can really see where the controversy is. I use Wikipedia not for my field but for other topics. But in terms of scientific knowledge and as regards things that can be objectified, or even for historical events that are no longer in the spotlight, I think it is fairly reliable.

A.P.: At the 2009 WWW Conference in Madrid, we celebrated the twentieth anniversary of the Web with Tim Berners-Lee and Vinton Cerf. Daniel Schwabe, who was a member of the original team of the TCP/IP, led by Vinton Cerf, told us that when Tim Berners-Lee presented his project at CERN more than twenty-five years ago and later in the United States for the first time, at the Technology Exhibition in San Antonio, nothing happened in the following eighteen months. So they presented the Web, and nothing happened for almost two years. Nobody did anything. No scientific authority responded, not from CERN or elsewhere. Nothing was done.

Bernardo, how come something like the Web, which twenty-five years later connects billions of people together, went unnoticed for a year and a half?

Can decisive knowledge go undetected?

Is it a problem of knowledge representation?

B.C.G.:   To a great extent, whether a scientific idea in technology thrives or not depends much on chance. Some very good technologies have not made any progress while worse ones seem to have flourished. What we are now doing at Oxford, for example, might be considered a total failure in five years' time, or perhaps the whole world will use it. We don't know. Many candidate technologies are developed; some make it and some don't. It is a combination of factors: how good the idea is, if there are people willing to develop it, the community that forms around it, luck, and also the quality of the technology, although in my opinion the last aspect only accounts for 30 percent.

The Web is not the only example of what you just said about good ideas going unnoticed at the beginning. When Douglas Engelbart and engineers at Xerox PARC invented the mouse and showed it to their company directors, the executives said, "What is this?" And when the same people at Xerox developed the first graphical interface, their people did not like it either. On their visits, Steve Jobs and Bill Gates saw it, and they immediately saw the potential. So basically, this has always happened.

A.P.:   On Google your name is associated with a very trendy term, the Semantic Web.

How would you define the Semantic Web in simple words so that we can understand it conceptually, and what is meant by this concept in Oxford's scientific circles?

B.C.G.:   The Semantic Web has no definition. It is still a very vague concept that encompasses many things, though it has drawn a lot of attention. If you think not about the idea itself but about what will come out of it, the specific side of it, it is a number of techniques, from knowledge representation to natural language processing and information retrieval. That is, it is a framework in which some techniques have stimulated research in certain areas with a particular application. The outcome of it will be a set of methods to process information in a smarter way. And that's coming from such techniques, and even from some older ideas about deductive databases, and so forth It will be a great framework, a large "basket" with lots of things in it.

How can you define that? It's very complicated; I think it has no definition. How will that be seen in the real world? There are already small applications where such things are working.

A.P.:   Should we see it as more like a tag cloud? Like a cloud with those semantic tags that we see on the Web?

B.C.G.:   It can be understood with a simple example. Imagine someone makes an inquiry, "I want to find all the science fiction books written by a particular author," and with that you gain access to a number of websites containing information, maybe Amazon or another type of website where you would find such things. If Amazon's ontology says that science fiction books and biographies are disjoint concepts—for example, that a biography cannot be a science fiction book—then an algorithm immediately "knows" that it has to ignore all the lists of biographies written by a particular author because that will an incorrect result.

A.P.:   In a debate on Internet governance that I attended in Madrid, during the question round, an executive in the audience said to Vinton Cerf and Tim Berners-Lee that he did not understand why they had let the Internet grow in such a "disordered" and "wild" way, almost without control.

Vinton Cerf replied, "We did it because both Tim and I continue to believe that the Internet should be an open place where no one has to ask permission of anyone to innovate."

Do you agree? Should the Web be an "open" place?

B.C.G.:   Those terms, "open" and "closed," are a bit blurred. In the end, there must be some type of regulation. Because with information. ... It all depends on the reach of the violation of people's privacy and data's confidentiality, both for individuals and for companies. If people start reporting breaches and there are scandals, then people will consider regulation. Until these things start happening ... we'll see ... we don't know.

A.P.:   Bernardo, you are cooperating from Oxford with the World Wide Web Consortium (W3C), the body that coordinates the development of standards and technologies for the Web. In the World Summit on the Information Society in Tunis in November 2005, Viviane Reding, the European Commissioner for the Information Society, said she did not understand why the regulatory agencies of the Web, such as the Internet Corporation for Assigned Names and Numbers (ICANN) and the W3C, were not directly answerable to "democratic governments." She did not understand that the bodies that make decisions on the progress of the Internet do not depend on the "political class." What do you think?

B.C.G.:   I do not understand why the political class should have a say in this. I think the W3C does an important job.

A.P.:   The Internet is used by billions of people, and it works perfectly. At the conference, we said, "What's this woman saying?"

B.C.G.:   I've taken part in standardization meetings and, in groups with twenty or more people—as was the case with the second version of OWL, with twenty-five or thirty participants—it is really hard to make decisions and move on, and there was already a very solid base, namely, the initial proposal that we put forward. It's a heavy burden. And you have to deal not only with technical issues but also with interpersonal relationships. It's complicated … and there are economic, commercial interests. In other words, it's really hard to reach consensus even on small things. And that happens with a group of twenty people, all of them technically skilled and working for leading companies and universities. Imagine what would happen if bureaucracy became stronger.

A.P.:   It's like what they say in design: a camel is a horse that was designed by a committee. The Web would end up becoming a camel instead of a horse.

B.C.G.:   In technology, things cannot be overly bureaucratized because changes are so rapid that it would be terrible to overly burden processes with further bureaucracy.

Bureaucratization confuses process with progress, and with technology issues one cannot afford such burdens because progress must be made. It is difficult enough to evolve in the current situation. We don't want to complicate it any further by creating more committees, implementing additional processes, filters, and so on. If we need a standard for people to start using it now … it currently takes us a year and a half or two. Imagine if it took ten years! By the time the standard was finished, nobody would use it. It would be totally outdated.

A.P.:   The mathematician Godfrey Harold Hardy once said,

A mathematician, like a painter or a poet, is a maker of patterns. … The mathematician's patterns, like the painter's or the poet's, must be beautiful; the ideas, like the colours or the words, must fit together in a harmonious way. Beauty is the first test: there is no permanent place in the world for ugly mathematics."[8]

Bernardo, do you think there is beauty in the algorithms of the Semantic Web?

B.C.G.:   I think we deal with ugly mathematics. And I think it should be so. For example, there are knowledge-representation formalisms in my field that have been designed by mathematicians. They always think about the elegance of the formalism and the properties that it should have to be easily manipulated. But when you compare that with applications, or reality, you realize those formalisms do not work the way they are. You have to modify them, and in doing so, obviously, the mathematics behind them "get dirty." Then those of us who try to prevent mathematics from getting too dirty see what we can do. In fact, if you try to prove theorems or whatever and you have a formalism that is not precise or is too difficult, then it is very difficult to prove anything.

You need to reach a compromise between what is needed in practice and the little "hacks" that we must do to formalisms and to the degree of dirtiness they have, because if they get too dirty it would be impossible to do anything with them.

A.P.:   So you don't believe in the beauty of equations? I saw one of your recent projects and the part with the mathematical equations is almost like a score, so beautiful. But I may be wrong because I am a neophyte. I did see beauty in those equations.

B.C.G.:   Maybe if a pure mathematician saw it, he would be outraged! When we do theory, we demonstrate, for instance, that an algorithm is suitable for a particular formalism, and so forth. We have to write proofs and deal with formalisms. One of the problems we usually have is that formalisms not designed by mathematicians are more difficult to handle, but we have no choice.

A.P.:   Bernardo, thank you very much for your time and for your words.

B.C.G.:   Thank you!

## Notes

1. The Semantic Web is an extension of the Web that provides a common framework for data to be used, stored, and retrieved. It was described by Tim Berners-Lee, James Hendler, and Ora Lassila, "The Semantic Web," *Scientific American,* May 2001, https://www-sop.inria.fr/acacia/cours/essi2006/Scientific%20American_%20 Feature%20Article_%20The%20Semantic%20Web_%20May%202001.pdf.

2. In artificial intelligence applications, knowledge representation entails providing information about the world to a computer in a form it can use to make decisions. See the Web page http://en.wikipedia.org/wiki/Knowledge_representation_and _reasoning.

3. Ontology in information science entails "a formal naming and definition of the types, properties, and interrelationships of the entities that really or fundamentally exist for a particular domain of discourse" (https://en.wikipedia.org/wiki/Ontology _(information_science)).

4. SNOMED is the initialism for Systematized Nomenclature of Medicine, a computer-processable collection of terms, codes, definitions, and synonyms to aid in human and veterinarian diagnosis.

5. A good introduction to knowledge representation, prepared by the MIT AI Lab, is Randall Davis, Howard Shrobe, and Peter Szolovits, "What Is a Knowledge Representation?," *AI Magazine* 14, no. 1 (1993): 17–33, http://groups.csail.mit.edu/medg/ftp/psz/k-rep.html.

6. Hal Abelson participates in dialogue 14.

7. Javier Echeverria participates in dialogue 28.

8. G. H. Hardy, *A Mathematician's Apology* (London: Cambridge University Press, 1967), 84–85.

# 16   The Emergence of a Nonbiological Intelligence

Michail Bletsas and Adolfo Plasencia

Michail Bletsas. Photograph by Adolfo Plasencia.

*I think we are on the point of seeing nonhuman intelligence, or, better put, nonbiological intelligence, in the twenty-first century, an intelligence that will probably be superior to human intelligence as we know it at present.*

*We have to learn to relate to more and more complex systems. I think we have to increase our ability to confront greater complexity yet further, to at least a few orders of magnitude.*
*—Michail Bletsas*

Michail Bletsas is Director of Computing at the MIT Media Lab and Director of the Network Computing Systems Group, where he is responsible for designing, installing, and maintaining the infrastructure that the laboratory uses to produce, create, move, convert, store, and consume its "bits."

He studied at the University of Thessaloniki, subsequently did postgraduate work in computer engineering at Boston University, worked for various wireless technology companies, and eventually joined the MIT Media Lab, where he helped create mesh networks technology (networks that do not need to deploy physical infrastructure in a zone to function).

Adolfo Plasencia:   Thank you for receiving me Michail, for this conversation.

Michail Bletsas:   My pleasure. Welcome to MIT Media Lab again!

A.P.:   Michail, you were born in Crete, a pivotal center of the Mediterranean. You studied at the University of Thessaloniki. Later you undertook postgraduate studies in computer engineering at Boston University. And later still, you arrived at the MIT Media Lab, founded by Nicholas Negroponte, also of Greek origin—whom you have worked with—and where you are now director of computing.

Is it by chance or has Greek vision got something to do with the Media Lab?

M.B.:   The fact that I ended up working at the Media Lab was something completely unrelated to my origin and Greek roots. I think I can safely say that my friendships, although they have a lot to do with who I am, did not play any significant part in my professional career. But I can say that Greeks fit easily into an atmosphere like that of the Media Lab. We don't require much structure. We like a bit of chaos. We prefer it. And I believe that is what the atmosphere of Media Lab reflects.

A.P.:   I'm interested in your opinions as director of computing at the Media Lab. Yours is a symbolic location for people who think about advanced technology. Research undertaken here associates computation with intelligence in many ways. In the names of the courses and projects, terms such as "Intelligence," "Thinking," "Intelligent Computation," "Things That Think," or "Ambient Intelligence" appear. However, not everyone agrees with associating computers with intelligence or thinking. For example, Roger Penrose said at a conference in Barcelona, "Computers cannot be intelligent in any way because they are not conscious. Perhaps someday, in some laboratory, someone will construct a conscious artifact, but it will certainly not be a computer."[1]

What do you think of Penrose's opinion?

M.B.:   I don't agree with him. I believe that human intelligence is something we have only recently begun to systematically analyze, and it's a very complex system. I would agree much more with Marvin Minsky, who influenced me a lot at the Media Lab. He claimed that to actually construct intelligence is not so difficult. It can be constructed using a very simple set of basic primitives. It's the massive connectivity among those primitives that gives rise to the complexity from which behavior stems. But it's a system we have not gone deeply enough into. The study of the brain is something that has only begun to gather pace in the last twenty years. We have recently made great strides in our attempts to understand the codification of human intelligence. That acceleration is also happening in technology, although we often tend to underestimate the goals that we may achieve in the medium-term future. We get very excited about our short-term achievements and sometimes are so disappointed when things don't go well that we completely lose track of what may happen in the long term.

I think we are on the point of seeing nonhuman intelligence, or, better put, nonbiological intelligence, in the twenty-first century, an intelligence that will probably be superior to human intelligence as we know it at present. We are part of evolution, no matter how you look at it. If past history is of any importance, the answer to whether it is going to emerge must be yes. People say that we are the most intelligent species on the planet and that we cannot improve any more. I tell them, Remember where we came from: remember the Neanderthals, remember Cro-Magnon people and the speed with which we have been able to advance through biological evolution. Now things are moving faster, and yes, I would call it non-carbon-based biological evolution. Biology will also evolve, not only intelligence. I believe that in the twenty-first century, we shall actually see advanced intelligence forms that are not *Homo sapiens*. This is the natural order of things. I don't see evolution stopping with us. There is nothing less natural than thinking that everything ends here with us.

A.P.:   There are people who say that the general paradigm of informatics is, again, "turning a corner," changing the direction of its evolution. I'm going to give some relevant examples:

In infrastructure, in the explosion of big data, in the server farms—by which Google with its complex software system Omega, and its new generation, Borg, which makes each data center with its hundreds of thousands of servers function together as if they were a single machine, just as Twitter does with its Mesos system. In computing: the "systematic" computer that repairs itself, which already exists at University College of London. In the online stores: iTunes and Google Play have sold billions of apps. In services: a single engineer at Facebook now provides support to a million users because the Internet has become gigantic, and so on.

Weren't these things and computer technologies almost unimaginable a short time ago?

M.B.:   The Internet, and its giant computation system is already the most complex system that human beings have ever constructed. We have to learn to relate to more and more complex systems. I think we have to increase our ability to confront greater complexity yet further, to at least a few orders of magnitude. We are now beginning to do that with our attempts to decipher how the human brain functions, something we are having increasing success with. The most important thing is that many questions that we were unable to answer before now have concrete answers.

Today we have enormous volumes of data, and undertaking their analysis is very important and useful for us. I don't think we should be put off tackling new challenges because of their complexity. The Internet has to be applied in many more contexts. The Internet of Things is now beginning, although we cannot yet say that it has been implemented.[2] My documents are still not connected! We are learning to understand that huge complexity, and, more important, we are learning to respond to very complex questions by simply gathering and analyzing huge amounts of data (big data). Doing this in the past would have required very costly equipment and a great deal of effort. Today—no.

A.P.:   Steve Jobs, before he died, said the era of the personal computer was over, but not everyone agrees.[3] Michael Dell and several well-off friends, among them several people from Microsoft, said that they disagreed with what Jobs said. Moreover, they wanted to refute that statement, wrong as far as they were concerned, by proving him wrong. They repurchased Dell Company, the former leading manufacturer of PCs, for $24,000 million. Intel also wished to prevent the end of the PC and considered the announcement, at least, premature.

What do you think? Is a slow death of information technology based on the PC taking place?

M.B.:   We have to separate questions of business from scientific questions. The important thing in business is not only what will happen but when it will happen. Michael Dell repurchased his great PC company. PCs continue to sell well, although their sales have reached a plateau. They have stagnated. The important thing to be aware of is that the profit margins of PCs are tiny now because it is a very competitive and mature market, whereas tablets are just arriving and so allow for a much greater profit.

Every time Apple sells a $400-plus iPad it earns a lot more money than Dell does when it sells a $1,000 portable computer. Michael Dell should be very careful when making predictions. He contradicted Jobs because there was a moment in the past when his company overtook Apple, but today Apple is the biggest and most profitable company in the market.

From a scientific perspective, we know that PCs are disappearing. Really, it's not natural for us to interact with them using the keyboard and mouse. Computers will be everywhere, we will have screens and reproduction surfaces everywhere, and we will interact with them through gestures and voice commands. They will look at us and respond to our movements. So yes, I can tell you that PCs will disappear in the form that we know them today, but this is going to take quite a long time. And there is still a multi-billion-dollar business there. Apple perhaps can afford to say, I can abandon that market because I sell other products that are much more profitable, but that's not the same as saying that Dell will no longer be able to sell PCs. Again, we have to separate issues of business from scientific arguments and predictions.

A.P.:   Nicholas Negroponte, with whom you have worked for many years, first launched the book Being Digital;[4] years later, in 2002, he published an article in *Wired* titled "Being Wireless," in which he compared the working of Wi-Fi to the action of frogs leaping from lily pad to lily pad on the surface of a pond ("lily pads and frogs" will transform the future of telecom).[5] And after came something you had a lot to do with: the mesh networks or wireless networks that do not need infrastructure to be deployed. Then came the mobile explosion, the ubiquitous access to the Internet using mobile devices: smart phones and tablets. I talked about this article in another conversation in this book with Howard Rheingold, and he calls this type of communication "horizontal and lateral."[6]

What do you call it? What do you think will be next in access technologies?

M.B.:   We have seen how wireless access technologies are converging. If we look at wireless in general, Wi-Fi and technologies using the Long-Term Evolution (LTE) standard are all very similar in their use of underlying technologies. One difference, however, is that LTE (4G and 5G) technologies use licensed spectrum and Wi-Fi uses unlicensed spectrum. There will be space for both, and we will see Wi-Fi continuing to grow. Over the years, telecommunications companies will shift toward Wi-Fi much more, even though they now see it as something they cannot use because of its unlicensed spectrum and their concern that they cannot guarantee quality of service. If you go back a few years, similar arguments were made against the use of packet switching and Internet Protocol–based communications when these started to replace traditional physical and virtual circuit switching.

Nevertheless, unlicensed access is the best means available for deeper penetration. Wi-Fi is going to play a very important role as an access technology.

We have an ever greater number of devices that are wireless. Now hardly anyone uses just Ethernet without good Wi-Fi access. As access technologies converge, we see that, at the purely technological level, LTE and Wi-Fi utilize very similar radio technologies. The main difference is a decision of business policy as regards the markets and the use of licensed or unlicensed spectrum. I think we will increasingly use devices that connect online using Wi-Fi. In any case, most of our devices today are wireless, and we will see them prevail over wired devices. Today we have certain things that are wired, but now the tablets, mobiles, the electrical appliances in the home are increasingly wireless, and we can't be creating cellular provider accounts for each one of them.

On the other hand, the value of having a network managed by a telecommunications company is something we cannot ignore. However, we will see part of that migrate toward the Wi-Fi side and Wi-Fi networks will obtain elements of the larger (wider area) networks. We will see a convergence between the two. The difference is that you will have control over the Wi-Fi part of the network and that, in turn, will be linked to the macro network, which today are two separate and mostly distinct parts. We are going to see all these parts converge, and most access will be wireless. Also, the first networks to be established in developing countries will directly be wireless, so bypassing the copper wire stage.

A.P:   You are a computation scientist but have ended up in computation applied to the Internet.

Do you think there is a before and after in computation sciences following the Internet explosion?

M.B.:   Here we can talk about two stages. As an information technology scientist, I could say that there is a "continuum," as the explosion of the Internet is accelerating several technological trends. Obviously, if you are outside, you are not going to recognize the concepts such as those that existed before the explosion of the Internet. What we are seeing now are the effects and the growing number of applications that those concepts have had in our daily lives. I would like to point out here that, after the Internet boom, we are beginning to formulate a more participative Internet, and we are realizing the potential of the Internet, which was always supposed to be a bidirectional medium from the beginning. That is the essence of what Tim O'Reilly formulated with his definition of Web 2.0, which tells people that they can also produce bits, so that there is no need for them to only be passive consumers. That is the most important difference in itself.

A.P.:   The networks and the Internet are another of your fields of research. The sequence we mentioned before for hardware has a parallel in the evolution of the network: the Web (Tim Berners-Lee), Web 2.0 (Tim O'Reilly), the Social Internet (social networks such as Facebook and Twitter), the Internet of Things (more than half of Internet traffic is no longer between humans), the Internet of Everything. ...

What would the definitive Web be like, if it were possible?

That interplanetary Internet that Vinton Cerf announced some time ago?

M.B.:   At the moment, we don't have much interplanetary activity, although the interplanetary Internet exists and will increasingly be seen. However, the most important thing is how the finely granular Internet is going to reach Earth. By that I mean that there are still thousands of millions of humans who are not connected, so there's a long way to go before we all are. This is one of my favorite problems and it is where I have undertaken most of my research.

A.P.:   The term is "finely granular Internet"?

M.B.:   It's not just a term, *connectivity everywhere*. There are still many people without connectivity to the Internet because they cannot afford to pay for it. We have to work honestly on the commitment to Internet being a fundamental human right. And until we make that come true, I'm going to continue to work on achieving it. I'm going to leave the interplanetary question to Vint. Right now we have several specific ways of connecting to the Internet—increasingly those are wireless ways—but we are going to witness ever more widely reaching methods over time.

I don't know exactly what they'll be like, but we will be more and more directly connected as humans and, of course, everything else will also be connected to the network.

A.P.:   Let's talk now about the evolution of the MIT Media Lab. After a glorious stage led by Nicholas Negroponte, in which you all brought to fruition the sentiment expressed by your friend, Alan Kay, "The best way to predict the future is to invent it,"[7] your laboratory grew and was extended with an impressive, transparent, and light-filled second building, which is now fully operational.

What computation and connection technologies have you deployed in the Media Lab this time?

M.B.:   The most interesting thing for me is that when one builds one's own laboratory, you notice more what goes on inside. The network of sensors that we have deployed and interconnected throughout the interior of the building allows us to know in real time where there is activity at each moment and so immerse in the laboratory. We can do all kinds of things: switch off the lights in places where there is no activity at that moment, so saving energy; control the air conditioning or refrigeration, for the same reason; tell people in which other place there is activity so that they go with the others and maintain their social activities. We can recognize and know who people are when they arrive, we can help and guide them to where they wish to go. If they want, their visit can be recorded, and we can send them it as a diary so they know whom they have spoken to, how much time they spent and with whom.

We don't have many fixed signs: most are touch-screen monitors with which one can interact, with maps and interactive catalogues of Media Lab activities. The building is much more interactive than the old one. At a more mundane level we have a wireless network that provides connectivity throughout the building. We have point-to-point real-time video capabilities, which are used a lot because of the continuous flow of events that occur here.

A.P.:   You are one of the inventors of mesh technologies for wireless networks. I imagine there will be a lot of that in Media Lab. ...

M.B.:   Yes, all the computation is ubiquitous and connected: coverage for smart phones, Wi-Fi, and so forth is very, very dense. That was a problem in some of the MIT buildings, but I believe we have solved it here quite well and to the point that we are providing the model for the others in terms of ubiquitous mobile connectivity and for places with a communication system and interactive screens with a lot of feedback.

A.P.:   Do you have your own "Internet of Things" in the building?

M.B.:   Yes, "Internet of Things" at the Media Lab means that all the thermostats are controlled, that the whole Media Lab is online with a whole range of flow sensors: a microphone network that does not record what you say but just records noise, so that we know where there is human activity and where not, always respecting privacy, obviously, at the same time. We also have little robots that go around acting a little like your "agent," all connected to the network. Another important thing we have is telepresence, as there is a lot of collaboration with companies throughout the world. For them and for us, it is important to what we used to call videoconferences and what these companies call telepresence. You can place your avatar in any place in the building and let it interact with anyone.

A.P.:   At the MIT Media Lab, almost every day, new artifacts and technologies are being invented that have never been seen before. Your capacity to amaze seems limitless. How do you do it?

Does it still surprise you as much as in the past that young people come here every year from all over the world to invent things in your building?

M.B.:   It doesn't surprise me at all. The lab's recipe is pretty simple at its base: find a corner at the best technical university in the world, bring in brilliant people that don't fit well in other places, give them freedom to pursue what they feel passionate about. Make them build things to demonstrate their ideas, and get feedback often from the lab's sponsors. Make the place fun, and repeat, year after year.

A.P.:   Is innovation something natural here?

M.B.:   Yes, it's what we do. When you bring so many intelligent people here and you give them creative freedom, that's what happens; innovation takes place day-to-day. It's the best thing about working here. Something comes up week after week.

A.P.:   Talk to me about the relationship between the MIT Media Lab and the companies that finance a huge amount of research and laboratories here. Hal Abelson says in another dialogue in this book that the companies that finance MIT and its laboratories cannot "redirect" research in line with their own company interests and, at times, the convincing arguments of the companies have to be resisted.[8]

What is the technological relationship like at the MIT Media Lab?

M.B.:   The Media Lab was a pioneer and was the first to go into this. Many companies came and became members of the laboratory because they were impressed with what they saw here and wanted to support it. They see the laboratory as a window to the future. We shouldn't discredit models of directed research that seek to solve very specific problems that require directed research. It's another way. Those models, of course, have their place at an institution like MIT. That research will always have a boost effect on undirected research, because specific problems have to be solved and so resources have to be made available and efforts focused on finding the solution. The problem is overcome by evolving. You need not only innovation but also evolution, incrementally improving concrete things. That is important.

The world does not advance only in quantum leaps, but at times you have to go to a specific place and pursue a specific goal intensively. Most often companies encourage this. Most often when a company finances you, it comes to you because it thinks you are more intelligent and capable than it is in certain things and it pays you to solve something specific. The companies, by definition, cannot tell you how the research must be done. That is good engineering, that is evolution. To be a good engineer, you have to know beforehand what you want to achieve. Here at the Media Lab we advance the other way around. We say, We are not going to go very deeply into this. We are going to try to create for you a window to the future and show you that there are ideas on what the future will be like. How do we get there, exactly? That's another question, but we tell them that we are going to get there.

A.P.:   Here you have strong relationships with communications companies. The boom in social networks, the boom in smart phones. This is all around you here.

How do you relate with the large companies of ubiquitous communication?

M.B.:   Telephone companies are very hierarchical in a top-down fashion. Control flows "from above." Innovation that comes from below is foreign to them. They don't understand the peer-to-peer effect. In Usenet culture, there used to be an ethical criterion: do not disperse, do not deviate, do not waste time. Those companies do not understand that. They don't understand "opening up," and in the long term they are going to regret that. Without this bottom-up movement, there is no innovation, because that's the direction that the best innovation takes.

A.P.:   What is the importance of the boom in social networks and of the competition between Google and Facebook for the Internet?

M.B.:   At this point in time, our lives would be poorer without Google than without Facebook. Google is useful in a much deeper sense than Facebook, because it satisfies our need to find relevant information for all aspects of our activity. Facebook, in contrast, although it also satisfies a deep need (communication), it does so in a relatively superficial manner.

A.P.:   Michail, many thanks for this dialogue.

M.B.:   Thanks to you too.

## Notes

1. Roger Penrose, in "El Universo no nació con el 'Big Bang,' sino con una expansión previa," said, "Consciousness can only be understood, and potentially be replicated, by a machine when we have a new physical theory that goes beyond quantum mechanics," Elmundo.es, August 10, 2006, http://www.elmundo.es/elmundo/2006/10/17/ciencia/1161080064.html.

2. Kevin Ashton, a British entrepreneur, is credited with coining the term "the Internet of Things" in 1999. The term refers to "things"—hardware, software, services—that collect data using existing technologies and flow it to other devices, where it can be used (https://en.wikipedia.org/wiki/Internet_of_Things).

3. Jason Hiner, "Steve Jobs proclaims the post-PC era has arrived," *TechRepublic*, June 2, 2010, http://tek.io/1BwHa9D.

4. Nicholas Negroponte, *Being Digital* (New York: Alfred a Knopf, 1995).

5. Nicholas Negroponte, "Being Wireless," *Wired Staff*, October 1, 2002, http://archive.wired.com/wired/archive/10.10/wireless.html.

6. Howard Rheingold took part in dialogue 30.

7. Alan Kay, "The best way to predict the future is to invent it." There are many formulations of this statement. Perhaps the earliest was Dennis Grabon's in his 1963 book, *Inventing the Future:* "We cannot predict the future, but we can invent it." Alan Kay's formulation, "The best way to predict the future is to invent it," Kay delivered in a public address in 1982, but he has said he began using it as early as 1971. Alan Kay, "Learning vs. Teaching with Educational Technologies," *EDUCOM Bulletin* 3/4, (Fall/Winter 1983), 17.

8. Hal Abelson took part in dialogue 14.

# 17 Remembering Our Future: The Frontier of Search Technologies

**Ricardo Baeza-Yates and Adolfo Plasencia**

Ricardo Baeza-Yates. Photograph by Adolfo Plasencia.

*[Control of the Internet] is not a matter of scale so much as a question of diversity.*

*Everyone's behavior has its own long tail.*
*—Ricardo Baeza-Yates*

Ricardo Baeza-Yates is Chief Technology Officer, NTENT, California, and part-time Professor in the Department of Information and Communication Technologies of Universitat Pompeu Fabra, Barcelona, Spain, and the Department of Computer Science, at the University of Chile. Formerly he was Vice President of Research at Yahoo! Labs, leading teams in Europe, the United States, and Latin America. He is a founding member of the Chilean Academy of Engineering.

His research interests include algorithms and data structures, information retrieval, web data mining, and data visualization. His contributions include algorithms for string search, such as the shift-or algorithm, and algorithms for fuzzy string searching, which inspired the bitap algorithm.

Among his publications are the second edition of *Modern Information Retrieval,* co-authored with Berthier Ribeiro-Neto (Addison-Wesley, 2011); the second edition of *Handbook of Algorithms and Data Structures,* co-authored with Gaston H. Gonnet (Addison-Wesley, 1991); and *Information Retrieval: Data Structures and Algorithms,* co-edited with W. Frakes (Prentice-Hall, 1992).

*Note:* This dialogue was done in 2014, and was revised by R.B.-Y. in 2016.

Adolfo Plasencia:   I've come to your Yahoo! lab in Barcelona to talk to you about technologies and the digital revolution, which is proving to be something like Pandora's box.

Ricardo Baeza-Yates:   Welcome!

A.P.:   Ricardo, I think the search technologies that you use can anticipate behaviors. In your research, you handle data from 700 million people at the same time. With your search technologies, you are able to "catch" and figure out what a user can do the day after tomorrow in a particular region of the world, with a specific culture and activity. Are we so predictable? Could you do something like "remembering the future" of what many people are going to do?

R.B.-Y.:   Your question has many facets. I'd rather divide my answer into two parts: one about the immediate future, that is, what you are going to do in the next few minutes, and another one about the future-future, that is, what you are going to do the rest of your life. They are two different futures.

I'll start with the immediate future. In this case technologies try to predict what a person is going to do. The answer about the immediate future is that, in part, it is true. We can predict the next app you will access on your smart phone. We have information on many mobile phones and on many apps. We can predict that you are going to access a specific application today, with a 90 percent success rate. We are all creatures of habit to a great extent—up to 80 percent, for instance. And that part of us can be predicted. So my answer is yes.

A.P.:   Then 80 percent of our behavior is determined?

R.B.-Y.:   Yes. Let me give you some figures, though it depends on the person. Some people have more "deterministic" behaviors than others. Some people are "strange." With strange people, we are likely to predict their behavior in 50 percent of cases. But there is also the long tail of people's behavior.

A.P.:   A long tail of human behavior?

R.B.-Y.:   Yes, everyone's behavior has its own long tail, a different long tail, and that is very important. We found this out in a study by Yahoo! in 2009, analyzing five sets of big data on music, searching, movies, and so forth. Basically, we saw that there was a long tail in everybody's behavior, not only strange people's behavior. On the other hand, we all are much like the rest of people; we all have to work, eat, and feed the cat. And there's a part of what you do—20 percent, or 50 percent if you are very "strange" or very different—that includes specific things that matter only to you—your hobbies, your worries, or your personal motivations.

A.P.:   Is it the intellectual dimension that makes us more unpredictable? Or not necessarily?

R.B.-Y.   Not necessarily. Maybe in your case, for example, your long tail includes a health problem, because you may be ill, something that has to do with your body, with your "animal" side. The long tail is made up of many specific problems, things that I want to do right away, such as: I'm very worried about something, I want to find information about it. That's something intellectual because I'm looking for information, but perhaps only you are interested right now in graphene and its future possibilities. Things that are common to everyone are easier to predict because we have a lot of data, and people are predictable in that respect. But it is very difficult to predict things that are related to the very specific part of you, because maybe you didn't even know these things ten minutes ago. Imagine that you think about a particular topic for the first time. It is impossible to predict that.

A.P.:   It's true.

R.B.-Y.:   Let me answer the second part, the future-future. Something that can be done and that we have actually achieved—and we won an award in 2010 for the best news analysis software. Let's say you have a very big news collection. You can find there what people think of the future. It is like analyzing wisdom. Some people share opinions about politics, economics, or the environment. Other people express opinions on conflicts or tensions between countries. They even do so for the conflicts that may arise over the next hundred years. Many people make assumptions about the future. In a collection of millions of pieces of news from the *New York Times,* one can find things many people speculate about, and that helps you understand what people think about the future. Among them there are also outstanding individuals whose decisions may influence the future. So there is a greater chance that what they say may actually come true, because they influence the present by expressing their opinion about future things. This facilitates what I call "looking into the future." As Alan Kay said, "The best way to predict the future is to invent it."[1]

A.P.: According to the neurologist Alvaro Pascual-Leone, at Harvard Medical School, neuroscientists already know that the brain is a hypothesis generator and that it is always making hypotheses.[2] We like imagining what will happen. Is that right?

R.B.-Y.: Exactly. It is a variant of the big questions we want to answer. Where do we come from? Where do we go? I think they have no answer, but people want to and have to live with such questions.

A.P.: In another conversation in this book, I asked Bernardo Cuenca, an ontology and Semantic Web researcher in the Department of Computer Science at Oxford, about his work: Can you briefly describe what you do here? He said (I paraphrase slightly): "We work on Semantic Web technologies to make the implicit explicit." In relation to that, do you think the Semantic Web—which apparently will be part of the Internet's mainstream in the near future—will make explicit the implicit in global knowledge?

R.B.-Y.: My answer is the same as that to your first question: Yes, in part. Perhaps the Semantic Web will be 20 percent successful in that, but it will fail in 80 percent, and not because the technology does not work. The technologies that can do the conversion already exist. The problem is that people, even though they know them, do not really want to use them. How many people, when editing a Word document, fill in the metadata, in other words, the explicit part? Nobody. They do not try to be clear about the day, the subject, what keywords they used before when they wrote about the same topic. If you use different words every time, explicitness does not work.

A.P.: Why do you think people do not want to make the most of such a fantastic opportunity?

R.B.-Y.: I think it has to do with our nature. It has to do with what is not deterministic. Let's see: How many people have everything organized on their computer? You can see it when you go to their home. Only the obsessive-compulsive are very orderly. But that's why many people say of them that they are not normal. For example, who has everything sorted out in different drawers so that they know exactly where things are? The Semantic Web is just that, having everything in the right place. But people are not like that; we like having a bit of a mess. Chaos is part of us. I think chaos makes life interesting. Imagine if everything were totally tidy, it would be very boring. There would be no problems; there would be … nothing.

A.P.: Human beings cope better with determinism than with uncertainty.

R.B.-Y.: I think we're more comfortable with determinism because that is exactly what we want to achieve. That's our goal, but we are chaotic. How often do you forget things? Imagine you could stop that and be methodical. If you were methodical in everything you did, your life would be less enjoyable. I don't think all people want to be methodical.

A.P.: Maybe it's because I can't change the "wiring" in my brain that easily.

R.B.-Y.: That's why I say it is part of our nature. Even if we want to, we may not be able to. An example is violence: we do not want to have wars, but we seem to fail all the time.

A.P.: Speaking about being willing to do things and being able to do them, and about technology: do you think we should use technology as we want to or as we are told to?

R.B.-Y.: That's a good question. I think we should use technology as we please. There are examples in technology—some are clearer than others—that we use for some things and not for what we are told to use it for. The Internet and the Web were not designed for what they are today. Many technologies are now being used for something other than what they were planned for. We take the best of technology and modify it to use what is best. But if someone designs a technology so that it is definitely the very best one for a particular purpose, we will surely use it the way we are told. For example, iTunes is for music and is designed for that. It is very good, and there is no dilemma between using it for what I want or for another purpose. But when you're faced with a dilemma, in the end, you will do what you want, if it is possible and legal, of course.

A.P.: Let's talk about cybernetics. Cybernetics is defined by Norbert Wiener as what in ancient Greece was known as a *kibernos,* a helmsman in charge of the ship's course.[3] In that sense, who is the *kibernos* of Internet cybernetics today?—if there is one.

R.B.-Y.: I think that, in Norbert Wiener's sense, there is no such figure.

A.P.: Is there at least a wheelhouse?

R.B.-Y.: Let's say, for example, that the United States must have its own wheelhouse or bridge for any Internet infrastructure in its territory, but it cannot control all of the Internet. Again, the answer is "in part." But in a global sense, there isn't a master helmsman. It is very difficult to exert control because it depends on many countries, companies, and ultimately on how people use technology. People adapt to the conditions in which they live, and the same goes for the Internet. The Internet is our own reflection. The virtual world is an illusion; it's just a reflection of the real world.

A.P.:   But a lot of people are worried about being under surveillance by the National Security Agency and other agencies. Apparently, some government agencies are so powerful that they can conduct mass surveillance over the Internet. Some countries have banned the Internet or its applications— unsuccessfully, though. Is the Internet so huge that not a single country can constrain it under its command? Is it bigger than what a vertical power can cover?

R.B.-Y.:   It is not a matter of scale so much as a question of diversity— diversity in the countries and languages involved. The Internet is very heterogeneous, so heterogeneous that nobody can control it.

A.P.:   Then is it digital diversity that has an endless dimension for the known powers?

R.B.-Y.:   Let me use a metaphor. If I have a thousand sheep, I can use a dog to guide them. But if I have a thousand different animals, I have a much more complex problem in controlling them. I don't need a dog; I need a thousand dogs, at least. That's the problem with diversity. The diversity of the problem is so complex that no one can control it, and then you have all the legal and political issues. In some places, although there is a very powerful entity that wants access to everything, it can't have full access. Some data only you can have access to.

A.P.:   Diversity in the sense of complexity?

R.B.-Y.:   Complexity comes from diversity. If everything were uniform, it would not be so complex.

A.P.:   I have another, somewhat provocative question, though not from your field. Michail Bletsas, director of computing for the MIT MediaLab, is sure that there will be nonbiological intelligence in this century, or at least an intelligence not based on *Homo sapiens*. Do you think that is possible? Is it feasible? Will it ever happen?

R.B.-Y.:   The answer depends on how we define intelligence. If intelligence means beating the best chess player, then yes; if intelligence means beating the best Jeopardy player (as was the case), then yes.

But the question is, is that intelligence or something else? In those two well-known examples, we find the same thing: the computer does it differently from a person. It's more brute force, massive computing; it involves much more data; and in the end, the computer performs better than a person. But is it because of greater computing capacity? I don't think so. Because a computer has more information capacity, more storage? Maybe. Perhaps the human brain stores much more information but we do not know how to access it so quickly.

If we stipulate it will be nonbiological intelligence, therefore a different intelligence, and that it must solve this type of problem, then my answer is yes. Of course, I think this kind of intelligence is not human intelligence; it's another kind of intelligence.

A.P.:   So you agree with someone who, in Barcelona—the city where we are now—said something that made me think about this a lot. Roger Penrose, after a lecture, said: "Maybe, in the future, at some lab, someone might make an intelligent machine but, of course, it will not be a computer." Do you agree with Penrose?

R.B.-Y.:   Not fully, because if it really is as smart as a person, it will not be a computer for sure, but maybe it'll be like a computer of the future, and we don't know what computers will be in the future.

A.P.:   Maybe a quantum computer, like the one described by Ignacio Cirac?[4]

R.B.-Y.:   A quantum computer, or something else entirely. We are constrained by what we know today. The best example for this is Asimov's famous story, "The Last Question" (1956), in which a huge computer stores all the knowledge in the world. Every now and then, they asked the gigantic machine how something intelligent could be done, how life could be created, and the computer said again and again: "I do not know (... there is insufficient data for a meaningful answer)." When the last human was about to die, he asked the question again. And the computer said the same thing: "I do not know." Then, after a very long time, the computer suddenly said: "Let there be light."

So, maybe yes, but I don't know what that computer is. So far, we use machine learning techniques to predict individuals' behavior, but basically, all we're doing is trying to predict data.

A.P.:   You are a computer scientist. Does intelligence—artificial or not—have more to do with the critical mass of complexity or with computing capacity?

R.B.-Y.:   I would say it is related to both things. The more complex the world is, the more computing power you need, and the problem is that, as humans, our computing power is limited.

A.P.:   With your big data technology, a few individuals can handle a huge, superhuman amount of information, something formerly impossible. And some research with big data is simulating a massive data recombination, because it is thought the brain might do this to create something new. Do you think that by recombining numerous data, you can come up with a creative process?

R.B.-Y.:   The process of creating things is not something I have delved into, but if research is analyzed—which is one of the creative processes we do here—there are two types of creation. One is like the great idea that you did not have a second ago and that you know once that second has elapsed. Once you know it, it seems obvious and trivial. You say "of course, it's obvious" (and then it is trivial because you already know it). They are great ideas, but this happens very rarely. I think there's a combination of factors that ultimately lead to thinking about that. It does not happen spontaneously. There is a process that allows you to find them. But you have to be thinking about a problem. You're thinking about it and suddenly it happens: "I've got it!" In the other creative style, innovations are quite frequently the result of transferring what you already know in one area to another one. In other words, there is something I want to solve, and by making a change I can do it. It is a process that does not come from *ingenie* (not *ingenuo*) but from having a comprehensive, overall view of things. The more global your gaze at the world, the better, but it is now more difficult to know all the knowledge that we have and all the technologies available to us.

People are able to innovate and create because they have a more global view of things, a holistic vision that allows them to lay bridges between subjects. Not everyone can do that. This takes us back to the problem of diversity and complexity. There is so much diversity and knowledge that we cannot have a holistic view. It is part of the problem. In the Renaissance, there were people who knew almost everything that was important in human knowledge at the time—people like Leonardo da Vinci, who knew mathematics, astronomy, geometry, physics. But today it is very difficult for a person to have such a comprehensive knowledge. That's part of today's complexity.

A.P.:   When you started to study computer science, did you imagine that the Internet would impregnate all of your scientific life?

R.B.-Y.:   No, but when I was a student the Internet was almost unknown.

A.P.:   A group of people, not too big, invented what we now call the Internet, in stages. Before, there was nothing like it, in computing and technical terms. Jon Postel, Vinton Cerf, Robert Kahn. Later, Tim Berners-Lee invented the Web, then Tim O'Reilly formulated Web 2.0. Some people made the infrastructure, the communication mechanisms, the computing; and others connected that with the public. Do you think this group of people (to name just a few) opened a huge Pandora's box?

R.B.-Y.:   I don't think so. I'm sure because both the Internet and the Web were designed for other purposes. Those who were thinking about it devised the Internet to transmit information, files. The first widespread use was email, something they had not imagined yet, but that was a natural outcome. The Web was also intended to organize information. Now it is a platform for things as powerful as social media. I don't think the earlier workers ever imagined the impact. People used it for something else; people used it for what they thought was more interesting.

A.P.:   People skipped the sequences of commands and the user guide. ...

R.B.-Y.:   But that's what we do all the time. Who reads manuals? Nobody. We are people. Maybe it's part of our wild side: I want to find out for myself how it works, and if I find out something else, great. I'm inventing something, I'm teasing technology. The best recent examples of innovation have to do with the use of technologies for purposes other than those for which the technologies were created. It happens quite a lot. You see people using something in a completely different way. They are called crazy but they are actually inventing something new.

A.P.:   Bill Aulet, the managing director of the Martin Trust Center for MIT Entrepreneurship, told me that entrepreneurship is not an algorithm, and apparently, success isn't either.

R.B.-Y.:   Sure. I agree with that. It is a creative process. Both things are creative processes.

A.P.:   You can never imagine that, out of something that you know, something you never imagined may arise. And that's not an automatic thing.

R.B.-Y.:   If I could turn it into an algorithm, if I could innovate automatically, it would no longer be innovation to me. If you can repeat it on an industrial scale, it is not innovation any longer. Innovation is something you cannot repeat.

A.P.:   It is the opposite of falsifying, in Popper's sense, a scientific experiment.

R.B.-Y.:   Exactly. Creating, designing, is an art, a craft. There is an artist behind it, a creator. If one could automatically paint as perfectly as Van Gogh, then we wouldn't consider Van Gogh to be such a great artist. We make massive something we believe to be special. Art is something special. If everyone did it, it would not be art. What about you, Adolfo, do you agree with this?

A.P.:   I agree.

And here is the big question about the Internet. From the Web Internet invented by Tim Berners-Lee, we moved on to Web 2.0. Tim O'Reilly formulated it in 2005, and it became social. Shortly after they said, that's it, the next one will be the Web 3.0 ... but not really, apparently. The Semantic Web, the mobile Internet (Mobile Web), the Internet of Things, the Internet of Everything emerged. Do you think that the future of the Internet can be predicted? Or is it impossible to imagine what direction the mainstream Internet will take and how it will evolve in the near future?

R.B.-Y.:   It's hard to predict. Five years ago my answer would have been different because I look at developments closely. One thing is clear: diversity is on the rise. The Internet does not consist only of connected computers. It includes smart phones, sensors, machines. Maybe we'll see "wired" people, robots and brains directly connected. The Internet will more and more reflect the diversity of people and the world. And that is even more complex. I think there will be different Internet worlds to be shared. Some talk about what will happen and predict that the whole Internet will be social. But there is a big contradiction here: how can something that is inside something eat that something? It's like telling your stomach to go and eat yourself.

A.P.:   Darwin said: we have found that evolution exists. But we do not know where it leads. Perhaps the evolution of the Internet could be Darwinian in that sense. We know it exists, how it works, that it generates growing complexity, but we do not know where it leads. Is it impossible to know, in the same way that Darwin did not know where evolution would take us?

R.B.-Y.:   Yes, exactly, and it has to do with the same thing, because the Internet is a collective creation by those who use it. More and more, it is a reflection of us. Currently, the Web 2.0 is a collective creation.[5] None of the people you mentioned earlier could imagine that Wikipedia would be what it currently is. Nobody imagined that one billion people would be connected to a single online social network.

If we could answer the question of where the Internet is heading, we would be answering the question of where we are going.

A.P.:   Where the human condition is going and where humankind is going?

R.B.-Y.:   Interestingly, although people say that this creates a kind of chaos and things that are negative, that is not true. We are reflecting what we are, the positive and the negative things that were already there before. The difference is that we can now see it. Now, with the Internet, we can amplify the world and it is far more transparent, and you can even find things that you did not know of before.

On the Internet, you can find out who did something. In real life, if someone does something wrong, it is much more complicated to know. The Internet as an amplifier also amplifies new possibilities. What is happening is that the Internet also allows us to modulate the evil in ourselves, and that is very important. I do not want to use words like "censorship" or "blocking"; I'd rather use the term "modulate," having control mechanisms to ensure freedom of expression, the democratization of the Internet and its diversity and, at the same time, exert some control over the misuse of technology in respect to issues that are important for a lot of people, such as privacy. These issues are being handled quite well, if one considers the global reach of the Internet.

A.P.:   So you're an optimist. You are an optimistic technologist.

R.B.-Y.:   I am an optimist technologist, but sometimes I am also a skeptic.

A.P.:   Well, we do not have all the answers. If you find the answer, please let me know, and I'll be here straightaway.

R.B.-Y.:   If I find the answer I don't think you will find me. I would have to hide!

A.P.:   Thank you very much, Ricardo. It's been a pleasure.

R.B.-Y.:   The same to you. Thank you, Adolfo.

## Notes

1. There are many formulations of this statement. Perhaps the earliest was Dennis Grabon's in his 1963 book, *Inventing the Future:* "We cannot predict the future, but we can invent it." Alan Kay's formulation, "The best way to predict the future is to invent it," Kay delivered in a public address in 1982, but he has said he began using it as early as 1971. Alan Kay, "Learning vs. Teaching with Educational Technologies," *EDUCOM Bulletin* 3/4, (Fall/Winter 1983), 17.

2. Alvaro Pascual-Leone participates in dialogue 24.

3. A discussion of *kibernos*, or governor, can be found in Niall Lucy, *Beyond Semiotics: Text, Culture and Technology* (London: Bloomsbury, 2001), 138, or Michael Shally-Jensen, ed., *Encyclopedia of Contemporary American Social Issues*, 4 vols. (ABC-CLIO, 2010), 1608, http://bit.ly/1rKHrxk. The latter gives examples of governors and

feedback systems in daily life, such as thermostats and robotic toys. Information about the environment, or feedback, is translated into a command for the system. A Roomba is another example.

4. Ignacio Cirac takes up this question in dialogue 1, "Quantum Physics Takes Free Will into Account."

5. Further discussion of Web 2.0 can be found in a post on Tim O'Reilly's website, "What Is Web 2.0: Design Patterns and Business Models for the Next Generation of Software," O'Reilly.com, September 30, 2005, http://oreilly.com/web2/archive/what-is-web-20.html.

# 18 The Challenge of the Open Dissemination of Knowledge, Distributed Intelligence, and Information Technology

Anne Margulies and Adolfo Plasencia

Anne Margulies. Photograph by Adolfo Plasencia.

*There is a struggle today to understand where to draw the line with technology. ... For our students, who have never known life not connected to technology, we are trying to figure out when technology is beneficial and when it's a disruption.*

*What we need to do as an element of university education is to make sure we are educating our students about the difference between data, information, and wisdom.*
*—Anne Margulies*

Anne Margulies is Vice President and University Chief Information Officer at Harvard University, responsible for information technology (IT) strategy, policies, and services that support the university's mission of teaching, learning, and research.

Before returning to Harvard—she was Assistant Provost and Executive Director for Information Systems in the late 1990s—Margulies was

Assistant Secretary for Information Technology and CIO for the Commonwealth of Massachusetts.

In 2002 she was founding Executive Director of MIT OpenCourseWare,
MIT's initiative to publish the teaching materials for its entire curriculum
openly and freely over the Internet. She has served on IT advisory boards at
MIT, Princeton, and Dartmouth and was also instrumental in the early formation of edX, a joint venture with MIT to expand access to education
through open online courses. She has received numerous awards recognizing her IT leadership accomplishments, including recognition in 2014 by
the Center for Digital Education as one of the Top 30 Technologists, Trans-
formers and Trailblazers and in 2010 as one of the Top 25 Doers, Dreamers
and Drivers by Government Technology.

Adolfo Plasencia:   Anne, you've come back to Harvard, where you've already
worked on issues related to information technology (IT). You headed
OpenCourseWare, MIT's bold initiative, and later you became Chief Information Officer for the Commonwealth of Massachusetts. Now you are Vice
President and Chief Information Officer (CIO) of Harvard University.

How would you describe the evolution of the relationship between people and information technologies, from both a professional and personal
viewpoint—and as seen from the perspective of those three stages in your
life?

Do you still stand up for the openness of knowledge, as you did with
OpenCourseWare?[1]

Anne Margulies:   Thank you very much for the question. I have been very,
very lucky and fortunate to have been able to serve in the three positions
you mentioned: at MIT, as CIO for the Commonwealth of Massachusetts,
and now here at Harvard. Through all three of those experiences I've been
able to witness firsthand how significant, how enormous the impact of IT
is on everybody's life. I'm going to state the obvious, which is that IT is
affecting people's lives in almost every way, while everyday technology
makes our work life and our everyday life ever more intermingled. But each
of my experiences was actually quite different. At MIT I learned how technology could be used to take educational materials created on MIT's campus and share them all over the world. So just up the river here in Cambridge,
from a very small office with an equally small staff, we were able to have a
truly dramatic impact by benefiting from the generosity of MIT faculty and
distributing educational materials around the world so they would help
millions and millions of people.

At the Commonwealth of Massachusetts I learned how technology can dramatically make government more efficient and, just as important, make lifesaving services accessible to citizens every day in areas such as public safety, health, and human services; without technology the government couldn't function and couldn't provide the services that citizens rely on every day.

As for Harvard, it's a vast, very special community that is extremely diverse, and there are tremendous pockets of innovation. So I'm seeing how we are just at the beginning of how technology is going to affect education. I believe that over the next decade technology could in fact reshape higher education and will certainly change what happens inside the classroom. We've been teaching inside classrooms in pretty much the same way for over 375 years, but I believe technology is at the point where there are going to be major changes in how we teach over the next decade.

A.P.:   In your opinion, what guidelines could be applied to integrate changing technologies into educational processes so that such processes can make the most of their potential?

How is innovation understood at Harvard, as seen from today's technological point of view?

A.M.:   There is an amazing amount of technological innovation happening here in Harvard. We recently did a survey to identify all the different types of technological innovation in teaching here, something that we called "innovative pedagogy." We discovered more than one hundred different examples of how technology is being used to improve research and, especially, teaching. To summarize what we learned from this experience, we found that technology is improving teaching here, first by helping us to create greater access to Harvard's faculty, so that students' access to faculty isn't limited only to their office hours or in other traditional ways. Faculty here are using lots of electronic ways to communicate and they are using a lot of collaborative tools so that discussions and contact with the professor occur between classes.

Second, many, many courses are being videotaped, so students are able to view these tapes before they go to class. In this way, faculty can use classroom time interactively so that students engage in active learning. A very significant impact that technologies are having on education here is in helping to support active learning by bringing the outside world into Harvard's classrooms. This means that when a faculty member wants to bring into the topic of the day an expert from the field or from a different culture or from elsewhere in the world, he or she can do so easily through videoconferencing.

Technology is also having a major impact by enabling Harvard to export its teaching to the outside world. Although we haven't done as much as MIT has with OpenCourseWare, we have a number of open videotaped lectures that are now published to the world, and people around the world can now get a glimpse of what happens inside Harvard's classrooms. So we can bring in the outside world and make the learning experience more active and engaged and also enable others to experience Harvard's outreach. Another very important way that technology is having an impact on teaching here is that Harvard is fortunate to have very rich collections of art, rare books, artifacts, and so on. With technology, we are able to digitize some of those collections. For example, we can take an ancient Chinese rubbing that has to be kept under glass in a museum here and digitize it so that faculty can provide access to such rubbings for the students in the classroom. So that's another very important area we are focusing on: using technology to help bring Harvard's collections into teaching.

A.P.:   At the beginning of your mission as CIO at Harvard, you spoke about the innovative vision that a meeting point between information technologies and advanced research should have. In the past, people talked about super-specialized, somewhat isolated research environments, whereas today we talk about hyperconnectivity.

How can we arrive at a favorable medium for progress in cutting-edge research to be as efficient as possible with support from IT and the Web?

A.M.:   I don't have a good answer to this because we are still struggling with the question. It's a struggle because both faculty and students have so many disparate demands and needs. If you think of the process of doing research and discovering and disseminating new knowledge, one of the ways of disseminating it is through teaching. We have a critical process that is interconnected. We have to be able to support that entire process, every aspect of which relies more and more on technology. So, for our part, we are currently investing in a high-performance computing center with truly enormous-scale computer clusters for very advanced scientific research. Elsewhere, our social scientists are also using extremely sophisticated research computing. Social scientists used to do their research by sending out surveys and by interviewing people. By contrast, social scientists today often do their research through massive data mining, processing data that's available through the Web, so social science research now also depends very significantly on high-performance computing.

The humanities, which often would rely on ancient texts and rare books, are now focusing on "digital humanities," so they need significant resources, all digitized and made available electronically. To ensure that knowledge creation and research, plus the teaching process, remain robust as they themselves create new knowledge, we have to support everything, from highly complex, very intensive scientific high-performance research computing to classroom websites. We also must be able to balance both our investments and our technologies across all of this and ensure that everything works together as seamlessly as possible in support of faculty.

A.P.:   As Tim O'Reilly says, this is the time of "network effects," which create turbulence but also great opportunities in the operation of complex organizations.[2]

How do these network effects affect Harvard University, which in so many ways is an organization that universities all over the world look up to?

A.M.:   In my estimation, one of the clearest examples of those network effects that I think Tim O'Reilly is talking about is that it's very disruptive at times because we really haven't figured out how to harness all of its positive aspects. I know that it is very difficult sometimes for our faculty when they are teaching and all their students have their heads down because they are looking at their laptops and we can't tell if they are shopping or looking at Facebook or are engaged in the classroom. A frequent topic of debate here is whether laptops should or should not be allowed in class in certain circumstances. For example, some of our faculty require that laptops be shut down and turned off in class so that there can be genuine interaction and discussion in the classroom. I think there is a struggle today to understand where to draw the line with technology. For this generation of students, who are referred to as "born digital"—students who have never known life not connected to technology—we are trying to figure out when technology is beneficial and when it's a disruption.

A.P.:   Ricardo Baeza-Yates, researcher and formerly Yahoo! Labs' Vice President of Research for Europe and Latin America, says we shouldn't yet talk about a "knowledge-based society"; he says that instead of "knowledge" we should say "information," and when we say "information" we often mean simply "data."[3]

Do you agree with Ricardo?

Regarding these three levels—data, information, and wisdom—how do you think these concepts should be addressed in relation to IT?

A.M.:   I agree with Mr. Baeza-Yates. I think these are the correct three levels, and I do think they are different from each other. I believe that what we need to do as an element of university education is to make sure we are educating our students about the difference between data, information, and wisdom. When students come to university, we teach them how to think critically and creatively, and hopefully this results in a degree of wisdom. But we need to make sure that this wisdom is based on quality information, both raw data and truthful information. Here again, one of the disruptions that technology has caused is that there are vast amounts of data and information available on the Web. The first place most of our students go to for research today is Google. But Google doesn't carry only quality information and data, so students have to learn a new skill: the capacity to distinguish between raw data and *accurate* raw data. Then they must learn how to turn accurate data into quality information and how to identify information that's accurate and of quality, so that what they assimilate through their research and through their studies can be transformed into wisdom.

A.P.:   The incredible growth of ubiquitous communication and its digital underpinnings is undeniable. It has resulted in the destruction of traditional classroom-centered teaching paradigms in favor of a "classroom without walls" and ubiquitous learning.

How can we arrive at the best of both worlds, combining the virtues of quality teaching and the ubiquitous learning paradigm?

A.M.:   We are having discussions at Harvard about this very topic. A very important part of Harvard education is the residential experience, living here on campus, being within the gates of Harvard. It's Harvard students interacting with Harvard faculty, Harvard's students interacting with each other in their houses and in their dormitories.

We believe that the residential experience is a very important part of our Harvard education. However, we don't think that Harvard's impact on education should be limited only to those who can come and experience Harvard here on campus.

This is something that's under discussion as we speak: with today's profound ubiquitous access to computing, how can Harvard students have a more global experience?

And how can students and educators from around the world participate as well in the Harvard community?

There are so many different models that other universities are implementing and there are many different experiments and models that

Harvard has implemented. Harvard has an incredible number of global programs with Harvard faculty and Harvard students, scattered around the planet. So there are lots of ways in which we can balance the virtues of a residential, on-campus experience with a more virtual one, even though we don't know yet exactly what the right model should be. Actually, I don't believe there'll ever be *one* right model. I think there will always be plenty of different models.

A.P.:   The explosive worldwide growth of social media has been massively boosted by the global success of Facebook, originally created at Harvard, as featured in the Oscar-winning film *The Social Network*. Unless you are familiar with the campus, you wouldn't know that the film wasn't filmed here except for one shot. As a work of fiction, reality is exaggerated in the film.

What do you think of the student microcosm in the movie, especially in terms of technological infrastructure?

A.M.:   I'm very glad you ask this question because people should understand that while it's a great movie, it doesn't accurately reflect student life here at Harvard. Students need to spend much more of their time in study than the film implies. They have a social life, but the social life of most Harvard students is not what you saw in that movie. As for the technology infrastructure, the movie was all about what happened here years ago. Then, when the movie was made, the technology infrastructure was very different from what it is today. What you see in the movie isn't a true reflection of how this infrastructure was at that time or what happened to it. I'm told that the Harvard network didn't crash completely the way it appeared to in the movie, although the Facebook prototype that was created here did in fact slow the network down significantly.

A.P.:   Anne, what do you think the future holds for IT?

Will we remain slaves to Moore's law and planned obsolescence?

Are you a "techno-optimist" as far as the future goes, or has experience made you more cautious?

A.M.:   I've been working in IT for more than thirty years, and it has always seemed as though IT overpromised and we underdelivered. There's always a lot of hype about what the next technology is going to be, so I'm always very cautious not to overstate what the impact of technology will be. However, after thirty years in IT I really believe that we are about to experience an inflection point.

I sense that we are on the verge of what is going to be a point in history where technology is going to make dramatic changes in higher education and dramatically improve access to education, improve its quality, and also improve how students are able to learn and faculty are able to teach. This advance will also accelerate how faculty undertake their research and discover new things.

Much of all this is going to happen by enabling much more collaboration and work across boundaries.

I think technology is really going to be able to abolish many of the boundaries that prevent people from having access to its potential, and also that technology will remove boundaries between academic disciplines. So I'm an optimist. I think this is an incredibly exciting time for technology and it's a particularly exciting time for technology and education.

A.P.:   Thank you so much, Anne.

A.M.:   And thanks to you too.

## Notes

1. Anne Margulies, "The OpenCourseWare Initiative: A New Model for Sharing," videotaped lecture, July 23, 2013, http://videolectures.net/mitworld_margulies _tocw. See also "About OCW" for a description of MIT's OpenCourseWare initiative (http://ocw.mit.edu/about).

2. Tim O'Reilly, "Network Effects in Data," *Radar*, October 27, 2008, http://radar .oreilly.com/2008/10/network-effects-in-data.html.

3. Ricardo Baeza-Yates explores this topic further in dialogue 17, "Remembering Our Future: The Frontier of Search Technologies."

# 19  Technology Is Something to Make the World a Better Place

## Tim O'Reilly and Adolfo Plasencia

Tim O'Reilly: Photograph by Adolfo Plasencia.

*Technology is not neutral. Technology has enormous power for good or evil and we, as consumers of technology, have to make the choices that will help to serve us and make the world a better place.*

*In the era of the network and the network as platform, network effects are what matter.*
*—Tim O'Reilly*

Tim O'Reilly is Founder and CEO of O'Reilly Media Inc. and a supporter of the free software and open-source movements. He is also a partner at O'Reilly AlphaTech Ventures, a founder and board member of Safari Books Online and Maker Media, and on the boards of Code for America and PeerJ.

Tim has a history of convening conversations that reshape the industry. In 1998, he organized the meeting where the term "open-source software" was agreed on, and helped the business world understand its importance. In 2004, with the "Web 2.0 Summit," he defined how Web 2.0 represented not only the resurgence of the Web after the dot-com bust but a new model for the computer industry based on big data, collective intelligence, and the Internet as a platform.

In 2009, with his "Gov 2.0 Summit," he framed a conversation about the modernization of government technology that has shaped policy and spawned initiatives at the federal, state, and local levels and around the world.

He has now turned his attention to implications of the on-demand economy and other technologies that are transforming the nature of work and the future shape of the business world.

Adolfo Plasencia:   Tim, you have long been known among information technology (IT) specialists as the best IT book editor in the world, but you have also done other outstandingly relevant things in the digital culture.

For instance, yours is the creation of the truly commercial website, described as "the surfing site of O'Reilly's Global Net" (Global Network Navigator.GNN), which was sold to America Online in September 1995. You created a great uproar on the net, in the middle of the euphoria over the dot-com boom, in year 2000, with your open letter to Jeff Bezos, in which you criticized him for his Amazon 1-click patent, a letter that gathered 10,000 support signatures in just four days.[1] You have belonged to the Internet Society Council and to the Electronic Frontier Foundation, together with John Perry Barlow and others. You started O'Reilly Media, which organizes seminars that position themselves more in the future than in the present. During one of those seminars, the term "Web 2.0" was coined. This term conceptually defines and identifies the second Internet generation. The word "editor" is not enough to explain what you do. On the net and in the mass media they systematically call you a guru, a word I am not sure you like.

I'd prefer you tell us how we could explain to the public *who* you are and *what* you do, as I don't think any profession has yet been invented with a name that fits your huge range of activities.

Tim O'Reilly:   That's a big list you gave me there! I would say, first of all, when I think about my company—let me talk about my company first, rather than just myself—we started out as a book publisher, and then we started doing online publishing and conferences, and somewhere along the line we realized that through all those vehicles we were actually doing something very, very different. The way we describe it to ourselves in our company mission statement is "changing the world by spreading the knowledge of innovators." We look for people who are creating disruptive innovations. These are usually people from outside the business mainstream, whether hackers or inventors, open-source software developers, or people who are technology innovators. Then we try to help other people follow in their footsteps, whether that's by documenting, in the form of books, information about their work, or bringing people together at conferences or other events, and also describing to the world what's important about it. So that's really our mission, and then everything we do as a business is just a vehicle for that mission.

As for myself, I would say that I started out as a writer living out a fabulous quote about writing that so shaped my thinking. Edwin Schlossberg said many years ago: "The skill of writing is to create a context in which other people can think."[2] And that's what I have always tried to do. I apply a sort of pattern recognition and then try to reframe issues in ways that help people understand them differently.

A very good example of this was when we originally brought together the group at the meeting that came to be known as the "Open Source Summit," because that was the meeting where the term open-source software was voted on and agreed on by these free-software leaders. Up to that time, a lot of these people had never met, and they didn't think of themselves as part of the same movement, and yet we were able to bring them together, and then they were able to see what they had in common and to embrace this new idea with a new name.

There's something really important in this story. Up to that time, there was a lot of focus on Linux and the Free Software Foundation's Gnu Project, but no one had really made the connection that the Internet was also based on free software. By bringing Apache, sendmail, and the domain name system into the same conceptual map as Linux I was able to help transform the narrative from one of rebellion against the old order of the industry into one that showed how open-source software was enabling the next stage of the industry. The stories we tell each other shape how we think.

I'm also an activist. I believe that the choices we make matter, that technology is not neutral. Technology has enormous power for good or evil and we, as consumers of technology, have to make the choices that will help it serve us and make the world a better place. So, like people who are activists in other areas, I am somebody who wants to make the world a better place.

A.P.:   Do you think it's possible to reach a permanent or durable consensus in a world with such a global diversity as that of open knowledge?

Do you think that by means of open collaborative processes on the Web, agreements can be reached and standards can be established and stabilized?

T.O.:   First of all, I believe that change is good. But there is a difference between permanent knowledge and standards. Standards are agreements, and in fact the most important standards are very, very simple. If you look at the Internet, the reason why it took off was because they agreed on as little as possible.

There's a fabulous concept, which is actually I believe originally in RFC 861, which was the request for proposal concerning the Transmission Control Protocol (TCP) by Jon Postel, who's one of the real fathers of the Internet. He wrote something that sounds a lot like the golden rule from the Bible: "Be conservative in what you put out and be liberal in what you accept from others." And that kind of simple rule of how do you connect, how do you connect well, was enormously powerful.

On the other hand, you had all of the big standards committees coming together around what was called the OSI stack, the open standards inter-connect. They specified all kinds of technical details for how everything would connect. And which of those two standards won? It was the simpler one. Therefore, I think that a lot of the details we fight over are very small. In fact, there's a wonderful poem by Rainer Maria Rilke in which he says something like, "What we fight with is so small and when we win, it makes us small. What we want is to be defeated decisively by successively greater things."

I really believe that the challenges we face are, How do we be open to a future? How do we allow new things to happen? And that means we won't agree on everything, but, for example, leaving room for disagreement while understanding where we need to agree in order to allow the future to happen, that's what is important.

I think I've been quoting too much here, but there is a wonderful passage from Laozi, the Chinese philosopher, who said: "Losing the way of life, men rely on goodness. Losing goodness, they rely on laws." And in that hierarchy I like to think that first of all, what we really want to find with open source or Internet standards is what Laozi would call "the way of life," which is: What's the real way that things work? What makes for optimum results? As I've often said, that's science; that's not religion, that's not politics. It's understanding what is effective. For example, when I've tried to work with Web 2.0, it's not about trying to advocate for a particular set of beliefs; it's trying to understand what is effective. And in that hierarchy there's a set of values. I believe that the values of open source are a lot about trusting people and relying on goodness. Wikipedia does that, and only when that fails do you want to go down and say, "You know, here there are some restrictions."

But I'd much rather appeal to the goodness in people and, even more than that, appeal to the way things work.

A.P.:   One of the complaints about the world of technology is that the industry forces user knowledge (and their intellectual work in the form of digital files) to become almost immediately obsolete. Their reason is that there is no other alternative but to constantly and rapidly update the contents because of Moore's law. People from a humanistic background, whom you know well, are especially unhappy, as your Harvard thesis dealt with the classic tension between the mythic and the rational in the dialogues of Plato.

Is this true? Is there no other alternative to this tyranny of constant upgrading and obsolescence?

Is digital nature really like that, or is a hoax perpetrated by the technology industry to keep users captive and hooked?

T.O.:   As I said earlier, I believe that change is natural and good. And I think there is a stress in modern life from the pace of change, and certainly there are people throughout history who have looked back to times when in theory, at least, the world was more stable and peaceful. If you want to be stable and peaceful, you can opt out of a technology society. There's nothing stopping you from doing so. However, I actually think that the excitement of having to come to grips with the future is a good thing. For me, it's a fabulous intellectual challenge. Every day there is something new; I'm always being surprised by the inventions of other people. It's a wonderful artistic world that we live in. I go back to speaking of art! The wonderful poet Wallace Stevens talked a lot about a future in which we're all trying to persuade each other of truths that are not necessarily even true but are aesthetic visions of possible futures.

There's a way that Moore's law certainly makes new things possible. But, I don't think that it changes the fact that we as human beings need to be responsive. When we stop responding, we grow old, we die. And I'd much rather be facing the challenges of Moore's law than, for example, the challenges of global warming, but we're going to have to do that too! We live in a dynamic system and we should be grateful that we have the ability to *respond*.

A.P.:   Let's carry on in the same vein, but now I am talking to O'Reilly the editor. I don't know whether you know about the Long Now Foundation, which approaches projects and reflections on a long-term basis, in some cases up to ten thousand years. The Library of Alexandria was created with a vision and a time horizon of up to seven thousand years. If it had not been deliberately destroyed in a fire, it might still be running today, with its mission and vision still in force.

Couldn't the digitally stored memory of humankind be approached using this same philosophy? Would the content usage and storage tools used in IT prevent this, or are we the ones who are incapable, in this digital era, of contemplating a cultural perspective of thousands of years, as the Egyptian, Greek, and Roman people of the Classical era did?

T.O.:   I just want to say that people like Brewster Kahle, of the Internet Archive, who in a way is really the modern heir to the librarians of Alexandria and who has worked with the Egyptian government to begin work there on a new digital Library of Alexandria, is one of my heroes. I've actually been a sponsor of events for the Long Now Foundation, and I work with them.

I believe that having a long-term perspective is very, very important.

As I said, we need to be responsive to our immediate environment, but a big part of that is thinking about consequences. We make choices, and often the consequences are not visible to us for a long time. I do think, on the one hand, though, that remembering everything is overrated. Part of what it is to be human is to forget, and if we had all accumulated knowledge, would we really use it? For example, even in the forgetting there is often new finding.

A very good example of that is the wonderful Classical statuary that we so admire, the white marble Greek statues. Originally they were all painted! So these visions that were remade in the Renaissance of what Classical antiquity was all about were wrong! But they were beautiful nonetheless. And they were beautiful because of what we had forgotten. It's like music. If it were all there, all the same, it wouldn't be music. I think that we live in a constant tension with memory and possibility, and we need to be sympathetic. Yes, it's sad when things are lost, but in some ways I think that one of the reasons we grow old is because we cannot forget.

Being able to forget and being able to start anew is an important part of being human.

A.P.: Tim, how could we briefly explain to a normal reader what the concept and essence of Web 2.0 is, the Web 2.0 that you formulated in a seminal paper in 2005?[3]

T.O.: Web 2.0 is a convenient name for a phenomenon that has been happening for quite some time. I've actually being talking about it probably for the better part of a decade. I just was calling it different things.

What I have been seeing is a shift, whereby the Internet is moving from being a sort of add-on to the PC to becoming the real platform.

In 2004 we decided to launch a conference that was really about a small idea, which was just that despite the dot-com bust, the Web wasn't over, that it was coming back. So we called it the "Web 2.0 Conference." But then, as we started to ask ourselves "What does that mean?," we started to explain what it meant in terms of this vision of the future Internet as platform, the Internet operating system, and really the rules for what made for success on this new platform. And there are probably four or five that I've talked about a lot. But the first one is that in the era of the network and the network as platform, network effects are what matter. So you could frame Web 2.0 as the creation of applications that use network effects, particularly harnessing their users and user contributions, in fact, harnessing collective intelligence, so that they get better the more people use them. That aspect of Web 2.0—that these are applications that connect people via computers to applications, and then harness the activity of all those users so that the applications improve—is, I think, the heart of Web 2.0.

There is a set of corollaries to that. For example, one is that you have to be open and networked, and so share data, so that the data sharing makes your application work well with others and become transparent and the like. However, in that openness are the seeds of the next closing. Because I really believe that technology goes through waves of open innovation and then consolidation and closing down.

One of the things that are really interesting to me about Web 2.0 is that when you have network effects, you actually have increasing returns to the players who first generate those network effects. Take eBay, for example. For a long time it was very hard for other players to enter the auction marketplace because once eBay had a critical mass of buyers and sellers, why would you want to go somewhere else? Similarly, if you look at Amazon, they've gotten stronger and stronger because they were really good at harnessing user contribution. We're seeing that with Google, where Google is just becoming better and better because they've done a better job than their competitors at harnessing their users to make their search results better, and to make their advertising network better.

This will likely lead to a period of consolidation, but of course, innovation will break out again in some new area.

A.P.: In a text you wrote, titled "The Open Source Paradigm Shift," you outlined the concept of paradigm, which Thomas Kuhn developed in his pioneering book, *The Structure of Scientific Revolutions.*[4]

According to WorldWideWebSize, at present, the indexed Web contains at least 4.71 billion pages. Now, a decade after your pioneering formulation of Web 2.0, we are clearly undergoing an enormous creative boom on a global scale of the network effects, which you have outlined, with technologies, concepts, and new meanings generated by social media (Wikipedia, Facebook, Flickr, Twitter, Linkedin, Reddit, Google+, Whatsapp, GitHub, the universe of mobile Internet apps; Kickstarter), which are all based on the concepts you formulated about the second generation in the meme of your seminal article, "What Is Web 2.0," published in September 2005.[5]

Bearing in mind the concept of your text that was mentioned at the beginning, could it be argued that during the decade of Web 2.0 there has been a paradigm shift in the Internet in line with or on the scale of those described by Kuhn?

T.O.: If by "paradigm shift," we mean a fundamental change in assumptions, absolutely. When I wrote that piece, we were still in the PC era. That was an era characterized by Microsoft Windows as a de facto standard. Commodity software was shipped every couple of years, and didn't change in the interim. We're now in an era where software changes many times a day, imperceptibly. New features are tested on users and changed based on the data that are gathered about whether and how they use them. It's all running on the Internet in one way or another, and the rules are completely different. The companies that were the first to understand the new rules gained enormous competitive advantage.

A.P.: Tim, do you think the Internet is going to evolve with a succession of new generations in the future of different kinds of Internet: the Semantic Web; the Mobile Web (Ubiquitous Internet); the Internet of Everything or the Internet of Things (which you say we should call "the Internet of Things and Humans"), and so on?

Will the definitive Web ever arrive, or can the same observation that Jimmy Wales applied to Wikipedia be applied to the Internet and the Web: "The Internet will always be something that is unfinished, incomplete"?[6]

T.O.:   Yes, of course: the Internet will continue to arrive, like all products of human culture.

A.P.:   Big data is one of the cutting-edge technologies of the day. As you have said, technology can be used for good or for bad. In Europe, a huge controversy has been stirred up concerning the widescale espionage carried out by the NSA, following Edward Snowden's revelations.

You have written a paper called "The Creep Factor: How to Think about Big Data and Privacy."[7]

However, with all that is going on, do you not think that, in certain circles, the expression "big data and privacy" has become an oxymoron?

T.O.:   Expectations of privacy have changed many times over the years. And usually, when new technologies move the goalposts in some way, it takes some time to establish what the new social norms ought to be. Often there are landmark legal cases decades after a new technology emerges. That will happen here too. In the meantime, though, I urge companies to be thoughtful about what they do with data they gather or purchase. Is the data being used on behalf of the customer or against the customer? That is the key question. When companies or governments use our data against us, we are rightfully outraged. When they use it to make things better for us, that's a powerful tool to increase our loyalty and satisfaction.

A.P.:   Since you have said that technology can also be used for the better, let's think positively:

I'm going to ask you three questions, which are really one:

Do you think the universe of the great "connected brain" of the Web and its technological evolution—which, as you have said, benefits from collective intelligence—could finally turn out to be the "cement" or binding agent that leads us back to a period of understanding human knowledge as a "whole"—as it was, for instance, in ancient Greece and Alexandria, as well as during the pre-Renaissance efforts in Bologna concerning the *trivium* and *quadrivium* = *Unus-Versus-Alia* = university?

Could the cybernetics of the Web work as a binding agent for breaking down barriers, which are sometimes unfathomable, referred to by C. P. Snow, which still separate the arts and humanities from the sciences and their useful applications?

In other words, do you see the global Web as an instrument for bringing together, or merging disciplines, which urge us, with a more highly interwoven knowledge, to progress toward a better human condition in the near future?

T.O.:   That's a tough question. On the one hand, there is no going back to the days when a learned individual could have read most of books that mattered, and perhaps own most of them in a library that would fit in a room or two. At the same time, in every era, there is some amount of canonical knowledge. STEM (or STEAM) is the new trivium today, and it's just as flawed. The most interesting people depart from the canon. All of our great cultural heroes are people who went beyond the boundaries of knowledge or practice in their day. Their artistic transgressions or scientific discoveries eventually became part of the new canon. So let's hear it for the explorers, the rule breakers, those who heed Ezra Pound's advice to "Make it new!"

**Notes**

1. O'Reilly and Associates, "An Open Letter to Jeff Bezos," O'Reilly.com, February 28, 2000, http://oreilly.com/amazon_patent/amazon_patent.comments.html. See also the compilation of links on the matter on "The Amazon patent controversy," O'Reilly.com, February 28, 2000, et seq., http://www.oreillynet.com/pub/a/oreilly/news/patent_archive.html.

2. Quoted in Edwin Schlossberg, "Leadership: An Interview with Tim O'Reilly," in *Beatiful Teams*, ed. Andrew Stellman and Jennifer Greene (Sebastopol: O'Reilly Media, Inc., 2009), 2.

3. Tim O'Reilly, "What Is Web 2.0: Design Patterns and Business Models for the Next Generation of Software," O'Reilly.com, September 30, 2005, http://oreilly.com/web2/archive/what-is-web-20.html.

4. Tim O'Reilly, "Open Source Paradigm Shift," O'Reilly.com, June 2004, http://www.oreillynet.com/pub/a/oreilly/tim/articles/paradigmshift_0504.html; Thomas Kuhn, *The Structure of Scientific Revolutions* (Chicago: University of Chicago Press, 1962).

5. Tim O'Reilly, "Web 2.0 Meme Map," O'Reilly.com, September 30, 2005, http://www.oreillynet.com/oreilly/tim/news/2005/09/30/graphics/figure1.jpg.

6. Jimmy Wales in recorded conversation with the author, in *Copyfight* Meeting, July 17, 2005, Barcelona, Spain.

7. Tim O'Reilly, "The Creep Factor: How to Think about Big Data and Privacy," O'Reilly.com, March 6, 2014, http://radar.oreilly.com/2014/03/the-creep-factor-how-to-think-about-big-data-and-privacy.html.

## 20 Encryption as a Human Right

David Casacuberta and Adolfo Plasencia

David Casacuberta. Photograph by Adolfo Plasencia.

*The right to privacy or intimacy is not the right to have secrets; it's the right to decide whom you want to share information with. It's a basic thing. … It has always been part of the definition of human beings and their dignity.*

*The right to electronic privacy should also be part of the declaration of universal rights and democratic constitutions.*

*—David Casacuberta*

David Casacuberta is Associate Professor of Philosophy of Science, the Universidad Autònoma de Barcelona (Spain). He earned a PhD in philosophy and a master's degree in cognitive sciences and language. His current research focuses on the cognitive and social impact of new media, especially with respect to privacy issues, e-learning, and social inclusion.

He co-edited (with Jordi Vallverdu) the *Handbook of Research on Synthetic Emotions and Sociable Robotics: New Applications in Affective Computing and Artificial Intelligence* (IGI Global, 2009) and is author of *La mente humana: Cinco enigmas y cien preguntas* (Océano, 2001), *Que es una emocion?* (Crítica, 2000), and *Creación e Colectiva. En Internet el creador es el público* (Gedisa, 2003).

Adolfo Plasencia:   David, you are a professor, a humanist, a scientist, a researcher. How do you like being introduced?

David Casacuberta:   As a professor. "Researcher and philosopher" is a bit too much. Today we all have to be researchers and publish papers in big journals as a duty, which is a bit absurd in the humanities.

A.P.:   One of the things that brought me here to talk to you is a text you wrote, "The Right to Encrypt."[1] In it you say that, according to Guibbard's 1992 human rights model, rights must be based on basic standards for human beings. After digitization and the deployment of the cyberspace, new technical and cultural contexts have emerged for those rights.

You wrote that a human right cannot exclusively stem from technical reasons., What would the nature be of the human rights that are now needed in a world ruled by digitization and coding, as Lawrence Lessig says,[2] and in the global Web or cyberspace in which we also live?

Should we adapt the rights we had or reformulate them?

Should we create new human rights to protect the new areas of human relationship brought along by technology, which did not exist before?

Should we enlarge the Universal Declaration of Human Rights to include new rights?

Maybe we have to create global human rights, which are different because human rights based on local laws are not suitable for certain purposes.

What do you think?

D.C.:   All this is at the very basis of the big problems we now have. First, what we should never, ever do is create dominant rules or standards, or a dominant policy. The rights of citizens should not be reduced. And yet not everyone seems to think this way. More and more people ask me why we keep on and on about privacy. They say it is an "old" issue, that privacy no longer makes sense because you are going to be spied on anyway. You can't help it. The NSA spies on you, they can use microphones, they read your email. People say, what's the point of privacy today? It's not worth discussing.

That's why I quoted Guibbard and the idea that behind human rights there must be laws, mental states, opinions, and emotions in humans. So if we want privacy, stay calm. If we don't want strangers to know things about us, that should be respected, because it is in our nature as human beings.

I do worry about something I keep hearing: young people who come and tell you, "What's wrong with Google knowing things about you and recommending things to you via 'your' gmail address? It's quite a good thing." Well, I don't think so!

I think we should all have our "vital" rights preserved, rights that are vital to us for emotional reasons, because we are human, we are mammals, and we work that way, and that needs to be respected.

Something is clear: with the advent of digital technologies, these rights are threatened by new problems that require new solutions. I think the simplest solution in the long run is to create new rights; that is why I wrote "The Right to Encrypt." It is a twenty-first-century right. I think it didn't make much sense to a lot of ordinary people even in the last century, in the 1980s. But things have changed dramatically. Today, the right to decide whom I allow to listen to my conversations makes more and more sense, because basically it is a right we have in the physical world. If I'm at home and I do not want anybody to see me, I draw the curtains. That way I feel safe. So I decide the level of information people can have about me.

A.P.:   When we talk about human rights, what is actually enforced is a law passed by a local parliament of a given country. Regulating human rights is a competency that has been transferred to each parliament, for each democratic country to pass such regulations and include them in its legal system. But this is not the best way to approach the problem, and the logic of the digital, immaterial world of the net is an absolutely global phenomenon.

I do not know if the scope of these new rights you're talking about should be directly global. It would be very complicated for 190 countries to agree on new human rights to be recognized by the UN and applied by their parliaments.

Should we start thinking about global citizenship and a scope in which these rights can work and have a global nature?

D.C.:   Yes. Absolutely, the view on these things must be comprehensive. I even think this should go beyond formal laws. We should consider some ethical foundations on which global agreement could be reached, as with the Universal Declaration of Human Rights.[3] I think it is essential to bring this up. Otherwise you see things like the NSA and the way it works. And things like "I can't spy on American citizens because they are protected by the American Constitution, but I can spy on the rest of the world because they are not."

A.P.:   Well, perhaps behind that logic, namely, "American citizens should not be spied on because they are a genuine democracy," is an attempt to ensure good legal protection for some governmental leaders.

D.C.:   Of course. We can see it right from the beginning of the Internet, the need for other types of legislation. We now talk about privacy, but the same happens with copyright violation or classic rights breaches as regards theft, but with digital media. What happens when you have a criminal who is stealing credit cards and using a computer on Bouvet Island or near Lake Baikal but controls it from another computer on Fiji, and the bank responsible for those cards is based in France, for example?[4] It gives way to such absurd problems that the whole thing needs to be reviewed. But no one seems to be willing to do so.

A.P.:   In "The Right to Encrypt" you refer to a concept by David Brin. In his 1998 book he said we were heading toward becoming a transparent society.[5] The transparent society is now a matter of fact. All our intellectual actions involving electronic media are recorded in all kinds of digital devices. Cameras record our images, Internet servers have a duplicate of our messages, email accounts are a source of information, and the same is true for everything we do with our cell phones.

Any digital instrument of this type is available to the network system, along with their locations and space-time coordinates. This information makes it possible to track users wherever they are, anytime, anywhere.

Do you think all this is inevitable? Some of the creators of the Internet thought that, in order to avoid the potential dangers they foresaw at the time, Internet access had to be anonymous. Tim Berners-Lee always argued that access had to be anonymous whenever possible. Presumed security advocates seem to thinks this is almost a heresy.

Do you agree with this idea of the creators of the Internet? Should it be public domain, and much closer to openness and freedom than to security?

What do you think?

D.C.:   I totally agree with them. It should always be open. But I am pessimistic. I don't think we will succeed, but it should be like that, absolutely. Cryptography can be a central element in that respect.

What finally allows you to anonymously gain access to the World Wide Web, and send messages without anybody reading them, is encryption. I am absolutely certain about that. But there are so many interests. ... Lawrence Lessig has already talked about this in his book, *Code and Other Laws of Cyberspace*.[6]

But piracy, terrorism, pedophilia, and so forth are increasingly used as excuses to try and turn the Internet into a space with more control, recordings, spying, surveillance ... don't you think so?

A.P.:   Cryptography is not new. Generals in Roman legions already wrote messages on the scalp of slaves, waited for their hair to grow, and then sent them to another location, where they had their hair cut so the message could be read. So apart from Caesar's cipher and other codes, there were many kinds of encryption.[7] And two thousand years ago, the Chinese general Sun-Tszu in *The Art of War* spoke about six different kinds of spies, so this is also part of the history of humankind.

But I think Brin's transparent society is not egalitarian. We are not all equal when it comes down to technological capabilities. We do not all have the same power. Some have an advantage; they have the capabilities and the means for very sophisticated electronic espionage. In other words, it is a "transparent" society only for those who are behind the mirror, as in the interrogation room, for those who are behind the glass. They run at an advantage. They watch without being watched. Am I right?

Besides, we take it for granted that everybody spies on everybody, but the question is, how much of this is transparent, and for whom in society? And in fact, this transparent society has multiple levels of transparency: the level of those in front or behind the mirror varies. In other words, it is transparent only for those who have the right means.

What do you think about this sort of security cynicism in governments in democratic countries? And what do you think about differences between the helpless and those who have all the advantages in the transparent society?

D.C.:   Well, this is the way things are. Technological changes have given way to enormous cultural changes, especially in recent times.

When Brin wrote about this, and when I wrote "The Right to Encrypt," there were not yet such huge inequalities technology-wise. There were CCTV cameras in the streets, and very few people had video cameras. We seem to have forgotten all this. At the time it was very clear that technology was more or less the same for almost everybody. In that period, HTML was the same for everyone, so I was able to have a website as cool as that of the CIA, for example. The only thing it had was links, photos. …

A.P.:   But we can already see that in a sentence paraphrased from Thomas Jefferson's Declaration of Independence, "All bits are created equal. It's not just a good idea. It Ought to be the LAW," has been applied to the digital world and the Web, where it is used symbolically in the defense of net neutrality.[8]

D.C.:   At the beginning, yes, but what is happening now? The Web 2.0, according to Bruce Sterling in his speech on the subject,[9] has apparently facilitated publishing, but that sort of hides the fact that you no longer have direct access to the code behind it. You work on programs that were done by other people, and they are in control.

Before Google, when people searched for information, we were all the same, but now Google and others have information about us that the rest don't have. The cell phone boom means that mobile phone companies have data and metadata on people's location and actions; we don't have access to that. So things have changed dramatically. Brin's transparent society is no longer possible because there is clearly a monopoly on people's private data, which was much better distributed before. In the past, one could be as effective as or more effective than the police or the government in obtaining and distributing information, but that has changed.

A.P.:   There is another subject in your texts: digital privacy. You speak about the need to redefine privacy. You have talked about the idea of "privacy as dignity."

But privacy is now threatened by many electronic media. It is not guaranteed even by the laws of formal democracies. According to you, the threats are also "a threat or attack against our dignity, whether we know it or not, whether it affects us psychologically or not."[10] Is this so?

Can we, in practice, do something as free citizens from democratic countries against the attacks unveiled by Snowden, for example? Or is this far beyond the actions of a citizen from a formal democracy in the Western world?

D.C.:   The essential thing is not to give up that right as citizens. It worries me to see some trends not only among ordinary people but also among politicians and journalists. *"Are you a terrorist? No. Then, what do you care if the NSA is listening?"* It's like a cultural stance. And that's a mistake because,

I insist, it is a matter of respect and dignity. If the NSA has a file about me, I don't think they will find much, that's true. But it doesn't matter. I don't want them to have it because it goes against a basic idea that I have about myself, about an autonomous person who has the basic right not to generate secrets. It would be a mistake to accept *"If you are not a criminal, you do not have to keep secrets."*

The right to privacy or intimacy is not the right to have secrets; it's the right to decide whom you want to share information with. It's a basic thing. ... It has always been part of the definition of human beings and their dignity.

Then we must stand up for that right and not give it up, even if they make it easy for us. That's where I see the next step that citizens will take. We are to think carefully about the famous sayings, "When something online is free, you're not the customer, you're the product," or "Anyone who gives you something for free is trading with you."[11] And be aware of it.

A.P.:   As a matter of fact, the "free" economy entails some traps.[12]

D.C.:   Definitely.

A.P.:   Would you say that to Chris Anderson?

D.C.:   Yes, of course I would. I think that part is not conscious. All these companies have a series of businesses, and it is unclear whether all the implications of those businesses for us are reasonable. And so people easily post information on those free spaces, and then there is always a price to pay. I am not talking about the future. I am talking about recent political cases of people talking nonsense on Twitter. They had to resign. They did not understand the medium in this basic way.

A.P.:   In a real-time electronic reality, everything is instantaneous ... we lose our midterm bearings and we don't even analyze the consequences of what we do with digital technologies. And the consequences can be very serious. We've seen it in the public world, in politics, and in other fields. But when you take an action with an electronic or digital medium, even people who are educated—perhaps not digitally—do not grasp how serious the consequences of a small electronic act can be.

D.C.:   Exactly. I think that is basic, it needs to be learned. And I think that cyber-anthropologist danah boyd explains it very well with her short, powerful sentences:

We're seeing an inversion of defaults when it comes to what's public and what's private. Historically, a conversation that you might have in the hallway is private by default, public through effort. It's private because no one bothers to share what's being said. The conversation may be made public if something worth spreading is said. Even though the conversation took place in a public setting, the conversation is private by default, public through effort. Conversely, when you engage online in equally public settings such as on someone's Facebook Wall, the conversation is public by default, private through effort. You actually have to think about making something private because, by default, it is going to be accessible to a much broader audience.[13]

and we continue working as if we were in the physical world. Now, you go to a job interview and the first thing they do is Google you.

A.P.:   We were saying that we do things that you think nobody will see and it turns out that everybody can see them. You didn't do your privacy settings on the Internet, and then you send something to someone you know and you are actually publicizing it in a football stadium.

D.C.:   Yes, you put it perfectly. On the Internet, "default" means that the whole world is watching you in the stadium, on a giant screen. You are telling the world quite a few things there. But we are at home with our curtains drawn, sitting in front of the computer. The door is closed. We think nobody can see us. It's a habit we need to face. We have to change our culture about it completely. Besides, we share more and more about ourselves, and there are more and more databases, and more cybernetics handling them, and all that stuff. The more entities there are willing to pay for such data, the greater the risks.

A.P.:   You wrote that in the world of mass electronic surveillance, one is guilty until proven innocent or until your innocence cannot be proved by yourself or by others. This sentence of yours is significantly earlier than Edward Snowden's revelations, so it was you who said it, not in connection with Wikileaks but before.

Were you already sure then that the situation would be as it is now?

Will the blame be put on those of us who cannot prove our innocence in the digital society?

D.C.:   Yes, because we saw it coming; we knew what things were making governments nervous at the time. In my experience as a "digital elder," having seen how all this arose has given me more of a perspective. I don't mean to say I'm smarter. I was there when it all started, and I saw how it developed. Obviously, governments did not understand a thing at the time. Back then, newspapers loved criminalizing the Internet. The Internet was a place full of child pornography, terrorism, information about bombs, and so forth. The headlines read: "The Web Threat." That's the way things were. A very few journalists were genuinely interested. The rest was witch hunting, pure McCarthyism.

A.P.:   Let's turn to another aspect of your article, "The Right to Encrypt." Let me remind you of some ideas and questions in relation to that central concept, "The space of digital communications or cyberspace should be considered a public space," and another premise, "The right to electronic privacy should also be part of the declaration of universal rights and democratic constitutions." This is more of a question than a premise, actually. Third, you defend the need to "establish cryptography as a right, since no ordinary ICT user has the power to protect their electronic communications other than with cryptography."[14]

Did what recently happened to the NSA strengthen that view?

D.C.:   Totally. There is something which is very interesting, talking about history. ... Originally governments, when they realized something was problematic, they would always prohibit it, right? That is what Clinton did with the Internet: he created a shield law to protect minors and avoid any kind of pornography on the Internet—basically, to eliminate it. Let me remind you that the definition of pornography in the United States is very strict. So we won that battle.

And what did governments learn in the first exercises of content control? That it was much better to pretend not to know and try to show that those contents did not exist than ban them. There's that famous phrase in *Time* by John Gilmore: "The Net interprets censorship as damage and routes around it."[15] In other words, content control is seen by the Internet as censorship, censorship as damage, and it routes content around to continue offering content. Therefore, universities and individuals decided to give shelter to documents that were censored by governments. They didn't care about organizations like ETA before or ISIS now; they don't care whether they are terrorist organizations or not. They believed that freedom of expression was important, that it was above everything else. No government has tried, for a very long time now, to stop a radical video because they know it is counterproductive.

A.P.:   And what have we learned from all this in two and half decades with the Web?

D.C.:   That silence is the best strategy. It even happens with cryptography. When I wrote the article, "governments" wanted to ban cryptography. They preferred citizens not to have any access to cryptography. Then they realized it was pointless because people were still using it. In fact, we used it to protest. And now what do they do? They don't take any notice. They don't mention it. They ignore it. And so people now do not use cryptography as they did before. But, as you rightly say, the existence of spy programs such as PRISM and other programs shows that we should all have cryptography and encrypt our mail, have contents in hard drives encrypted, and so on, and should trust these mathematical tools rather than "governments" that are not really willing to do their job.

A.P.:   We still say that cryptography prevents enforcement authorities from spying on our electronic communications. You wrote that a few years ago and you still think so. You said, "With a PGP [Pretty Good Privacy]-encrypted message, a police officer or an intelligence agent can only shrug their shoulders because there is no human way to decipher what it says." But some "powers" also have other methods. For example, some time ago, four companies offering fully encrypted email services, such as SilentCircle, had to close down because the method used by U.S. agencies, with U.S. laws, was to enable a judge with a "secret court" and force these companies to either deliver the complete cryptographic key to decrypt all telecommunications with customers or close down. All four companies closed down.

Then is cryptography still impossible to decipher for these people?

D.C.:   PGP is, because it is based on this idea of autonomy. You generate the key and you are the only holder of that key. Besides, there is a whole community of hackers that makes sure that the program has no back door. But of course, very few people use PGP now. In that sense, it is secure, but only at a theoretical level; in practice we know what happens: the NSA has back doors to cryptographic systems and other computer tools, and they promote the manufacturing of computers with a back door (accessible to them, of course). So in theory yes, in practice no.

A.P.:   Then do you think that the alibi of collective security versus individual freedom to encrypt will finally prevail in the near future?—because the right to encrypt is an individual right and an individual act. However, the alibi of collective protection against possible terrorist attacks is in fact collective protection used by state agencies or governments to do what they do.

D.C.:   I'm afraid so.

A.P.:   As I was saying, this "collective security" alibi alleged by U.S. law is actually based on individual freedom. ...

Will this alibi defeat our individual freedom to encrypt?

D.C.:   Yes, I think so. These are some common arguments used that are in line with the NSA: "Thanks to this program, we managed to arrest many terrorists; thanks to this program we managed to find Osama bin Laden. Thanks to this program, blah blah blah, blah blah blah." They are very powerful arguments. Another science fiction writer has argued about this quite well and has reflected about this a lot. His name is Neal Stephenson. In one of his talks, he analyzed the idea of people's changing nature.[16] He said that humans are very fickle; depending on the situation, we have different models of rights. For example, if you live in a wealthy neighborhood, you don't want security cameras and police walking around in the area, you love your privacy. But if you live in a neighborhood with frequent muggings and robberies, you want more cameras, more surveillance, and more police control. It is almost inevitable. Then it just takes a few alarms going off (even artificially) for people to accept it and say: "How many people die in a terrorist attack and how many die in a traffic accident?" We are culturally in that dilemma. Certain things generate alarm almost automatically. And if they are properly exploited, then people inevitably buy this collective security stuff. I think there is no easy solution.

A.P.:   Thank you very much, David. It has been a pleasure. We'll keep in touch, and we will see if we can get further confused about this.

D.C.:   That's unstoppable, I'm sure. Thanks to you; my pleasure.

**Notes**

1. David Cascuberta, "El derecho a cifrar" [The right to encrypt], Academia.edu, 2003, http://www.academia.edu/4534267/El_derecho_a_cifrar.

2. Lawrence Lessig, "Introduction," *Free Software, Free Society: Selected Essays of Richard M. Stallman*, ed. Joshua Gay (Boston: GNU Press of the Free Software Foundation, 2002), v, https://www.gnu.org/philosophy/fsfs/rms-essays.pdf.

3. The Universal Declaration of Human Rights was adopted by the UN General Assembly on December 10, 1948, in response to the atrocities of World War II. It is the most translated document in the world. The document sets out human rights in a number of spheres—civic, political, spiritual—and the duty of the individual to society.

4. Bouvet Island is an isolated island in the South Atlantic from which numerous cyberattacks have been launched; Lake Baikal in southern Siberia is also the source of many cyberattacks. David Casacuberta implies these attacks are carried out with impunity and without consequences for the attackers. See "Secret NWO Island

Headquarters Discovered at 'The Last Place On Earth'?," http://allnewspipeline.com/NWO_Secret_Island_HQ_Discovered.php.

5. David Brin, *The Transparent Society* (New York: Perseus Books, 1998).

6. Lawrence Lessig, *Code and Other Laws of Cyberspace* (New York: Basic Books, 1999).

7. The Caesar cipher is a substitution cipher that was used by Julius Caesar in his personal correspondence (https://en.wikipedia.org/wiki/Caesar_cipher).

8. The principle of net neutrality is a network-design paradigm and is one of the founding principles of the Internet as we know it. According to this principle, the net should transmit/transport all the bits as equals, without prioritizing or discriminating them in any way, regardless of the specific information they are part of. In current practice, this means that broadband internet providers must remain completely separate from the information that is sent through their networks. This ensures that the Internet is the same for any user without distinction, that is, maintaining neutral communication based on three key pillars: non-discrimination, continuous interconnection and equal access for all.

9. Bruce Sterling, "What Bruce Sterling Actually Said About Web 2.0 at Webstock 09," *Wired*, March 1, 2009, https://www.wired.com/2009/03/what-bruce-ster.

10. David Casacuberta, "El derecho a cifrar."

11. See Jonathan Zittrain, "Meme Patrol: 'When Something Online Is Free, You're Not the Customer, You're the Product,'" blog post, *The Future of the Internet and How to Stop It*, March 21, 2012, http://bit.ly/1QGU5r9.

12. Chris Anderson, "Free! Why $0.00 Is the Future of Business," *Wired*, February 25, 2008, http://archive.wired.com/techbiz/it/magazine/16-03/ff_free.

13. danah boyd, "Making Sense of Privacy and Publicity," SXSW, Austin, Texas, March 13, 2010, http://www.danah.org/papers/talks/2010/SXSW2010.html.

14. ICT here stands for information and communications technology.

15. As the Internet pioneer John Gilmore put it, "The Net interprets censorship as damage and routes around it." See Philp Elmer-Dewitt, "First Nation in Cyberspace," *Time International,* December 6, 1994, http://www.chemie.fu-berlin.de/outerspace/internet-article.html.

16. Neal Stephenson, "A Talk with Neal Stephenson," n.d., Cryptonomicon.com, http://www.cryptonomicon.com/chat.html.

## 21   Order in Cyberspace Can Only Be Maintained with a Combination of Ethics and Technology

John Perry Barlow and Adolfo Plasencia

John Perry Barlow. Photograph by Rafael De Luis.

*When I discovered cyberspace, I realized that the frontier was bound only by the human imagination. And that we would never run out of frontiers as long as human beings could dream.*

*The real question is, how do we maintain a sense of ourselves in an unbounded environment, and at the same time provide sufficient tolerance for other points of view, so that we are not constantly trying to suppress them?*
*—John Perry Barlow*

John Perry Barlow is a poet, essayist, and cyberlibertarian political activist. He is also a former lyricist for the Grateful Dead.

In 1986 Barlow joined The WELL online community. In 1990 he founded, along with fellow digital rights activists John Gilmore and Mitch Kapor, the Electronic Frontier Foundation (EFF). As a founder of EFF, Barlow helped publicize the Secret Service raid on Steve Jackson Games. He subsequently became a Fellow at Harvard University's Berkman Center for Internet and Society.

Barlow is the author of the famous "A Declaration of the Independence of Cyberspace," which was written in response to the enactment of the Communications Decency Act in 1996 as the EFF saw the law as a threat to the independence and sovereignty of cyberspace. He argued that a cyberspace legal order should reflect the ethical deliberation of the community instead of the coercive power that characterizes real-space governance. He strives to promote different rules for cyberspace, much closer to the common universal dreams and values for human beings, with no other moral or cultural distinction than those that gather around the common values that all humanity shares, in the tradition of Alexis de Tocqueville, Jefferson, Washington, Mill, Madison, and Brandeis.

*Note:* The following conversation took place at the encounter about digital culture, "Copyfight," in July 2005 in Barcelona, and was later supplemented at the "Powerful Ideas Summit," held in Valencia in January 2007. It was revised in June 2015.

Adolfo Plasencia:   In 1996, at the World Economic Forum in Davos, you read the famous "Declaration of Independence of Cyberspace."[1] Would you change anything after the experience of the last two decades?

John Perry Barlow:   I did not intend to make it into a declaration of intentions but rather a declaration that would describe the present moment. And to clarify that in spite of all the attempts that the industrial period has made to establish its sovereignty over cyberspace, it has always failed.

If I had to make any change, the only thing I would change would be to make it less arrogant.

A.P.:   It seems there are many powers that would like to govern cyberspace.

John, is cyberspace governable or not?

J.P.B.:   That was what I meant to say with the "Declaration of Independence," which was that cyberspace is naturally immune to rules by conventional means, partly because there are so many forces that would like to rule it and they cancel one another, and partly because I believe that law requires access to the physical body. As such, order in cyberspace can only be achieved by some combination of ethics and technology.

A.P.:   You published an article that first appeared in *Wired*, "The Economy of ideas," which is considered one of the founding texts of free culture on the Internet.[2] In that text you say, "Everything you think you know about intellectual property is wrong."

Are so many people wrong about this?

J.P.B.:   Anyone who uses the term "intellectual property" has the wrong idea.

Property, by its nature, is something you can take away from somebody else. It's something that one can have possession of. I can possess an idea only if I keep it to myself. But the second I reveal to another, no one possesses it. And everyone can. And if it's a good idea, it would spread naturally. And I will still maintain that idea in my head, and nobody will have taken it away from me as it's spread. And not only will it not become less valuable, because it's spread, it will become more valuable. This makes it fundamentally different from real property.

If I have a large diamond, it's very valuable. If everybody on the planet has a large diamond, it's not valued at all. Precisely the opposite is true of an idea.

A.P.:   John, you are one of the main driving forces behind the Electronic Frontier Foundation, which sounds fascinating. Is the electronic frontier a fixed or mobile frontier, or is it a horizon that we can never reach?

J.P.B.:   Frontiers, by their nature, move. And part of the reason I wanted to name it that was because my family came to America and then for three hundred years stayed on the frontier as it moved. Suddenly, I found myself at the end of physical frontiers.

But when I discovered cyberspace, I realized that the frontier was bounded only by the human imagination, and that we would never run out of frontiers as long as human beings could dream. There will always be the same questions asked by frontiers reviving themselves anew with each change in technology.

A.P.:   It is likely that there are companies, corporations, or perhaps leaders who would like cyberspace to be an official and commercial environment. But there are also many citizens, ordinary people, who would like a free and open cyberspace.

What do you feel is the current situation, or how do you see the future for this articulation between the official, commercial Internet and the free, open Internet?

J.P.B.:   I don't think that commerce and freedom are inconsistent. I believe in free markets, and in fact, one of the things that I did that is less well known was to help create something called the Commercial Internet Exchange, in 1990.

A.P.:   John, what is today's "economy of ideas" like?

Is it a capitalist or socialist economy?

Is it a true market economy?

And should it be supervised or not?

J.P.B.:   It's both capitalist and, I think, in a very pure way, Marxist—because it's impractical for people to be expected to share physical property, but it is practical to share the work of the mind. If I have a Mercedes-Benz, I don't want everybody to be able to drive it. It's practical to ask people to share the fruits of thought. It's capitalist only in the sense that a very different kind of investment is required to create wealth in cyberspace. But it's not the kind of investment the Marxist notion of "bourgeoisie" would make in order to ensure its power. Because I could be a poor graduate student at Stanford University, without any capital at all, and create one of the most powerful institutions on the planet in the form of Google.

A.P.:   Not long ago, John, you said in an interview that you do not believe that the digital divide exists. Many say that ever since the revolution on the Web has been going on, in the same time in which this revolution has been taking place, the differences between the rich and the poor have done nothing but increase.

Is it possible that the weapon of the Internet revolution is useless in narrowing the gap between those who have more and those who have less on the planet?

J.P.B.:   I think the differences between the rich and the poor have increased, but for reasons that have nothing to do with internet. If it has happened, it is because the rich have become more greedy, more covetous, and more active in appropriating richness as a response to these changes. For example, you can see it in the United States. The government has eliminated any limit in terms of taxes to enrichment and, in alliance with the richest and the largest corporations, it has eliminated all restrictions on monopolies. As a result, the United States is becoming an oligarchy.

A.P.:   I mentioned it because for many people, this revolution was a great hope for the poorest on the planet, a possible way to get out of their misery.

J.P.B.:   I spend a lot of time in Africa. It is very rare not to find a cybercafé in any village or city of Africa. And the first slow beginnings of a starting new economy are forming rapidly in Africa.[3] Starting with crime—which is how often a new economy starts anyway!

A.P.:   And what happens to that great hope?

J.P.B.:   Eventually even criminals discover that it is more profitable to earn money legally. But in order to do so, they first have to have the education, the instruments, and the contacts.

A.P.:   We are now in Spain. You live in the United States, but you travel all over the world. Is the European vision that we have of cyberspace of the Web different from the one people have in the United States or other places?

Do you think the local culture of cybercitizens affects cyberspace culture, or that when we act as such, we all act in the same way?

J.P.B.:   Had you asked me this same question only five years ago I would have answered that there exists only one global culture of Internet. And somehow, it still does exist. But what I saw as global then was in reality a strong injection of the United States' values at the beginnings in that global culture: the belief in freedom of speech, the belief in intimacy and privacy, the belief that there should be a limited government. But as other cultures connect to the Internet, for example, those of the Arab world, their values are being injected into the Web.

The real question is, how do we maintain a sense of ourselves in an unbounded environment, and at the same time provide sufficient tolerance for other points of view, so that we are not constantly trying to suppress them?

I am optimistic enough to think that once we realize that it's very difficult for us to suppress other peoples' cultures, and once local cultures gather more energy among themselves, by reaching or having the largest place in which to communicate, they will solve this problem. For instance, take a look at Catalonian culture: during the print period, only so many books could be published in Catalan because the market would only support so many. Now there is no such limit.

A.P.:   Excellent! That's everything!

Thank you very much for meeting me today and for your time. I hope to see you soon in my city, Valencia.

J.P.B.:   I hope so too!

## Notes

1. John Perry Barlow, "A Declaration of the Independence of Cyberspace," Electronic Frontier Foundation, February 8, 1996, https://projects.eff.org/~barlow/Declaration-Final.html.

2. John Barlow, "The Economy of Ideas," *Wired*, March 1, 1994, http://archive.wired.com/wired/archive/2.03/economy.ideas.html.

3. John Perry Barlow, "Africa Rising: Everything You Know about Africa Is Wrong," *Wired*, January 1998, http://archive.wired.com/wired/archive/6.01/barlow.html.

## 22   The Free Software Paradigm and the Hacker Ethic

Richard Stallman and Adolfo Plasencia

*Technology without the influence of ethics is likely to do harm.*

*It's absurd to ask whether ethics is profitable or not; because it is a prior responsibility, an obligation towards others. Choosing ethics as a guide only when it is profitable is really the same as not having any ethics.*
—*Richard Stallman*

Richard Stallman is Founder and President of the Free Software Foundation, a software developer, and software freedom activist.

### Some considerations

### 1.   Richard Stallman, according to Lawrence Lessig[1]
Every generation has its philosopher—a writer or an artist who captures the imagination of a time. Sometimes these philosophers are recognized as such; often it takes generations before the connection is made real. But, recognized or not, a time gets marked by the people who speak its ideals, whether in the whisper of a poem or the blast of a political movement.

Our generation has a philosopher. He is not an artist, or a professional writer.

He is a programmer. Richard Stallman began his work in the labs of MIT, as a programmer and architect building operating system software. He has built his career on a stage of public life, as a programmer and an architect founding a movement for freedom in a world increasingly defined by "code."

"Code" is the technology that makes computers run. Whether inscribed in software or burned in hardware, it is the collection of instructions, first written in words, that directs the functionality of machines. These machines—computers—increasingly define and control our life. They

Richard Stallman. Photographs by Adolfo Plasencia.

determine how phones connect, and what runs on TV. They decide whether video can be streamed across a broadband link to a computer. They control what a computer reports back to its manufacturer.

These machines run us. Code runs these machines.

What control should we have over this code? What understanding? What freedom should there be to match the control it enables? What power?

These questions have been the challenge of Stallman's life. Through his works and his words, he has pushed us to see the importance of keeping code "free." Not free in the sense that code writers don't get paid, but free in the sense that the control coders build is transparent to all, and that anyone has the right to take that control, and modify it as he or she sees fit. This is "free software"; "free software" is one answer to a world built in code.

(*Note*: Lawrence Lessig, introduction to the book *Free Software, Free Society: Selected Essays of Richard M. Stallman*, 2nd ed., GNU.Org, copyright c 2002 Free Software Foundation, Inc. This foreword was originally published in 2002 as the introduction to the first edition. This version is part of *Free Software, Free Society: Selected Essays of Richard M. Stallman*, 2nd ed. (Boston: GNU Press of the Free Software Foundation, 2010). Verbatim copying and distribution of this entire chapter are permitted worldwide, without royalty, in any medium, provided this notice is preserved.)

## 2.  Who is Richard Stallman?

"Who am I?" was the first question in Richard M. Stallman's initial announcement[2] of his GNU Project, an email beginning "Free Unix!" Sent on September 27, 1983, from his computer at the MIT Artificial Intelligence Lab, it stated the plan to develop the revolutionary GNU (GNU's Not Unix) operating system, and asked people to join in.[3]

It continued: "Why must I write GNU?" His answer was: "I consider that the golden rule requires that if I like a program I must share it with other people who like it. I cannot in good conscience sign a nondisclosure agreement or a software license agreement. So that I can continue to use computers without violating my principles, I have decided to put together a sufficient body of free software so that I will be able to get along without any software that is not free."

Here's the updated definition of free software:[4]

Free software means that the users have the freedom to run, copy, distribute, study, change and improve the software. A program is free software if the program's users have the four essential freedoms:

• The freedom to run the program as you wish, for any purpose (freedom 0).

• The freedom to study how the program works, and change it so it does your computing as you wish (freedom 1). Access to the source code is a precondition for this.

• The freedom to redistribute copies so you can help your neighbor (freedom 2).

• The freedom to distribute copies of your modified versions to others. By doing this you can give the whole community a chance to benefit from your changes (freedom 3). Access to the source code is a precondition for this.

A program is free software if it adequately gives users all of these freedoms. Freedom to distribute means you are free to redistribute copies, either with or without modifications, either gratis or charging a fee for distribution, to anyone anywhere. Being free to do these things means (among other things) that you do not have to ask or pay for permission to do so.

### 3. An extraordinary life, an extraordinary biography

Stallman was born in 1953 in New York City. He was interested in computers at a young age; when still a pre-teen at a summer camp, he read manuals for the IBM 709410 and wrote assembler programs for it on paper. From 1967 to 1969, Stallman attended a Columbia University Saturday program for high school students. He was also a volunteer laboratory assistant in the Biology Department at Rockefeller University. Although he was interested in mathematics and physics, his teaching professor at Rockefeller thought he showed promise as a biologist.

His first experience with actual computers was at the IBM New York Scientific Center when he was in high school. Then IBM hired him for the summer of 1970, following his senior year of high school, to write a numerical analysis program in Fortran He completed the task after a couple of weeks ("I swore that I would never write in FORTRAN again, because I despised it as a language compared with other languages") and spent the rest of the summer writing a text editor in APL.

As a first-year student at Harvard University in fall 1970, Stallman was known for his strong performance in Math. Stallman graduated from Harvard magna cum laude, earning an AB in physics in 1974. His time at Harvard was not happy, as he did not have even one date, but at least he felt less of an outcast: "For the first time in my life, I felt I had found a home at Harvard."[5]

In 1971, near the end of his first year at Harvard, he became a programmer at the MIT Artificial Intelligence Lab, and therefore part of the hacker community. There he was usually known by his initials, RMS (which was the name of his computer accounts).[6]

Stallman considered continuing at Harvard in physics, but instead he decided to enroll as a graduate student at MIT. He dropped out of his doctorate in physics after one year, and chose to focus on his programming at the MIT AI Laboratory.

In 1975 and 1977, Stallman did a couple of AI projects with Professor Gerry Sussman; their 1977 paper established the technique of dependency-directed backtracking for constraint satisfaction problems (subsequently called "truth maintenance"). To this day, it is still the most general and powerful form of intelligent backtracking. Stallman likes to say, "This is what enables the computer not to crash when you tell it two contradictory things."

As a system hacker (*) in MIT's AI laboratory, Stallman's job was to improve the operating system software. He worked on projects such as TECO, Emacs for ITS, and the Lisp machine operating system. He became an ardent critic of restricted computer access in the lab, which at that time was funded primarily by the Defense Advanced Research Projects Agency.[7] When MIT's Laboratory for Computer Science (LCS) installed passwords in 1977, Stallman found a way to decrypt the passwords. He sent users messages containing their decoded password, with a suggestion they change it to the empty string (that is, no password) as a way to resist the pressure for "security." Around 20 percent of the users followed his advice at the time. Although passwords ultimately prevailed, Stallman boasted of the success of his campaign against passwords for many years afterward.[8]

In the late 1970s and early 1980s, the "hacker community"[9] that Stallman thrived in began to fall apart. To prevent software from being used on their competitors' computers, most manufacturers stopped distributing source code and began using copyright and restrictive "end user license agreements" to limit or prohibit copying and redistribution. Such proprietary software had existed before, but at this time it became the norm. This shift in the legal characteristics of software was facilitated by the U.S. Copyright Act of 1976,[10] as stated by Stallman's MIT colleague, Brewster Kahle.[11]

Around 1978, Xerox donated a laser printer (Xerox 9700, known as the "Dover") to the AI Lab but without the source code for the software that controlled it. Stallman had modified the software controlling the lab's previous graphical printer (the XGP, Xerographic Printer) so it electronically messaged a user when the person's job finished printing, and messaged all logged-in users waiting for print jobs when the printer had a problem. Being blocked by lack of source code from adding these features to the new printer was a major inconvenience; as the printer was on a different floor from most of the users, it tended to stay jammed or empty of paper.

When Stallman was visiting Carnegie Mellon University, he approached a researcher there who (he had heard) had a copy of the source code, to ask for a copy, following the hacker community's usual practices. He was stunned when the researcher said he had promised not to share copies of the source code. Reflecting on the episode afterward, Stallman recognized that the researcher had betrayed his MIT colleagues; indeed, he had betrayed the whole world, as the evil Chinese emperor Zao Zao spoke of doing in the famous novel, *The Three Kingdoms*. This comparison led Stallman to conclude that it is wrong, evil, to block or forbid people from redistributing or changing software.

Richard Greenblatt,[12] a fellow AI Lab hacker, founded Lisp Machines Inc. (LMI) to market Lisp machines, which he and Tom Knight[13] had designed at the lab. Greenblatt rejected outside investment, believing that the proceeds from the construction and sale of a few machines could finance the growth of the company. In contrast, the other hackers felt that the venture capital-funded approach was better. As no agreement could be reached, hackers from the latter camp founded Symbolics,[14] with the aid of Russ Noftsker,[15] a former AI Lab administrator who had left to found other companies. Symbolics recruited most of the remaining hackers, including the notable hacker Bill Gosper,[16] who then left the AI Lab. Symbolics also forced Greenblatt to resign, by citing MIT policies. While both companies delivered proprietary software, Stallman believed that LMI, unlike Symbolics, had tried to avoid hurting the lab's community. In 1982, Symbolics stopped contributing changes to MIT's Lisp machine operating system and made AI Lab users choose between the MIT version (available to LMI and to Symbolics) and the Symbolics version (not available to LMI). From March 1982 to the end of 1983, Stallman worked (alone for most of that time) to reimplement the Symbolics fixes and improvements for the MIT version so that Symbolics would not gain by its act of aggression.

Stallman believes that software users should have control over the software, which requires the four freedoms that make the program free: to run it, to study and change it, and to redistribute it unchanged or modified. Freedom to study and change it requires having the program in the form of source code. Stallman maintains that it is antisocial and unjust to deny users any of these freedoms.[17]

The phrase "software wants to be free" is often incorrectly attributed to him, and Stallman argues that this is a misstatement of his philosophy.[18] He argues that freedom is vital for the sake of users and society as a moral value, and not merely for pragmatic reasons such as possibly developing technically superior software. Moral arguments can persuade people who are not specialists in technology, and are not particularly concerned with technical advantages.

In January 1984, Stallman quit his job at MIT to work full-time on the GNU Project,[19] which he had announced in September 1983, and today serves as president of the Free Software Foundation, founded in October

1985. He still features, as Visiting Scientist, as a member of the Computer Science and Artificial Intelligence Laboratory (MIT CSAIL).

(*) About Hacker, according to Richard Stallman in his article "On Hacking,"[20] "The use of 'hacker' to mean 'security breaker' is a confusion on the part of the mass media. We hackers refuse to recognize that meaning, and continue using the word to mean someone who loves to program, someone who enjoys playful cleverness, or the combination of the two."

At the MIT AI Lab, a "system hacker" was someone who did a lot of development or maintenance of the operating system. Doing that was Stallman's job at the AI Lab, so the term was his unofficial job title.

*Note:* Richard Stallman is referred to as "RMS" below, as he prefers. Richard Stallman revised the texts of this dialogue in November, 2015.

Adolfo Plasencia:    Hello, Richard, nice to meet you.

Richard Stallman:    Hello, Adolfo.

A.P.:    Richard, let's talk about free software first as it's at the very heart of your thinking. Free software is no longer just a matter for programmers; it now involves society in general. For some time now, in your writings and in what you say, you've been associating it with a philosophy strongly focused on ethics. I would like you to tell me about your idea of free software and ethics.

RMS:    Free software is software that respects users' freedom, particularly their freedom to cooperate with others. Otherwise it is "proprietary," in which case the software is available only for you to run; besides, the developer usually imposes conditions that keep users divided and helpless. You are helpless because you have no access to the source code. You cannot change anything or know for sure what the program does, or if it has malicious functionalities —something that's not uncommon at all. And you feel isolated because you can't share copies with your neighbor. To me, this seems so unethical. ...

So, in order to have that freedom, I had to completely write another operating system, a free one.

A.P.:    Do you know the book by philosopher Pekka Himanen,[21] *The Hacker Ethic,*[22] with contributions by Linus Torvalds[23] and Manuel Castells?[24]

RMS:    Yes, but I don't entirely agree with what he says. Hacking to me is a kind of activity you do for fun (sometimes with other motives as well), using your intelligence to challenge yourself or to amuse others. I think that hacking does not imply a specific ethic. Our ethical principles apply to hacking, just as they do to the rest of life.

A.P.:   The book makes a distinction between "crackers" and "hackers."

RMS:   Yes, the term "crackers" refers to people who break the security of a computer system. This can be done for a whole variety of purposes, destructive or not, depending on the goal or what you do afterward. I'm not saying that breaking the security of something is necessarily bad, but it can be bad.

A.P.:   Hacking can also be done just to outsmart the people who built the system. Hacking is like a competition. ...[25]

RMS:   Yes, sometimes. Anyway, you can be a hacker in any area of life, not necessarily involving computers. For example, sometimes I eat with chopsticks. Normally, you would use two chopsticks; using three sticks in one hand is a kind of hack; using four sticks with just one hand is a better hack. When I told a friend about my three-stick hack, he showed me how skillful he was, using four sticks with just one hand to pick up two morsels at once. He topped my hack. This has nothing to do with computers or security, but it's fun and breaks the "rules," and requires using your intelligence—so it is hacking.

A.P.:   If you tell me that hacking can apply to other fields, can you imagine, for example, intellectuals and writers doing it? It's hard to imagine. ... How can this apply to other fields?

RMS:   Well, for example, several composers have written music structured as a palindrome.[26] There is a fourteenth-century piece titled "My end is my beginning and my beginning is my end," which sounds so beautiful.

A.P.:   The piece you mean is by Guillaume de Machaut. But the most famous palindrome in music is the one by Johann Sebastian Bach. He included a crab canon, or retrograde canon, in his 1747 composition, "Musical Offering."

RMS:   Then, they were hacking.

A.P.:   What about John Cage's composition, 4'33"?

RMS:   That's an empty composition, with no notes—the musician literally plays nothing. This illustrates the kind of hack which is a joke or prank whose brunt falls on a particular target: in this case, the audience, which expected a nontrivial composition of musical interest. We hackers would say that the piece is "a hack on the audience."

A.P.:   Have you seen the film *Amadeus*, directed by Milos Forman, about Mozart? There is a scene in which Mozart, age twenty-one, is at a party. He plays a divertimento for his audience, making fun of all the famous musicians of his time: "Let me play like Handel," he says, placing his hands on the piano but with his back turned on it—and he sounds even better than Handel. "Now let me play like Johann Sebastian Bach...." Is that hacking?

RMS:   It sounds that way, though I did not see that film. In the 1960s and 1970s, sometimes, someone would try to figure out a new way of doing a particular software job with fewer code instructions than before. This was a hackers' competition. In those years, programming was often done in machine code, so these clever ideas were occasionally useful too.

A.P.:   Do you think MIT is a place of freedom for those who work in the field of technology?

RMS:   I don't know about MIT as a whole, but the AI Lab at MIT certainly was one in the 1960s and 1970s. It was a very special place, where attention was paid to what people could contribute rather than to rules about who was authorized.

A.P.:   I think you know that Tim Berners-Lee, who now works at MIT CSAIL, was working at CERN, the European Organization for Nuclear Physics, when he invented the Web. CERN is a European institution financed by European governments. Anything discovered there is always published as freely accessible information, with a description and the names of those who have participated in the research to bring about the discovery, and obviously, it is free for humankind. This concept ... do you think it is possible to come closer to it and spread its application to the rest of the world? Do you think that that the philosophy of science is regressing or progressing in today's world?

RMS:   For sure it is possible to spread it, but companies often oppose it. I think companies already have too much power and their power should be stopped, but I do not know how. "Corporatocracy" is not good.

A.P.:   Do you think that this concept of public availability, which already applied at CERN, was a good starting point for the scientific world in relation to your own ideas about proprietary software?

RMS:   Sure it was—though free software entails more than just "availability." CERN is an example of a principle that was one of the inspirations for free software. When I worked at a scientific laboratory in the 1970s, the spirit of scientific cooperation was highly valued at the MIT AI Lab. Sharing our software with others was an example of scientific cooperation. Nowadays, when companies fund research, they often lock it up.

Sometimes funding from companies makes bad science. For example, when studies of the effects of drugs are financed by drug companies, they report fewer negative effects. We must stop allowing companies to do this testing; instead we must tax them to support state-funded tests. Companies even suppress research results, as we now know Exxon did in the 1970s and 1980s after its scientists recognized the danger of global heating.

My focus is not a technical matter, it is a matter of ethics, and it relates to other ethical issues.

A.P.:  I know, that's why I'm bringing this up now. Although it shouldn't be, it is quite exceptional for a technologist to have such views. There are many technologists who do not talk about society, or about ethics, as you do.

RMS:  How silly of them! They are apparently not public-spirited citizens. Their attitude is dangerous, because technology without the influence of ethics is likely to do harm.

A.P.:  But you know there is a current now in the business world which says that ethics can be profitable, and there is a tendency to talk about the culture of business ethics.

RMS:  It's absurd to ask whether ethics is profitable or not, because it is a prior responsibility, an obligation towards others. Choosing ethics as a guide only when it is profitable is really the same as not having any ethics.

A.P.:  There's another thing I would like to ask you: years ago you came up with the idea of the four freedoms of free software. Do you think they are still relevant? Or do we need some additional ones?

RMS:  The issue, overall, has not changed. Free software means that the users control the program, and that requires the four freedoms. If the users don't control the program, the program controls the users, and the developer controls the program. That is a system of unjust power, so nonfree software is an injustice.

We still use copyleft[27] to defend freedom for all users of our program. A program is free software because it carries a free license, placed on it by its copyright holders. Some free licenses permit putting the code into a nonfree program. Some free licenses, which we call "copyleft" licenses, don't permit that: they say that all versions of the program must be distributed only as free software. The GNU General Public License (GNU GPL) is the premier copyleft license.[28]

But there's a change in our approach to certain cases. For example, if a program is useful on Internet servers, someone might make an improved version to run on a server and never release the improved version to anyone. If this happens often, the development of the released version can be held back because these improvements are not available for it.

We now have a license, the GNU Affero General Public License (Affero GPL for short), which requires making the source code of the modified version available for the users of the service to download.[29]

A.P.:   Before working on the GNU Project in 1984, you sent an email in September 1983 that became famous. Why did you send it?

RMS:   I sent it to announce the development of the GNU system and invite people to become involved, to recruit collaborators. I did not want to have to write the whole system myself.

A.P.:   Another thing: do you think that, now that IBM and other companies have finally opted for free software, they have done so because they believe in it, or is it just a marketing strategy?

RMS:   I suppose it is a marketing strategy, but that does not mean that these actions are wrong. The same companies do good things and bad things at the same time. I criticize the bad things and I praise the good things, regardless of their motives.

For example, IBM contributes to the development of free software, that's a good practice; they promote the use of a free system, that's a good practice. But they also develop proprietary programs: that's a bad practice. To have the most influence on them, we need to judge each separate behavior for what it is, rather than lumping them together into a simplistic overall "grade" for a large company.

A.P.:   You travel to Europe quite a lot. What do you think about Europe and Spain and what is happening there as far the free software movement and people's willingness to participate are concerned? What's your impression?

RMS:   This varies across Europe. I think Spain is the most enthusiastic country in Europe as regards free software. Other countries too, such as France and Germany. ... However, some countries, such as Italy and Portugal, are far less enthusiastic. Spain is the leader anyway.

A.P.:   In Europe the opposition to software patents is relatively high. Some time ago, in April 2003, a group of high-ranking scientists petitioned the European Parliament to prevent the patenting of algorithms and software ideas. In April 2005 in Spain there was a demonstration, with participation by all Spanish universities, asking Parliament to reject software patents. Via Eurolinux,[30] more than 400,000 people signed the petition on the Internet against patents. In the end, the draft directive on software patents did not succeed because on July 6, 2005, following intense debate, the European Parliament turned down the software patents directive with 648 votes

against, 14 in favor, and 18 abstentions. It was indeed a historic vote. Una-
nimity was almost total against software patents. But the threat continues.
Although Article 52 of the European Patent Convention specifically
excludes them, years later there are still political and economic pressures
from the industry, and there are now lobbies or "patent trolls," as they are
known, pushing for Europe to finally accept software patents in law. What
is your opinion about the concept of software patents?

RMS:  Computational idea patents restrict the use of a computer. If they
are allowed, they directly restrict what you can do on your computer. They
are a restriction imposed on all computer users, and this directly affects at
least half of Europe's citizens. It is not just an economic issue, it is also a
matter of justice. It is not fair to restrict in this way what can be done on
your own computer. But it does not make sense either as an economic
policy, because computational idea patents are obstacles to innovation, to
software development.

A work of software can be very large, and combine thousands of ideas.
And if each idea can be patented, then developing or using a large program
is dangerous: you could be sued. You can see it in your word processing
program … it has hundreds of features. Every feature in the program embodies at least one idea, but could include many ideas.

One functionality can be seen in several different ways. Let's use graphic
designs as an analogy. If you draw a rectangle, one of the sides is the base.
If there is a patent covering using a side as a base, then you could be sued
over the side which is the base of your rectangle. But if we turn the rectangle 45 degrees, it looks like a diamond, with a point instead of a straight line
as the base. So if there is a patent for a point as the base of a figure, you
could also be sued under that patent. In turn, the rectangle is just a small
part in a very large drawing, and there may be patents on ideas found in
other parts of the drawing. You could potentially face many lawsuits for just
one drawing.

Computational idea patents work like that. Combining hundreds of
functionalities entails thousands of ideas, each of which you could be sued
for, so what can you do? How do you deal with that?

A.P.:  In one of your talks, you explain that there was once a time when
each patent covered the whole design of a physical product.

RMS:  Yes, two centuries ago that was more or less the case.

A.P.:  It's not the same thing with software. Can you explain the
difference?

RMS:   A patent is not about a specific product, at least not directly. It is a monopoly imposed by the state on implementing an idea—a feature, a method, a concept, a technique. If a product is complex, it combines many ideas, and each idea is susceptible to being patented.

How many ideas are there in each product? This varies from field to field. For example, in the pharmaceutical field there are relatively few ideas integrated into each product. Twenty or so years ago, a patent on a drug covered only that one drug—its entire chemical formula. Patents were not issued on aspects or parts of a drug. When someone invented a new drug, no previous patents could cover the drug, so that inventor could obtain the sole patent covering the newly invented drug.

Things have changed. Even in the area of drugs, it no longer works that way because patents are much broader and more general. Instead of covering an entire molecule, they may now cover parts of molecules, or ways that parts of a drug can function. If you develop a new medicine today, it might already be covered by someone else's broad patent, or several such patents.

Software is very different from pharmaceuticals. It is hard to combine dozens of different ideas in one pharmaceutical. It is easy to combine thousands of ideas in one program, because software is mathematics. In physical engineering you are building physical systems consisting of physical elements, and they don't have definitions. Thus, you always have to cope with the perversity of matter. You can try to model or predict what matter will do, but if it doesn't behave as you expected, tough luck. The matter is what counts, not the model.

A.P.:   Yes, in the end, what the matter does, the developments, can be good or bad, right?

RMS:   Yes.

A.P.:   But we shouldn't always be pessimistic. There are good things, too.

RMS:   Surprises may occasionally lead to a discovery that is useful for something else, but they are generally a problem for the specific project you're working on.

A software program is made up of mathematical elements, which have definitions. Each element always does what it is supposed to do; the challenge of design is in combining them. This makes it much easier to make a design that does what you want it to do. So we can build much larger structures than in other fields.

A software program can easily have a million elements in its structure. That wouldn't be a huge program. However, a physical structure with a million design elements would be a megaproject.

A.P.:   A car, for example?

RMS:   No, no; the physical structure of a car is not that complex.

A.P.:   It would have to be something a lot bigger.

RMS:   Exactly. With one million design elements, it would have to be a colossal project, something on the border of the possible. Maybe a jetliner. Note that multiple identical parts count only once in the complexity of the design.

But in software, that many design elements is not unusual—a small team can write that in a few years. Software is easier to design than physical systems, so we programmers can make systems with elements, implementing more ideas.

The formula "one product, one patent" applied, more or less, a few centuries ago. Today, in all fields of physical engineering or computer science, there may be many possibly patented ideas in one product. Many of those patents will exist, and give someone a way to sue whoever developed that product. It makes gridlock.

I'll give you an example in computing. Some time ago, a lawyer studied the code of one large program, the Linux kernel. He found 283 U.S. patents, each prohibiting a computing operation done within the Linux code. The code of Linux was only one quarter of 1 percent of the entire GNU/Linux system.[31] By multiplication we can make a rough estimate of around 100,000 computational idea patents, each prohibiting one or more ideas implemented somewhere in the GNU/Linux system. Here you see how computational idea patents create gridlock. Applying the patent system to software does not promote progress at all; rather it hinders progress in our field.

A.P.:   What could we do to make people understand this?

RMS:   We make an analogy between programs and other large written works, such as symphonies or novels. A novel implements many literary ideas. Each page may implement two or three literary ideas: there are ideas about events, and ideas about symbols or metaphors. Stylistic ideas may be implemented too. In the whole novel there may be thousands of such ideas.

For instance, one idea is a love scene with a woman on a balcony. That's a fairly old idea, so it would not be patentable today. Now imagine a love scene with a woman on a balcony and she is singing. Perhaps there is no public evidence or record of such a scene in the past, so someone might patent the idea. From that moment, anybody writing a novel including a love scene with a woman singing on a balcony could be sued for it. Every time you turn the page, you could find ideas that someone may have patented, someone who can sue the author. That would be an absurd system— producing gridlock comparable to what happens in software today.[32]

A.P.:   But all fields in knowledge make progress because they rely on what was done before. Am I right?

RMS:   Absolutely. But beware! Software and knowledge are not the same thing! "Knowledge" refers to ideas, not the works they are implemented in.

I'm talking about the use of ideas in computer programs, in writing software code. Each program must combine many ideas, perhaps thousands of different ideas, in one program. Overlooking this distinction may conceal the problem of gridlock.

A.P.:   Sure.

RMS:   It is not a matter of knowledge alone; it is a matter of using our knowledge to develop software to be used.

There is a huge difference between copyright and patents. If you write a novel, you needn't be afraid of copyright, because you would know you did not copy it. Also, the details in your novel will not be the same as those in another novel. But it's easy to infringe a patent without knowing. You simply couldn't tell whether an idea you use is patented because there are so many patents, and reading even one patent is hard work.

For example, in 2009 the European Patent Office had already granted 50,000 software patents, which is against the European Patent Treaty, but they have been granted despite the treaty.[33] With 50,000 patents, you can't possibly know them all. Imagine if there were 50,000 literary patents; you wouldn't know how to write your novel without being sued.

A.P.:   Yes. I have noticed that computer science students in European universities worry about this. They worry about their work as future developers if such things continue. And some of their teachers and lecturers are very close to the world of proprietary software.

RMS:   I suppose they are taught proprietary software programs in their lectures. That is "unethical." No university (or school of any level) should make students dependent in this way.

A.P.:   Yes, dependency on proprietary software.

RMS:   Exactly. See "Why Schools Should Exclusively Use Free Software."[34]

A.P.:   And that dependence will continue throughout their professional life.

RMS:   Absolutely. It is not impossible to escape, though, in the same way that it is not impossible to quit tobacco or alcohol. But inducing such dependence in students is a bad thing, and I think schools should stop teaching it.

If students want to make this a better world ... they may want to find an ethical job, in which case they will not find it in the field of proprietary software. But they can find ethical work in the software field.

Almost all paid work in the field of programming consists of developing software for private use by clients. It is not that different to do it ethically; just respect the client's freedom by delivering the code to the client under a free license.

If you want to be paid to develop free software to release, it's not easy, but it's not impossible. Many do get such a job, but it's not automatic. Your job hunting must be tenacious. It might even take a few years.

If you develop a very interesting free program, it will help you a lot, because you will then have a reputation as a great programmer; when jobs are offered, you are very likely to get one, as you would already be in contact with other people in the community and that will certainly help you.

A.P.:   Thank you, Richard.

RMS:   Thank you.

## Notes

1. Lawrence Lessig is the Roy L. Furman Professor of Law and Leadership at Harvard Law School, and the former director of the Edmond J. Safra Center for Ethics at Harvard University. Prior to rejoining the Harvard faculty, Lessig was a professor at Stanford Law School, where he founded the school's Center for Internet and Society, and at the University of Chicago. He founded Creative Commons, a nonprofit organization devoted to expanding the range of creative works available for others to build upon and to share legally. He is a former board member of the Free Software Foundation and Software Freedom Law Center. Lessig has long been known to be a supporter of net neutrality. He is author of *Code and Other Laws of Cyberspace* (2000), *The Future of Ideas* (2001), *Remix: Making Art and Commerce Thrive in the Hybrid Economy* (2008), *Republic, Lost: How Money Corrupts Congress—and a Plan to Stop It* (2011), *Republic, Lost: The Corruption of Equality and the Steps to End It* (2015), and more.

2. Initial announcement, GNU operating system (the original announcement of the GNU Project, posted by Richard Stallman on September 27, 1983), https://www.gnu .org/gnu/initial-announcement.en.html.

3. Richard Stallman, "The GNU Project," n.d., https://www.gnu.org/gnu/ thegnuproject.en.html. An overview of the GNU system may be found at "GNU's Not Unix," https://www.gnu.org/gnu/gnu-history.en.html.

4. GNU, "What Is Free Software? The Free Software Definition," last modified January 1, 2016, http://www.gnu.org/philosophy/free-sw.en.html.

5. Michael Gross, "Richard Stallman: High School Misfit, Symbol of Free Software, MacArthur-Certified Genius," interview, 1999, MGross.com, http://mgross.com/ writing/books/the-more-things-change/bonus-chapters/richard-stallman-hig h-school-misfit-symbol-of-free-software-macarthur-certified-genius.

6. Richard Stallman, "Humorous Bio," in *The Hacker's Dictionary*, 1st ed. (1983). See also *The New Hacker's Dictionary*, 3rd ed., ed. Eric S. Raymond (Cambridge, MA: MIT Press, 1996).

7. C. Arvind Kumar, ed., "Richard Stallman," in *Welcome to the "Free" World: A Free Software Initiative* (Hyderabad: Indian Universities Press, 2011), 90–91.

8. Steven Levi, *Hackers: Heroes of the Computer Revolution* (Cambridge, MA: O'Reilly Media, 2010).

9. GNU, "The Hacker Community and Ethics: An Interview with Richard M. Stallman, 2002," last modified April 12, 2014, http://www.gnu.org/philosophy/rms -hack.html.

10. The Copyright Act of 1976 is a US copyright law and remains the primary basis of copyright law in the United States, as amended by several later enacted copyright provisions: https://en.wikipedia.org/wiki/Copyright_Act_of_1976.

11. Brewster Kahle, colleague in MIT CSAIL of Stallman, graduated from MIT in 1982, and afterward founded the Internet Archive, which he continues to direct. In 2001, he implemented the Wayback Machine, which allows public access to the World Wide Web archive that the Internet Archive has been gathering since 1996. See "50 Ways That MIT Has Transformed Computing /31, 31," Internet Archive (1996), http://www.csail.mit.edu/node/2223.

12. Richard D. Greenblatt is an American computer programmer. Along with Bill Gosper, he may be considered to have founded the hacker community and holds a place of distinction in the Lisp and the MIT AI Lab communities: https://en .wikipedia.org/wiki/Richard_Greenblatt_(programmer).

13. Tom Knight is an American synthetic biologist and computer engineer, who was formerly a senior research scientist at the MIT Computer Science and Artificial Intel- ligence Laboratory: http://en.wikipedia.org/wiki/Tom_Knight_(scientist).

14. Symbolics, a spinoff from the MIT AI Lab: http://en.wikipedia.org/wiki/Symbolics.

15. Russell Noftsker is an American entrepreneur who notably founded Symbolics, and was its first chairman and president: https://en.wikipedia.org/wiki/Russell_Noftsker.

16. Bill Gosper is an American mathematician and programmer with Richard Greenblatt, he may be considered to have founded the hacker community: https://en.wikipedia.org/wiki/Bill_Gosper.

17. GNU, "Free Software Is Even More Important Now," last modified September 1, 2015, http://www.gnu.org/philosophy/free-software-even-more-important.en.html.

18. Free Software Is Even More Important Now: tps://www.gnu.org/philosophy/free-software-even-more-important.en.html.

19. GNU, "The GNU Project," last modified December 1, 2015, https://www.gnu.org/gnu/thegnuproject.en.html.

20. Richard Stallman, "On Hacking," Stallman.org, n.d., https://stallman.org/articles/on-hacking.html.

21. Pekka Himanen is a philosopher and currently teaches as a Professor at the Helsinki Institute of Information Technology. He previously was Director of the Berkeley Center for the Information Society, http://www.globaldignity.org/people/pekka-himanen.

22. Pekka Himanen, *The Hacker Ethic and the Spirit of the Information Age,* with a preface by Linus Torvalds and an epilogue by Manuel Castells (New York: Random House, 2001).

23. Linus Torvalds is a Finnish software engineer who is the creator and, for a long time, principal developer, of the Linux kernel, which became the kernel for operating systems (and many distributions of each) such as GNU and years later Android and Chrome OS: https://en.wikipedia.org/wiki/Linus_Torvalds.

24. Manuel Castells is Professor of Sociology, Open University of Catalonia (UOC), in Barcelona. He is also University Professor and the Wallis Annenberg Chair Professor of Communication Technology and Society at the Annenberg School of Communication, University of Southern California: http://www.manuelcastells.info/en/curriculum-vitae.

25. Richard Stallman, "On Hacking," Stallman.org, n.d., https://stallman.org/articles/on-hacking.html.

26. A palindrome is a sequence of characters, sometimes in the form of words or, which reads the same forward or backward ("Able was I ere I saw Elba").

27. GNU, "What Is Copyleft?," http://www.gnu.org/licenses/copyleft.en.html.

28. GNU, "GNU General Public License," version 3, June 29, 2007, http://www.gnu .org/licenses/gpl-3.0.en.html.

29. GNU Affero General Public License v3 (AGPL-3.0), https://tldrlegal.com/license/ gnu-affero-general-public-license-v3-(agpl-3.0).

30. Internet Archive, "Eurolinux Alliance: Petition for a software patent free Europe," http://web.archive.org/web/20051227032443/petition.eurolinux.org/index _html?NO_COOKIE=true.

31. GNU, "Linux and the GNU System," last modified January 1, 2016, http:// www.gnu.org/gnu/linux-and-gnu.en.html.

32. GNU, "Software Patents and Literary Patents," last modified April 12, 2014, http://www.gnu.org/philosophy/software-literary-patents.en.html.

33. Giuseppe Scelatto, coordinator, et al., *Study on the Quality of the Patent System in Europe*, DG Markt Patqual, March 2011, 158, http://ec.europa.eu/internal_market/ indprop/docs/patent/patqual02032011_en.pdf.

34. GNU, "Why Schools Should Exclusively Use Free Software," last modified January 16, 2016, http://www.gnu.org/education/edu-schools.en.html.

# III   Intelligence

What is intelligence? How did it come about? Where does it reside? These questions echo those we ask about human existence: Where did we come from? Where are we going? In short, what is our place in the universe? Surely if we could find the answers to all these questions, we would understand the meaning of our existence as a species, and knowing that would unlock the answers to countless other questions.

Questions about what intelligence is and why some human beings seem to possess more of it than others, or what differences there are between mind, brain, and consciousness (are they synonyms?), are clearly relevant to practical aspects of life. They interest not only philosophers and scientists. As Aristotle said, "Intelligence consists not only of knowledge but also the skill to apply knowledge in practice." Intelligence continues to be a mystery, and the more we know, the more the mystery seems to grow. It's a topic of such magnitude that both science and the humanities, as applied disciplines, are currently trying to break it down, divide it into its constituent parts, to analyze its workings.

For that project to succeed, we first must agree on how the question is to be formulated. When we talk of intelligence, what are we talking about? That which emanates from the human brain? What does human intelligence have in common with the type of intelligence found in the anthill? How are the multiple intelligences that Howard Gardner talked about to be accounted for?[1] And how do we tackle possible nonbiological or artificial intelligences, computational or otherwise? Clearly, the more we dissect and analyze the forms intelligence can take, the more problematic the term itself becomes.

Advances in neuroscience, computing, engineering, and nanotechnology contribute to the sense that the concept of intelligence itself may need to be reformulated. Not since 1956, when the information technologist

John McCarthy coined the term *artificial intelligence* (AI), have we heard such hype about the concept as we do today.

The development of natural and artificial intelligence was the focus of a 2011 symposium, "Brains, Minds and Machines," held as part of MIT's 150th anniversary celebration. At a session hosted by Harvard University's Steven Pinker, Marvin Minsky, a cofounder of the AI laboratory at MIT and a pioneer of AI and robotics, was asked why the AI project had not progressed more during the 1960s. Minsky replied that something had gone wrong: though robotics had advanced to the point that machines could "play basketball or soccer or dance or make funny faces at you," they were not in any real way "smarter" than their creators and playmates. Noam Chomsky, who participated in the same session, expressed skepticism about the possibility of machine intelligence and the behaviorist approach he associated with it. Chomsky has long criticized this approach, identified with Harvard's B. F. Skinner and his book, *Verbal Behavior*.[2] In that book, Skinner tried to explain linguistic skills using behaviorist principles, emphasizing the historical associations between stimuli and responses in animal behavior by using probabilistic analysis, a statistical method that predicts the future in accordance with past trends. The statistical methods of AI are currently much more sophisticated than those Skinner used, but this has not deterred Chomsky from reminding us how fundamentally wrong he believes these approaches are. Neither does he share the neuroscience field's emphasis on brain "wiring" or "connectomics" and the possibility of reverse engineering it for the purpose of machine intelligence. According to Chomsky, "You're never going to find them [linguistic competences] if you look for strengthening of synaptic connections or field properties, and so on."[3]

Chomsky's remarks suggest that what interests him are the root questions of what language and intelligence are, where they reside, and how they function. In a subsequent *Atlantic Monthly* interview, Chomsky described the intersection between the new technologies of big data and the development of software based on AI.[4] Chomsky criticizes this data-driven turn in AI, which has now become a large-scale industry, and the massive use of computerized statistical techniques (including those we now refer to as machine learning and deep learning). This path may produce successful engineering applications, Chomsky argues, but it will make it more difficult "to obtain true knowledge on the nature of intelligent beings."

Chomsky's remarks about AI at the "Brains, Minds and Machines" symposium did not sit well with everyone, particularly those in the software and computing industries. Peter Norvig, director of research at Google, wrote that what Chomsky "dislikes [about] statistical models is that they tend to make linguistics an empirical science (a science about how people actually use language) rather than a mathematical science (an investigation of the mathematical properties of models of formal language)."[5] According to Norvig, we "now know that language is like that as well: languages are complex, random, contingent biological processes that are subject to the whims of evolution and cultural change. What constitutes a language is not an eternal ideal form, represented by the settings of a small number of parameters, but rather is the contingent outcome of complex processes." If languages are contingent, Norvig argues, the most appropriate way to analyze them is by employing probabilistic models because people have to constantly understand uncertainty, ambiguity, the "noise" of the discourse of others. This is why the mind, according to Norvig, must be using something like probabilistic reasoning. This, therefore, is a "strong" debate.

AI has inspired more than its share of controversy. A recent debate, discussed in this book, concerns the possible threats and serious ethical problems that general-purpose AI poses, for example when applied to certain robotic systems. In January 2015, several leading researchers and scientists signed an open letter, published by the Future of Life Institute (FLI), on general-purpose AI, and then, six months later, on July 28, 2015, a second open letter on specific-purpose AI applied to autonomous weapons.[6] The latter was signed by Chomsky, the MIT professor and Nobel Prize winner Frank Wilczek, the philosopher and cognitive scientist Daniel Dennett, the physicist Stephen Hawking, and leaders from the technology and business worlds, including Elon Musk, CEO of Tesla and director of SpaceX, and Steve Wozniak, the cofounder of Apple, along with many other scientists; interviews with several of them appear in this book. If the first letter warned of the possible danger of a badly directed general-purpose AI, the second warned of an "AI arms race" that could be disastrous for mankind and urged the United Nations to consider prohibiting "offensive autonomous arms," which are today still outside "significant" human control. This letter pointed out that despite AI's huge potential for benefiting humanity, the risks of autonomous machines are real. The two letters also generated important responses, like that of MIT professor and robotics pioneer Rodney Brooks, who argued in an essay that AI is a tool, not a threat.[7]

The conversations in Part III reflect on aspects of intelligence in different fields and disciplines. The topics covered include the neurophysiology of the brain, affective computing, brain-machine interfaces, robot cognition, collaborative innovation, and the wisdom of crowds. My hope is that they will inspire your own thoughts about what intelligence is and where it might reside, now and in the future.

## Notes

1. Howard Gardner, *Frames of Mind: The Theory of Multiple Intelligences* (New York: Basic Books, 2011).

2. B. F. Skinner, *Verbal Behavior* (Brattleboro, VT: Echo Point Books and Media, 2014).

3. Quoted in Yarden Katz, "Noam Chomsky on Where Artificial Intelligence Went Wrong," *The Atlantic*, November 1, 2012, http://www.theatlantic.com/technology/ archive/2012/11/noam-chomsky-on-where-artificial-intelligence-went-wrong/ 261637.

4. Ibid.

5. Peter Norvig, "On Chomsky and the Two Cultures of Statistical Learning," Norvig.com, n.d., http://norvig.com/chomsky.html.

6. Future of Life Institute (FLI), "Research Priorities for Robust and Beneficial Artificial Intelligence: An Open Letter," last modified January 5, 2016, http:// futureoflife.org/ai-open-letter, and FLI, "Autonomous weapons: An open letter from AI & robotics researchers," last modified January 7, 2016, http://futureoflife.org/ope n-letter-autonomous-weapons.

7. Rodney Brooks, "Artificial Intelligence Is a Tool, Not a Threat," blog post, *Rethinking Robotics*, November 10, 2014, http://www.rethinkrobotics.com/blog/artificia l-intelligence-tool-threat.

## 23 "Affective Computing" Is Not an Oxymoron

### Rosalind W. Picard and Adolfo Plasencia

*I think that emotions are capable of being fully understood, although it may not be in this lifetime. We're starting to understand pieces of them: I am now wearing a sensor that measures signals that correspond to some kinds of feelings I have, although it doesn't correspond to all of them.*

*We are trying to invent the future: we bring to it certain values of what kind of future we want to make.*
—*Rosalind W. Picard*

Rosalind W. Picard is Founder and Director of the Affective Computing Group at the MIT Media Lab and Codirector of the Media Lab's Advancing Wellbeing initiative. She has cofounded two businesses, Empatica, Inc., which creates wearable sensors and analytics to improve health, and Affectiva, Inc., which delivers technology to help measure and communicate emotion.

Picard holds master's and doctoral degrees in electrical engineering and computer science from MIT. She started her career as a member of the technical staff at AT&T Bell Laboratories, designing VLSI chips for digital signal processing and developing new algorithms for image compression. In 1991 she joined the MIT Media Lab faculty. She is the author of *Affective Computing* (MIT Press, 1990) and is a founding member of the IEEE Technical Committee on Wearable Information Systems, which helped launch the field of wearable computing. She is an active inventor, holding multiple patents.

Adolfo Plasencia:   The expression "affective computing" is something of a contradiction, as it may seem as though we are asking mathematics to teach us emotions. Mathematics—and computation is pure mathematics—is said to provide us with beautiful, even elegant equations, but it cannot be asked to provide us with emotions.

Is the expression "affective computing" an oxymoron?

Rosalind W. Picard. Photographs by Adolfo Plasencia.

Rosalind Picard:   When I originally defined "affective computing," I said it was "computing that relates to, arises from, or deliberately influences emotion and other affective phenomena." Thus computers might be given skills to handle emotion—perhaps responding differently to help a frustrated user versus a pleased user—without the computer itself having emotions or becoming emotional.

Where it becomes controversial is when you cross that line, so that computers are not simply helping somebody by responding more intelligently to their emotions but have mechanisms that make the computer appear emotional.

I remember one time after a talk I gave in Spain, somebody came up to me who was really angry. He had steam coming out of his ears and could hardly speak. My talk was translated into several languages, and one of the translators made me sound as if computers can have emotions just as we have. He was right to be upset about that. However, it was a mistranslation: I wasn't saying that, nor do I think that it is possible, given what we know. Today I try to be very careful to distinguish that while we give computers some emotional abilities, they are quite different from human emotional abilities. Computers do not have emotional experiences as we do. They may compute in ways that express or simulate emotion, but they do not feel the way we do.

So the way I think of it, affective computing is not an oxymoron: computers are processing affective information. However, I think it can be misunderstood in a way that could be quite disturbing, worse than an oxymoron.

A.P.:   Your Affective Computing Research Group is trying to overcome or eliminate the divide between information systems and human emotions.

What characterizes that divide between machines and people?

Is what you are doing a kind of computational science "infected" or "tainted" by emotions?

R.P.:   I think the gap that most drove me to want to work in this area is the one where I felt that people's feelings are really important, they really matter, and they were being ignored by computers. You interact with a computer, then it frustrates you and it keeps acting like you are happy, and you're not. And that's a recipe for making you even more frustrated! One guy took a gun and shot his computer through the monitor once, and multiple times through the hard drive! Another man, a chef, got so mad at how his computer responded to him that he threw it in the deep-fat fryer. So I saw that how computers were behaving was unintelligent, and if we're going to make computers intelligent, then they have to start by being smart about responding to human emotions.

So the main gap I was interested in when I started in this area was to make computers more emotionally intelligent.

While we are starting to close this gap by giving computers much better abilities to see the emotions of their users, there is still a huge gap between what computers can do and what people can do. The biggest gap seems to be our ability to feel, to experience emotion, which is not something computers have. We can give computers lots of information about human feelings, tell them about our experiences, and instruct them in best practices for how to better treat people who are happy, bored, frustrated, or angry. But the best we can hope to do is not even as good as teaching a man what it feels like when a woman has a baby. While a man will probably never be able to experience childbirth, he can liken its pain to some other experiences he has had. He can respond better to a woman's pain by making this comparison. A computer can also learn descriptions of feelings such as pain, and it may even be better than some men at knowing what to say to you when you are in pain. But the computer cannot truly experience or understand pain as a human can; it has some fundamental differences that we do not know how to bridge.

A.P.:  You often use the expression "emotional communication" in your texts.[1] Perhaps you agree with the physiologist Claude Bernard, who, at the end of the nineteenth century, said that feeling is the origin of everything.

Do you agree?

Should the relationship between machines and human beings also be emotional, or do you believe there may be communication with humans that is not emotional?

R.P.:  The relationship is emotional, whether we want it to be or not. It's appealing to a scientist, and to me, trained as an engineer, to think we could put a piece of communication out like a pure mathematical expression and somehow allow it not to be tainted by emotion or feeling. However, what I came to learn is that even a purely logical and completely neutral piece of information is received by one or more persons, each of whom has emotions, and it might be received positively by one and negatively by another, even though it's a neutral piece of information. For example, if you say the word *presence* out of context, some people in a good mood are more likely to hear the word *presents* (gifts), or if you say the word *band*, some people in a bad mood are more likely to hear the word *banned*. The listener's mood colors even neutral information negatively or positively.

So that led me to understand that all communication has a sender and a receiver—think of Claude Shannon and information theory—and whether or not the sender puts emotion into it, the receiver can receive it as if it contained emotion.

So emotion becomes a part of all communication even if the content of the communication might not explicitly have emotion in it.

A.P.:   The Turing test was a test proposed by Alan Turing to demonstrate the existence of intelligence in a machine. In your opinion, which kind of test (or Turing test type) would have to be passed by a machine in order to call it a machine with emotions and feelings?

R.P.:   I wrote in my original book, *Affective Computing,* a bit about how we might test for different emotional abilities.[2]

Since then, the thing that I think is most important is the relationship that evolves when two people are together. A relationship also evolves when a person is interacting with a machine or with a robot over time. In all of the versions of the Turing test I've seen, they test short-term interaction (although the new movie *Ex Machina* performs a one-week test, which is a first!).

When you interact with a person over time, you might treat that person formally the first time you meet him or her, but then that loosens up over time: your language changes, your rapport changes. Your willingness to self-disclose more about your feelings usually increases over time, and all of these aspects of emotional intelligence, of showing empathy and responding and "reading between the lines," as we say, tend to change if an entity is truly emotionally intelligent. The classic Turing test cannot get at that because it uses just a short period of time. In a brief window of interaction, any entity can just "act emotional," and it is not sufficient to convey skill emotional skill or intelligence. A false empathy might be revealed only when you repeatedly interact with it and see that it is inconsistent and does not grow with a relationship.

A.P.:   Roger Penrose was recently in Barcelona delivering a lecture and said that perhaps someday, somebody in some lab may create an intelligent machine, but it will certainly not be a computer, because … "I believe that intelligence cannot be simulated using algorithmic processes, that is, by means of a computer."[3]

Penrose thinks that mind and brain are two separable entities and that computers cannot be intelligent in the same way as a human being, as, according to him, formal systems of sequential instructions will never give a machine the capacity to understand and arrive at the truth.

Do you believe that in order to understand and for there to be a conscience in some place or in some machine, emotions have to exist?

R.P.:    This is a huge question. The question of consciousness for most people is not just logical, like something you would write on a blackboard. It is not simply the ability to ask and answer, "What is going on? This is what is going on." While that is a part of consciousness, consciousness also involves feelings that are ineffable. While some can be described, such as a feeling of being, a feeling of knowing, a feeling of pain or hunger, and a feeling of significance, there is still a sense that there is more. We anticipate that feeling involves a lot of different components that we think are rooted in our physiology and in our chemistry, but even that may not be sufficient to describe them fully.

It's really interesting to look at when some aspects of these feelings are absent in some patients with certain kinds of damage. There are people who can't feel pain or can't feel their bodies, and people who can't tell that their internal feelings are changing until they change in unusually extreme ways. We've worked with some of those people, and there is no doubt to me that they are conscious. Parts of their feeling systems may be damaged, or maybe they are simply different in their ability to communicate or introspect. I think it is fine that some people's conscious abilities operate differently than others'—it adds diversity that may be good for all of us—and we should not diminish their humanity. Some fascinating studies have shown that even when an electroencephalograph of the scalp is "flat" there can still be signals generated deep in the emotion centers of the brain, which show up in electrodermal sensors on the wrist. These patients tend to survive. We are clearly dealing with a very complex system, and, like the blind scientists and the elephant, we have only felt the trunk, the tail, the legs, and the ears. We have a lot more to learn.

While my thinking may change as more scientific findings are discovered, right now I believe that emotional experience is a core part of a normal functioning conscious system.

A.P.:    In another conversation I had recently with the Harvard neurophysiologist Alvaro Pascual-Leone, he said that he agrees that the mind and the intelligence are emergent properties of the brain.[4]

Do you think that emotions are also emergent properties of the brain?

R.P.:    With all due respect for the great scientists who use the term "emergent," I think it's a cop-out. It's a single word for saying, "We don't really understand it." When you call something emergent you're saying, "It's there but we don't really know what it is or how it came into being."

I think that emotions are capable of being fully understood, although it may not be in this lifetime. We're starting to understand pieces of them: I am now wearing a sensor that measures signals that correspond to some kinds of feelings I have, although it doesn't correspond to all of them. So, as we start to learn about these pieces, we start to see that they're not merely an emergent phenomenon; they actually have components that can be measured. But am I then saying emotions are fully reducible to stuff we can measure? I won't go that far. There might be more to them, things we don't know how to measure or that we may not know how to measure for another century or more. It may be a long time before we understand, maybe in the next world that we understand, as now we see through a glass dimly. But right now, I'd say that they are not simply emergent. There is real understanding to be sought.

A.P.:  At the Affective Computing lab which you are in charge of here at the Media Lab, your present is like the rest of us imagine our future. But what does the future look like from the affective computing perspective? How do you see the future, here in your laboratory?

R.P.:  We are trying to invent the future: we bring to it certain values of what kind of future we want to make. One of the things that is in our minds constantly is, "What would we like that future to be?" I'd say, at present, what we would really like is to enable technology to be in a partnership with people to help people have much better experiences, much better well-being, much better understanding of their own emotions, much better ability to manage and regulate their own emotions. I am especially interested in helping people who have disabilities and difficulty regulating and comprehending emotions—in helping people who want to communicate those emotions. For example, we work with people who are nonspeaking, or people who just maybe have a hard time putting words and numbers around their feelings. They need to be better understood, and technology can help be a kind of affective prosthesis or tool to help them do better.

We do not want to falsely claim that we can reduce feelings to words and numbers. We make approximations with the technology, just like words or music approximate feelings. None of these replaces true feelings; rather, they are channels of their communication. It is my hope that affective technologies will be used with people in a way that always respects and honors human dignity and freedom. Our aim is to show that respect by helping emotions to be much better understood. The ultimate purpose is much greater than simply advancing science; it is advancing the worth of all people, no matter what their level of disability or ability.

A.P.:   Thank you so much, Rosalind, and we hope to see you again soon and, if possible, again in Spain and Valencia.

Thank you!

R.P.:   Thank you. It's been a pleasure to talk with you, and I look forward to my next trip to Spain.

## Notes

1. See, for example, Rosalind W. Picard, "Future Affective Technology for Autism and Emotion Communication," *Philosophical Transactions of the Royal Society B: Biological Sciences* 364, no. 1535 (December 2009).

2. Rosalind W. Picard, *Affective Computing* (Cambridge, MA: MIT Press, 1997).

3. See Roger Penrose, *The Road to Reality: A Complete Guide to the Laws of the Universe* (New York: Vintage, repr. 2007); and M. A. Sabadell, "Roger Penrose: 'La física podría ayudarnos a entender la conciencia,'" *MUY Interesante* 225 (February 2000), http://www.muyinteresante.es/ciencia/articulo/roger-penrose.

4. Alvaro Pascual-Leone participates in dialogue 24.

# 24  Mind, Brain, and Behavior

Alvaro Pascual-Leone and Adolfo Plasencia

Alvaro Pascual-Leone. Photograph courtesy of A.P.-L.

*The mind and what we are are a consequence of the brain and its structure. But the brain and its structure are also a consequence of the mind.*

*The brain is approximately 2 percent of a human being's weight. Nevertheless, it consumes 20 percent of the energy used by the human body. Why does it need so much energy?*

*I am a convinced believer in quantum physics, where chance ceases to exist.*
*—Alvaro Pascual-Leone*

Alvaro Pascual-Leone is Professor of Neurology and an Associate Dean for Clinical and Translational Research at Harvard Medical School. He is the Director of the Division of Cognitive Neurology and of the Berenson-Allen Center for Noninvasive Brain Stimulation at the Beth Israel Deaconess Medical Center and is a practicing behavioral neurologist.

Dr. Pascual-Leone received his medical degree in 1984 and his doctorate in neurophysiology in 1985, both from Albert-Ludwigs University in Freiburg, Germany. Following an internship in Medicine at Staedtisches Klinikum Karlsruhe in Germany and a residency in internal medicine at Hospital Universitario de Valencia in Spain, he completed a neurology residency at the University of Minnesota, and then trained in clinical neurophysiology and human motor control at the University of Minnesota and the National Institutes of Health. He joined Harvard Medical School and the Beth Israel Deaconess Medical Center in 1997, after several years at the Cajal Institute of the Spanish Research Council.

The overarching goal of Dr. Pascual-Leone's research is understanding the mechanisms that control brain plasticity and brain network dynamics across the life span to be able to modify them for the patient's optimal behavioral outcome, prevent age-related cognitive decline, reduce the risk for dementia, and minimize the impact of developmental disorders such as autism. He is a world leader in research and development, clinical application, and the teaching of noninvasive brain stimulation.

Dr. Pascual-Leone is the recipient of several international honors and awards, including the Ramón y Cajal Award in Neuroscience (2006), the Norman Geschwind Prize in Behavioral Neurology from the American Academy of Neurology (2001), and the Friedrich Wilhelm Bessel Research Award from the Alexander von Humboldt Foundation.

Adolfo Plasencia:    Alvaro, in life, do we have to learn and unlearn?

Alvaro Pascual-Leone:    Undoubtedly! I reckon that is one of the greatest teachings of Eric Kandel's marvelous work, and part of why he won the Nobel Prize. Learning is no longer just understanding the mechanisms of memory in the sense of how new memories are created but also how memories are forgotten, which memories should be forgotten, and, since Luria and many others, we now know that those who are unlucky enough never to forget, as is normal, suffer enormously from having an excessive memory. It is very important to forget.

A.P.:    You've done a lot of research using transcranial magnetic stimulation (TMS), a noninvasive method for stimulating the brain's neurons, and you and your colleagues are trying to improve the situation of the ill or at least reduce the suffering from illnesses such as migraines, Parkinson's disease, depression, and many more.[1]

Can you explain what TMS technology is?

A.P.-L.:    First we have to ask ourselves what the brain is.

Generally speaking, we think of it as a complicated chemical organ inside the cranium. It uses different chemical substances to do different things. But above all, it's electrical! The brain has as many neurons as there are stars in the Milky Way, and each of them is supported by and connected with thousands of other cells, which in turn are generators of electricity. The connections are like electrical cables. So, although it's true that there are chemical substances in the brain and in the transmission of information between neurons, the brain is above all an electrical organ.

Hence the question: Is it possible to use electricity to modify the brain's electricity and thus its activity?

The idea of using electricity to modify the brain's activity is old, but it's very difficult to do, because the skin and the cranium are wonderful insulators. If you want to put electricity inside, without opening the cranium and skin, it can hurt a lot! Even so, there are techniques still in use, such as electroconvulsive therapy for cases of depression. The patient, however, has to be anaesthetized, and a fairly strong current has to be applied.

Alternatively, you can open the cranium and directly stimulate the brain, as the pioneering work of Wilder Penfield clearly showed in Montreal in the last century. We can learn a huge amount about the organization of the brain thanks to this type of intervention, and I wanted to do the same, but without having to open the cranium and without being limited to patients needing surgical operations.

TMS is a technique that allows us to use a magnetic field to "make the bridge" between the current applied outside the cranium and the current that is induced in the brain. The bridge is created by the magnetic field, according to Faraday's laws of induction. You pass a current through a coil and the current flow generates a magnetic field perpendicular to that coil. That magnetic field, as time passes, will induce a secondary current in any conductor suitably directed toward it. If you then position a coil of electric cable—copper, for example—and discharge a sufficiently strong electrical impulse, a perpendicular magnetic field is created; the magnetic field contrary to the electrical one can indeed penetrate skin and bone without problem and on arrival in the brain a current will be induced, as the brain is an electrical organ and a great conductor. That's the principle.

Clearly, the difficulty is always in the details. What we want to do is stimulate a very specific point of the brain, and so coils of an appropriate geometry have to be designed to create well-focused electrical fields. We also need to know where to apply that electrical field in relation to the brain of the individual whose activity we want to modify or modulate.

All these technological problems can be solved and have mostly been solved over the last twenty years. Currently, we have very reliable techniques that allow us to stimulate areas of the brain the size of a fingertip or a pinhead, and to stimulate a specific point on the cerebral cortex in the convexity of the brain.

A.P.:   Modify here means to improve, doesn't it?

A.P.-L.:   That's another important point. Often when we think about what our brain does, when, for example, our hand is at rest, we know there are neurons that allow us to move the hand, and we assume that if we are not moving our hand, these cells are silent and doing nothing. But it's not true! The neurons that move the hand are discharging 50, 60, 80 times per second but don't cause movement because to cause movement, two things have to happen: those cells must increase their firing rate to about 110 times per second, and they must connect with specific other cells for a set purpose, and even modify the cell activity that normally exercises an inhibiting effect. It is always networks of synchronized cerebral activity that cause movement, a certain thought, and so on

We know we can modify activity in the neural networks in a controlled manner using, for example, transcranial stimulation. However, the consequences of that modification may be beneficial or harmful to the individual. So modifying doesn't always mean improving. The challenge is to modify in such a way that the result is truly beneficial for the individual, and the acute effect may be different from the long-term effect.

A.P.:   Let's move on to a mystery, namely, the functioning of the mind in the brain. Do you remember the verses of Calderón de la Barca:

"¿Qué es la vida? un frenesí.
¿Qué es la vida?, una ilusión,
una sombra, una ficción.
y el mayor bien es pequeño;
Que toda la vida es sueño,
y los sueños, sueños son."

(What is life? A frenzy.
What is life? An illusion,
A shadow, a fiction,
And the greatest good is small;
For all of life is a dream,
And dreams are only dreams.)

I quote those verses of Calderón because of what you said about the brain creating its own reality in its interior, it being a formulator of hypotheses, a generator of expectations.

Why are you convinced of that?

Does the brain generate its own fictions, its own illusions?

A.P.-L.:   Absolutely! There are a lot of data in cognitive neuroscience and there is a lot of common knowledge about the fact that the brain is really a generator of hypotheses. The brain only sees what it's looking for.

A.P.:   "Look" is not the same as "see," is it?

A.P.-L.:   No, it's not the same. There is a concept we call "inattentional blindness," according to which if you're attending to one thing and something else happens, you may not see it. The experiments I'm talking about are dramatic; for example, a study was undertaken where participants were asked to watch a group of six to seven people passing a basketball to each other and to count the passes. What nobody noticed was that someone dressed as a gorilla appeared on the court, turned, and passed! The participants didn't see it! People are literally blind to what they do not have to look for. This is inattentional blindness. This type of cognitive experimentation shows that our expectations determine what we see and experience.

A.P.:   The brain concentrates on what it likes, and what it does not like, it rejects.

A.P.-L.:   It doesn't necessarily focus on what it likes, but it fixes on something, on what it's seeking, what it wants, what it likes or sometimes doesn't like, but it has a series of expectations. The thing is, the brain does not face a situation without creating an expectation to compare it with. When reality doesn't match that expectation, surprise and discord are generated. Mostly it fits. Everybody knows that! We all know that none of us turns out to be who we really are!

A.P.:   But Alvaro, when we talk about the brain, aren't we talking about ourselves?

A.P.-L.:   Ah. That goes right to the heart of the question, whether the brain, the mind, and the soul are different things or not!

A.P.:   And if will has got something to do with it? The brain goes its own way and drags us with it. Are we its slaves?

A.P.-L.:   That's a very difficult question!

On the one hand I am a convinced believer in quantum physics, where chance ceases to exist. The mind and what we are are a consequence of the brain and its structure. But the brain and its structure are also a consequence of the mind. According to Daniel Dennett, "the Mind, the Soul, etc. are emergent properties of the brain." That doesn't mean that the soul and the mind don't exist as different realities. My godfather, Antonio Ferraz, is a philosopher, and one of his mentors, Xavier Zubiri, cites the example of watching television and says that the viewers who are watching see us and generate a hypothesis of what we are, and so we are a reality for them. Really, however, they are not seeing us but seeing phosphorescent or liquid glass points, pixels! However, their brain does see us! So there are two realities, both correct and both being captured at the same time. Just as in the verses of Calderón, the brain has an image of an inner life, of several dreams that are real and distinct from the hypotheses contrasted with outer life. From a neurological and physiological perspective, what is inside is important; one's dreams are important because if those dreams are a consequence of and generators of other realities, that changes your brain. If you dream of greatness, you will be greater. If you have positive dreams, you will enjoy better health, be more hopeful. If you think about what you're thinking you have to be careful because those thoughts are going to change your brain! And changes in the brain will change your reality. William James advanced such ideas when he said, "Believe that life is worth living, and your belief will help create the fact."[2]

A.P.:   Be careful with your desires; they may come true!

A.P.-L.:   Exactly! And therefore, it's important to realize that everything we do, think, experience changes our brain, even though we may not like the change.

A.P.:   You mentioned at a conference an experiment in which a group of randomly selected children at a school were sent and received the message that they were the best, the cleverest and, indeed, in the end, they were. It has also been shown that two individuals exposed to the same virus can infect each other or not, depending on the extent to which they are convinced they are going to become ill.

A.P.-L.:   What is really important in all this is to realize that when one thinks in a certain way or when one has an attitude or motivation, this induces changes of the nervous system, which in turn induce changes in health, in resistance to illnesses, or in success in daily life. This isn't magic; it's to the result of the activity of certain physiologically demonstrable connections in the nervous system. That's why, when you believe you are going to recover, when you are aware of the situation you're in from a health perspective and you're sure the help you receive is the best, you get better.

The circuits that have to do with "feeling safe" generate a certain activity that sets substances free and make the leukocytes, the cells that defend us from illnesses, stronger. So there is an axis, a connection between certain cerebral networks and the immune system. That fails when, for example, you lose a loved one and see no reason to continue to live. And when there is no reason to continue to live, you're more likely to become ill. Why is that? After all, we are constantly exposed to illnesses. We are literally continuously generating cancer. And yet most of us have sufficiently strong defense systems to kill off those cancer cells as they appear. Why, then, do we not always defend ourselves? Well, using the terminology of Susan Sontag, it's because we are "cancerizing"; that is, it's not the external agent that causes the illness: you have to have a bacterium that gives you the infection, but even that is not enough. Apart from that, there's a series of host factors in ourselves that make us more or less prone to contracting illness.

Those factors are governed by our brain, dreams, thoughts, internal motivation, beliefs, and well-being. It is therefore very important to control the brain's expectation chains, and this is possible because the cerebral networks of control are trainable.

A.P.:    Alvaro, another issue that concerns me is will.

Is it us that decides, or not?

Does our brain activate certain parts, certain circuits, beyond the will?

Does free will override external stimuli?

A.P.-L.:    Well, volition and free will pose problems in ethics and physiology, philosophy, psychology. As a physiologist, what I say is that from the physiological perspective, there's no evidence that there is a change in the patterns of cerebral activity when facing decision making. What we do see is compatible with something happening and our brain generating an explanation. Rather than making a decision, it comes up with an explanation.

We still cannot identify the cerebral activity associated with intention, with decision making. It is as if cerebral activity was changing at random or by environmental influences, and our brain was continuously incorporating explanations of what was happening and generating a story from unrelated events.

The fact that physiologically we still have not found, and perhaps never will be able to see, evidence of free will and voluntary decision making doesn't mean that those things don't exist. All of us, I believe, know from experience that we can make decisions, that there is an "I" separate from the connections involved in the movement of a hand. The challenge, from the physiological perspective, is how to prove it.

What we do know is that we can change the knowledge of the decision we have made. For example, in one of the experiments we did, we said to the participants: You'll see a light, and when the light comes on I want you, yourself, to make a decision, on whether to move your right hand or the left hand. Once the light goes off I want you to stick firmly to that decision until the light comes on a second time and then execute whatever you had decided."

We showed that if you apply cerebral stimulation on certain frontal parts of the brain at a specific moment, the participant doesn't decide, but "you decide" which hand to move for him, right or left, depending on where the stimulation is applied. We know this because the physiological pattern is fixed, so we know that we have done it. However, if we do it soon after the participant has had to make his or her decision, it is as if that decision had not settled in sufficiently, and the participant says to you, "This time I was very quick to move my hand!" and is convinced that she has made the decision, when actually you have induced the movement and she is inter- preting it as something of her own volition.

A.P.:   They interpret it a posteriori, you mean?

A.P.-L.:   In this case, yes, but they are wrong. Clearly, our nervous system has an enormous capacity for interpreting things that happen to us, beyond our control, in order to make us think that we control them. But that doesn't mean that situations where the brain may really control initiative don't exist. By the brain, I mean "I."

A.P.:   We also have to talk about intelligence.

Bertrand Russell wrote: "The difference between mind and brain is not a difference of quality but a difference of arrangement," that is, of the organization of the parts of the brain.[3] And for Marvin Minsky, the con- struction of a principle of intelligence can be made by starting with a fairly simple set of basic principles. It is the connectivity among them that gen- erates the resulting complexity from which behavior arises. What do you think?

A.P.-L.:   I think that when we speak of intelligence, we are often referring to a fluid intelligence or what my uncle, Juan Pascual-Leone, calls "mental capacity." We don't refer to the activity of a specific neural network but to the capacity to activate neural networks that pass activity from one network to the other for a certain purpose—the motor of activity in neural networks. The great challenge for a neurologist or a neurophysicist is to discover what the neurobiological substrate of that capacity is that allows one person to have more and another person less and how we can—if we can—increase it.

A.P.:   Alvaro, don't these technologies and means with which you are capable of penetrating and modifying the functioning of the human brain sometimes make you neurologists feel dizzy?

Are you going to be able to model the human soul, or am I going too far?

A.P.-L.:   The reason why I dedicate myself to this field is to increase skills that may be transferable so as to help alleviate suffering. No, I don't feel dizzy. I have a feeling of true happiness. When we find something that really helps reduce the symptoms and disorders of a sufferer, first, it is a huge honor that we are allowed to experimentally apply these techniques until we discover how they work and how to better apply them, and second, it is the best reward for work that one can receive. That's why we do it, basically. It not only fills you with very deep satisfaction when you realize that what you have dreamed of doing is possible, it also is very surprising to learn that things you never even caught a glimpse of are possible. In neurology and neuroscience, the field is moving so quickly that there is a whole mix of feelings.

As for the human soul, here we are entering the religious field. The question of whether there is a transcendental being we may call the soul is not physiology, it's a question of metaphysics and personal religious belief.

A.P.:   In 2006 the neurologist Marcus E. Raichle published in *Science* an article titled "The Brain's Dark Energy."[4]

What do you think of what he said and that theory?

A.P.-L.:   It's a huge problem. Let me give you some figures to give you an idea of the magnitude of the problem. The brain is approximately 2 percent of a human being's weight. Nevertheless, it consumes 20 percent of the energy used by the human body. Shocking! Why does it need so much energy? The traditional answer is that we relate to the outside world through the brain. That's no doubt true, but it's not the main factor for explaining the huge energy consumption of the brain.

First, most of the energy consumed by the brain has nothing to do with information about the world that enters the brain through the senses but with what it wishes to inhibit. The brain consumes an enormous amount of energy-inhibiting information that enters through the senses. Obviously, this begs the question: Why does the brain allow so much information in. Why not filter it? We don't know why, but it doesn't filter it. It inhibits it later, and thus generates a reality. I mean, to see a tree, the brain doesn't capture every leaf and branch and from each detail build up a tree. We don't "see" the leaves and branches but this "new" tree created from the original tree.

However, even including that energy consumption used for inhibition, the brain needs only 1 to 2 percent of body energy for its relationship with the outside world. We still don't know why the other 18 or 19 percent of the body's entire energy consumption is dedicated to this organ. It's more, much more, than your heart, liver, pancreas, or all of the other organs put together use. It's a huge quantity. What's it for? Marcus calls it the "dark energy" of the brain. Currently, one of the most influential hypotheses is that it has something to do with the self-reference effect, the concept of I, the conscience. But I don't believe that. It may have something to do with it but if it does, it is by accident. From the evolutionary perspective, the brain is not like that. Chance made us self-aware, but the brain was designed to control and regulate the entire organism. I believe that that energy is "dark" only because we still don't know for certain what the brain does with it, but things will change. We'll find out what it is used for. I have the impression that the brain uses it for supporting a critical and fundamental role in organistically regulating the homeostasis of the individual's organism, that is, health, in its widest sense.

A.P.:    The plasticity of the brain has become the holy grail of neuroscience. Is that true, or am I exaggerating?

A.P.-L.:    The nervous system is plastic, not in the sense that you can "plug in" the plasticity and use it but in the sense that it is changing constantly, and that change may develop into a benefit or harm to a specific individual. The real challenge is to sufficiently understand that plasticity in order to guide it, to increase certain changes and reduce others for the benefit of the individual at a given moment.

For a long time in my laboratory we have dedicated ourselves to the question of how to measure and guide the plastic capacity of the brain. That has produced very interesting results and we have learned a lot. Something of particular interest that has to do with what we were talking about before is that the brain is dedicated not only to its relationship with the external world but also to its internal world, to monitoring the internal organs of the body itself.

If when I relate to the external world my brain changes, does it change when it's interacting with my insides, with my pancreas, for example? And if my pancreas changes because I have diabetes or pancreatitis, does this also change my brain? And if my brain changes, does this have an effect on my pancreas again? In the end, what this means is that the brain is healthier in a healthier body, but also, simultaneously, the body is healthier with a healthier brain. So the impact of the brain is two-directional and implies that if you maintain a full and flexible cognitive function throughout your life, you will be healthier. Or expressed another way: part of the reason why we end up having health problems beyond our control may have something to do with a loss of cerebral plasticity as time passes. So, perhaps plasticity really is the holy grail. At my center, we have been searching for it for years, and increasingly we are studying cerebral plasticity in relation to internal means, to the organismic relationship with the body and to the concept of whether cerebral health, in the sense of optimal cerebral function, translates into better control of overall health.

The challenge we face is to understand these relationships between brain and mind, on the one hand, and brain and the body on the other, so as to develop interventions that can modify, restore, or optimize the cerebral plasticity of each individual. We've been developing methods to measure in vivo the cerebral plastic capacity of humans. It's an enormous challenge because that has an effect on the well-being of each individual, of all humans. It has a lot to do with preventive action for everyone—for my father, for me, for my wife, for my kids—regardless of the disease being treated or prevented.

That's the challenge we face, a challenge so huge that I see it as the main challenge for the rest of my academic life.

A.P.:    Wonderful. Thanks a lot, Alvaro.

A.P.-L.:    A pleasure, Adolfo.

## Notes

1. T. Wagner, A. Valero-Cabre, and A. Pascual-Leone, "Noninvasive Human Brain Stimulation," *Annual Review of Biomedical Engineering* 9 (2007): 527–565.

2. William James, *The Will to Believe and Other Essays in Popular Philosophy* (Auckland: The Floating Press, 2010), 80. [From a 1912 edition.]

3. Bertrand Russell, "Preface," in *Bertrand Russell's Best* (London: George Allen & Unwin, 1958), 5. [From a 1912 edition.]

4. Marcus E. Raichle, "The Brain's Dark Energy," *Science* 314, no. 5803 (November 2006): 1249–1250.

## 25  MIT Collaborative Innovation: It Takes >2 to Tango

**Israel Ruiz and Adolfo Plasencia**

Israel Ruiz. Photograph by Adolfo Plasencia.

*The combination we call mens et manus, the "mind and hand" of MIT's Technology-Enhanced Active Learning project, the projects and experience are what has always characterized education at the MIT.*

*Being guided by short-term results does not generate innovation. … Creative freedom is the long-term nexus. Therefore, we make sure that there is time for creating. It's part of our economic model.*
*—Israel Ruiz*

Israel Ruiz is Executive Vice President and Treasurer of MIT—the institution's Chief Financial Officer—and an officer of the MIT Corporation. He holds a master's degree from the MIT Sloan School of Management and a degree in industrial engineering from the Polytechnic University of Catalonia. Before joining MIT he worked as an engineer at Hewlett-Packard and at Nissan Automotive.

Adolfo Plasencia:   Israel, thank you for agreeing to this conversation.

Israel Ruiz:   Delighted.

A.P.:   One of the most outstanding characteristics of MIT is its ability to remain at the forefront of disruptive innovation, as well as cutting-edge research and, obviously, excellence in training.

From your viewpoint, and given your experience as MIT's Financial Director and Treasurer, what are the basic practical mechanisms and the finance and resourcing models, in both the short and the long term, that you need to ensure that continuity?

I.R.:   I believe that the mechanisms themselves are well founded because, when we finance the organization's operations, which today amount to $3.2 billion a year, of which 50 percent are earmarked for research, what we are doing is ensuring people's stability. We try to guarantee professors' and researchers' stability and through them their work groups', as in the final analysis it is they who make possible the innovation and disruption you mentioned, and the excellence in education we are renowned for. The essential function of our financial planning, therefore, is to try to ensure and safeguard as much as possible the stability of people and not so much of projects, because projects change and are financed in another way, normally with federal funding. Our funds are mainly aimed at people.

A.P.:   Let's talk about balancing the growth of your institution. Some time ago, I received a message signed by the provost, Martin Schmidt, and you inviting the MIT community to the debate forums on the extension to the northeast of your campus, in the Kendall Square area, as part of the MIT 2030 framework.[1]

How is the rate of growth at the MIT thought out, and how do you achieve the correct balance in economic growth?

I.R.:   One of the things we are known for is our twenty-year development plans. We began the 2030 project, curiously enough, during the economic crisis of 2008. That was perfect timing for thinking about the long term, because we weren't going to build in the short term, were we? So we developed the framework with a twenty-year vision, and now we are building next to Kendall Square, on the East Campus of MIT and the Kendall MBTA area, where we hope to combine physical space and people by adapting the infrastructures of the campus to people's needs.

How do we ensure economic sustainability? Well, by maintaining the number of professors that we employ. What works well at MIT is a model based and centered on team dynamics. We have around a thousand professors, 1,018, to be precise, and we know that each work team is going to grow at its own pace; that is, when the professor enters, he or she is going to organize and develop a work team together with his or her students. If the professor is successful with his or her team, it will grow until it reaches a predetermined limit. We understand perfectly well how this model operates, although the mechanism is obviously different, depending on whether it involves mathematics, economics, or quantum physics. But we are absolutely clear about the model.

We ensure MIT's growth by sustaining the rate of professor replacement and by controlling the growth in the number of professors, which, over the last ten years, has been very little, less than 1 percent a year. That's what our economic stability is based on. We don't grow more than we can. We don't contract more people than we can afford. What we do is adapt laboratories, classes, and infrastructures and see where the projects and the professional lives of our professors, researchers, and educators are going. Then we try to adapt the physical model, obviously involving a much longer timescale, to that.

Now we are trying to promote the maximum interchange of innovation between large companies, small companies, associated and unassociated research centers, and the schools and centers of the MIT campus. For example, we are going to set up a five-year project within this framework called the nanofacility of the future, for working with nanomaterials. It is a service center that involves more than seven hundred MIT professors. Of the 1,018 professors at MIT, seven hundred are working on issues related to nanomaterials, and we are going to create the nanofacility for them.

A.P.:    I am also interested in "MIT's envisioning of the future that does not yet exist." In the MIT 2030 plan, you try to envision the future development of the institution, its campus, and the broad outlines of its activity over the forthcoming twenty years.

Tim Berners-Lee, who worked at MIT CSAIL in Building 32 of the Ray and Maria Stata Center on the campus, is an example of what I mean. Twenty-five years ago he invented the Web, which is now used by more than two billion people all over the world. That was unimaginable twenty years ago. With your horizon of 2030, could something like that happen again, something of great consequence that does not exist now?

How do you go about tackling that long-term period toward a reality that is as yet unknown, where most things that you're going to dedicate your time to don't even exist at the moment?

I.R.:   You're absolutely right. Let me give you some key ideas. The 2030 plan we don't in fact call a plan. We call it a "framework," and it is a working framework. We specifically call it that because a plan with a two-decade vision would almost certainly be a mistake. So, when we are drawing up a working framework, we develop guidelines. We try to define several principles by which we are going to be governed over that period. And those principles obey the vision, the culture, and how MIT imagines that future.

The key idea is flexibility: flexibility of space and of ideas to ensure the existence of spaces that connect people. Tim's project, which gave rise to the Internet, had a clear connectivity principle: it was intended to connect people. It could not be guaranteed, nor could the current massive use of the Internet be predicted, but it did have that clear principle. And just as that project had it, so do many others. In fact, the social networks of today are another giant example of connectivity.

We are trying to develop buildings and infrastructures that "connect," that promote connectivity between people. We're clear about that. Our building work is going to closely accompany all that. Another point on the way we undertake things is that we never commission pharaonic buildings; we build incremental ones that give us room to maneuver.

But first several principles must be drawn up on which everyone agrees. The email you received was precisely to invite people to participate in that process, to give their opinion or to complain about what is wrong, and to tell us what it is they want from those spaces. From there we will embark on a process of synthesis. This is part of our daily routine. Every process that we face we tackle in this way.

A.P.:   Let's talk about adaptation. Darwin said that it is not the most intelligent who survive in changing conditions but those that best adapt themselves to those changes. The technological revolution of recent decades has brought about almost unbelievable changes. For the last five decades, MIT has continued to be the best university in the world in engineering and technology in all the rankings. This suggests you have adapted very well to changes in technology.

But what mechanism do you employ for adapting to an ever-changing reality, like that of the last half century?

I.R.: I think we have three mechanisms. Everyone at MIT will tell you something different, but I'll give you my opinion.

The first mechanism is that, in an institution of cutting-edge education where research is the central part of that education, we receive an influx of new talent on a grand scale every year. It brings in students, who come in their thousands. We have students who have just started university as well as postgraduate students who come with their needs, but also their very up-to-date way of looking at things. The professors have a longer time cycle with a much longer-term view of their careers. When they enter, tension is created with the students, who are perhaps far less expert but who bring with them much fresher and more up-to-date ideas, although perhaps of a short life span. This tension induces ideas and generates adaptation of what interests the professor or what he or she will be interested in in the future. That dialogue is fundamental to our model. And it is also fundamental that the dialogue is on familiar terms, not as father to son but as mentor to student, between two equals.

A.P.: Peer to peer, you mean?

I.R.: Exactly, peer to peer. I have always liked that in an institution: that an idea, even though it may be from a freshman, a first-year novice, will always be listened to from the other side of the table. We do, in fact, have many examples of ideas generated by first-year students that have really advanced knowledge in the project group of their professor. It's one way that a source of adaptation is guaranteed.

The second mechanism is that every year, even during the economic crisis, we contract fifty professors. Of the 1,018 professors that we have, fifty new ones enter every year. After seven years fewer than half remain. That's how the process of tenure operates in the United States. And those years are the hardest ones in their entire professional career. Their work, merit, and impact are evaluated. Over those years those fifty professors from all over the world inform and adapt their departments to everything concerning the new trends they bring with them.

And the third mechanism, perhaps the most important, is that all this is done in a very chaotic manner; the organization does not interfere. It occurs on an individual basis. The individuals go their own way, stating where it is they want to go, as the department that manages them needs that guideline. Each head of department, who normally has a professional cycle of five years—professors rotate in that post—undergoes an evaluation every two years by a visiting committee. It's like an advisory council. Every department has one. The committee is renewed every two years and is answerable to the corporation governing MIT.

The main aim of this visiting committee is to assess the state of a department in terms of professors, students, and discipline and to ensure that the department's future strategic plan is up-to-date and is well adapted to needs and related industries. This is guaranteed by the makeup of those committees, which include academics, industrialists, and corporations. They have no direct relationship with MIT but come to us and, over a few days, give us their critical opinion of how we are operating. We don't really want to know what we are doing right but what we could be doing, not what we are doing. This provides us with a perfect opportunity every two years to assess how we are currently doing and if we are heading in the right direction.

A.P.:  Something that I think is innate at MIT is exploring new forms of education. It's part of the institution's DNA.

In your EmTech España paper, delivered in Valencia, you spoke about MOOCs, the massive open online courses, in which MIT was a pioneer, with its MITX initiative, which it later developed, in collaboration with Harvard University, into the edX initiative.

How did you approach the economic model so that it would be sustainable? How did you do it, bearing in mind that, in a MOOC, dozens and even hundreds of thousands of students have registered? Do you think that MOOCs are just another incremental step, or are they a disruptive innovation that will change the economy and traditional approach to education?

I.R.:  We're still at the beginning, aren't we? I think that a priori, MOOCs have the ingredients and possibilities to be a disruptive educational technology, either those that now exist or are yet to be seen. Today we are still at the outset, still setting them up, but I see MOOCs as an opportunity for our mission, which has always been adapted to advancing knowledge, disseminating that knowledge and educating students in areas of science and technology associated with design, architecture, the humanities, and management.

On the one hand, there's this combination we call *mens et manus*, the "mind and hand" of MIT's Technology-Enhanced Active Learning project, with the projects and experience that are typical of education at MIT.[2] However, this education is still difficult to achieve online. On the other hand, we believe there is opportunity to experiment and see what can be done online today, and how we can improve access to information and knowledge.

MOOCs today present a unique opportunity to be able to educate more people and better on our campus. They can have much more of an impact on society, and, at the same time, educate many more people throughout the world, on a different level perhaps, and without the attendance element, but with content that we personally believe is really very useful for a lot of people. That was the main reason for creating them.

Moreover, in terms of the economic model, MOOCs could be totally disruptive if they turn out to be massive and more than 100,000 people join in, even if it's only to learn and not to receive a qualification. The person who is being educated does not necessarily have to pay. Take, for example, a company that wants to educate its engineers who have been working in the same place for ten years, and it is in the engineers' interest because they are distributed worldwide and the company wants to educate them well and efficiently. This may benefit our processes, as perhaps it will provide us with a sustainable model and in that way provide total access without any restriction to every student in the world.

A.P.:   That would make your campus the entire world.

I.R.:   Yes, the entire planet could be our campus, as far as information and knowledge are concerned, although not in terms of mentorship and apprenticeship, of teamwork and learning certain complex skills. There are some things that can only be done on our physical campus, in contact with the kind of professor we have and participating in the kind of student groups we have. This is always going to be the case. At present, technology still does not allow us to have that degree of interaction, does it? In a sense, that is why you and I are here now undertaking this dialogue. We could have done it on Skype, but it would certainly have been a different type of interview.

A.P.:   That's true.

I.R.:   We know the economic sustainability of that kind of training is going to be difficult to achieve if we overload the student with great economic expense. That is not going to be our model because we want the widest access possible. There is the possibility that the student pays us if he or she wants a certificate, but there are also other opportunities for people who only want to participate in an educational training platform, and perhaps this MOOC platform is the answer for them.

A.P.:   You have to try. …

I.R.:   Yes, everything has to be tried. This is also part of our DNA, don't you think?

A.P.:   Trial and error. ...

I.R.:   Exactly.

A.P.:   You said in a conference paper that thinking about things in the short term is a mistake. Do you believe that to be so?

I.R.:   Yes. What I mean is not that that thinking in the short term is an error. In fact, thinking in the short term is more than necessary, but one has to think in the short term within a context of a long-term idea. There have to be long-term guidelines.

Being guided by short-term results does not generate either the disruptive innovation or certain other characteristics that we are seeking to pursue. The short term tends to be a much more incremental and less disruptive process, so context is essential. Dealing with the short term and being up-to-date is necessary. You fall apart if don't handle the short term well.

A.P.:   Yes, but combining it with freedom is very difficult.

How does creative freedom operate at MIT?

I.R.:   That's a tricky question indeed. I firmly believe that creative freedom is the long-term nexus. One thing we make sure of is that there is time for creating, without limits or intrusions. This has worked well for a number of decades now. We believe in it and continue to operate in that way. It's part of our economic model.

A.P.:   Another dialogue in this book was conducted with Hal Abelson, one of the most widely respected professors and humanists at MIT. The dialogue is titled "Pillars of MIT: Innovation, Radical Meritocracy, and Open Knowledge."[3] There is a sentence in the description of the MIT 2030 framework that I am interested in, as it echoes Abelson's definition: you want "to keep the innovation engine running well into the twenty-first century."

How do you think that Abelson's definition translates specifically in your day-to-day activities?

Is it perhaps that the driving force that moves MIT is an engine whose fuel is innovation? And if it is, what is that engine like, and how does it work? Is it an "anxious" motor, like a Ferrari, a steady engine, or one that at times idles?

I.R.:   Hal's a genius at synthesizing things. I think the principles that he describes connect very well with MIT, which, in the final analysis, safeguards people. They are the ones who ensure the principles that Hal defines as the pillars of the institution, although each one of us may express it in a different way. The search for truth, sharing and opening up knowledge—all this forms part of the character of the people who make up our community.

We look, of course, for meritocratic value in people, but above all in relation to ideas. Meritocracy and also the best ideas gain through the process of many people collaborating.

I would say that our driving force or engine, as well as being "anxious," has other, different constants than a good engine. We need an anxious motor to be able to go fast and really innovate in things that may be radically different. We need it to be steady for activities that require a longer-term process.

A.P.:   Innovation would be one of those principles, I imagine.

I.R.:   Innovation is a main ingredient. It may be said that it is the gas for our engine. However, I think that what really drives us is our impact on the world. That may be an innovation, a social advance, or a positive change in the world. That, for me, is a fundamental factor through which the engine and the people of MIT move. Clearly, there is a cause-and-effect relationship between innovating and making an impact.

A.P.:   Another idea from the MIT 2030 framework is that of "an iterative, inclusive and intelligent process." This reminds me of what the neuroscientist Alvaro Pascual-Leone said in his dialogue about how the human brain operates: the brain is changing itself all the time.[4]

Does this happen at MIT?

And also, how is it that a huge and complex body like yours behaves as a single "intelligent entity"?

I.R.:   That too is a difficult question, although perhaps it may not seem so at first sight. We start from the assumption that MIT is a community made up of intelligent individuals. That is true, but the product of that community does not automatically have to be that of an "intelligent community." I'll give you an example. You could have many ants, each one doing its own thing, and you would have a community. When the ants get together, align themselves and work together, they create architectural structures that are so complex we do not even know exactly how they work. This process of community in which we talk of iterations, the inclusiveness of the process linked to that of your earlier question, to merit, is to a certain extent the secret of leadership on whatever scale. The question is how to direct intelligent bodies, each one innovating and pursuing challenges in its own discipline, and how to channel and galvanize that in order to institutionally achieve and finally fulfill our ten-, twenty-, or thirty-year challenges.

A.P.:   Again, what you said about ants reminds me of another metaphor that is mentioned in the dialogue with the philosopher of science Javier Echeverria.[5] It is that of "anthill intelligence." As he pointed out, there is not enough DNA in each ant to be able to organize the immensity of the whole anthill. It's a mystery. It is unknown what, but there has to be something that establishes and maintains that order and ensures the intelligent functioning of the entire anthill for the survival of the community. This is one of the mysteries of intelligence that we are trying to reflect on in this book and is directly related to what you have said.

I.R.:   It's exactly that. It is the same metaphor. The behavior of the intelligent group is something that requires a lot of guidance, discussion, and dialogue. And for that reason, in the end, maintaining ourselves through these principles of discussion, transparency, and dialogue provides us with the most important competitive advantages.

A.P.:   You mean that just as the intelligence of the anthill is a mystery, you also combine pragmatism and mystery?

I.R.:   Yes, pragmatism; and also mystery, it's true. Obviously, it's very complex. If I had to really analyze how we move our community, well, all I can tell you is that it involves a lot of work and time, as well as enthusiasm for speaking and listening, sharing, not making senseless decisions, believing. …

A.P.:   And believing in it.

I.R.:   Yes, believing firmly in it, of course.

A.P.:   Thank you very much, Israel. Thanks for your time and ideas.

I.R.:   It's a pleasure.

**Notes**

1. The MIT 2030 themes, framework, and related material can be found on this website: http://web.mit.edu/mit2030/framework.html.

2. MIT's TEAL project—Technology-Enhanced Active Learning—is described on this website: http://web.mit.edu/edtech/casestudies/teal.html.

3. Hal Abelson participates in dialogue 14.

4. Alvaro Pascual-Leone participates in dialogue 24.

5. Javier Echeverria participates in dialogue 28.

# 26 Mind over Matter: Brain-Machine Interfaces

Jose M. Carmena and Adolfo Plasencia

Jose M. Carmena. Photograph courtesy of J.M.C.

*Imagine a piece of technology that allows you to control an apparatus simply by thinking about it. Lots of people, it turns out, have dreamed of just such a system, which for decades has fired the imaginations of scientists, engineers, and science fiction authors. It's easy to see why: By transforming thought into action, a brain-machine interface could let paralyzed people control devices like wheelchairs, prosthetic limbs, or computers.*

*I am pretty sure, however, that neurotechnology and experimental paradigms based on brain-machine interfaces will play an important role in the quest for understanding how the brain makes the mind.*

—*Jose M. Carmena*

Jose M. Carmena is a Professor of Electrical Engineering and Neuroscience at the University of California, Berkeley, and Co-Director of the Center for Neural Engineering and Prostheses at UC Berkeley and UCSF. His research program in neural engineering and systems neuroscience is aimed at understanding the neural basis of sensorimotor learning and control and at building the science and engineering base that will allow the creation of reliable neuroprosthetic systems for the severely disabled.

Carmena received his BS and MS degrees in electrical engineering from the Polytechnic University of Valencia, Spain, and the University of Valencia, Spain. Following those, he received his MS degree in artificial intelligence and PhD degree in robotics, both from the University of Edinburgh, Scotland. From 2002 to 2005 he was a Postdoctoral Fellow at the Department of Neurobiology and the Center for Neuroengineering at Duke University. Among the awards he has received are the IEEE Engineering in Medicine and Biology Society Early Career Achievement Award (2011), the Aspen Brain Forum Prize in Neurotechnology (2010), a National Science Foundation CAREER Award (2010), the Alfred P. Sloan Research Fellowship (2009) and the UC Berkeley Hellman Faculty Award (2007).

Adolfo Plasencia:   Jose, thanks for taking time out of your hectic schedule to speak to me.

Jose Carmena:   It is my pleasure, Adolfo.

A.P.:   I can see from your CV that first you studied electronic engineering in Valencia, Spain, then artificial intelligence (AI) and robotics in Edinburgh, UK, and finally neurosciences at Duke University in the United States.

Now you are at the University of California, Berkeley, which is I think one of the best places in the world for studying these things. Along your career path, after electronic engineering came AI, robotics, and neuroscience, all hybrid disciplines. According to the website of the Neuroscience Institute at Berkeley, which you belong to, your scientific interests focus particularly on brain-machine interfaces in motor control and in neural engineering. After this long journey and all your training, which discipline is the dominant one in your lab, and the one you are the most passionate about?

J.C.:   Neuroscience, without a doubt. Basically, the brain is the principal part of all research we do in the laboratory and what I'm most passionate about! Despite not being the discipline I first trained in, neuroscience has become a common nexus for all other disciplines we work on.

A.P.: You haven't followed this sequence of hybrid disciplines just by chance, though. Was all this journey and the accumulation of learning necessary to research and focus on what you wanted?

J.C.: It's true. My interest for science started with my dad, who is an amazing hybrid of a clinician and a scientist. He has been a constant role model throughout my life. My interest in artificial intelligence and robotics came from movies like *Blade Runner* and *2001: A Space Odyssey*.

Around the time I was in high school I started reading magazines such as *Scientific American*. The Web didn't exist yet, and my window to the scientific world came from these sources. I specifically recall reading a fascinating article about the research on insect-like robots that Rodney Brooks was doing at MIT around 1990. This article changed my life. It introduced me to the amazing world of AI and robotics, interests that grew throughout college, to the point that by the time I was done with my electrical engineering degree, I had decided to pursue postgraduate studies in AI, which I did at Edinburgh. Halfway through my doctorate, in 1999, I came across an article by Miguel Nicolelis, which was the first demonstration of a brain-machine interface.[1] This article blew my mind and triggered a change in direction toward the study of natural intelligence instead of the artificial; it steered me toward the brain and the clinical applications of combining engineering with neuroscience. When I finished my PhD I was lucky to be offered a postdoctoral position by Nicolelis to study how the primate brain learns to control a brain-machine interface, and I didn't think twice about it.

A.P.: Yes, but where is the center of gravity that balances this multidisciplinary work to allow it be as efficient as possible?

Is this center a fixed point; is it a mobile point that moves with the job; or is its nucleus in a different place in each case?

J.C.: This center of gravity actually defines the field we work in, neural engineering, which can basically be defined as the fusion of neuroscience with the main engineering disciplines. In my laboratory in Berkeley we have students who come from different PhD programs (neuroscience, electrical engineering, bioengineering, others), which means they have to learn from other areas of knowledge out of their comfort zone right from the first day.

A.P.: Rodney Brooks, who was director of the MIT Computer Science and Artificial Intelligence Laboratory (CSAIL), has hinted that after many years of running the laboratory, he thinks it is better to leave the big questions about AI for the next generation of brilliant young scientists. I get the impression that for him, this science, AI, is moving too slowly in relation to the expectations he had of it.

However, I see everyone saying the opposite, that science and technology are moving really fast.

What do you think?

Do you feel that advances in AI are happening very quickly, very slowly, or does it depend?

J.C.:   In AI there are the "hard problems," such as vision, natural language, and perception, that we haven't solved yet in the natural domain. These are problems that advance very slowly. But there are also clear examples of success in AI, of really useful tools that make our life easier and become indispensable, and that we carry on our smart phones, providing instantaneous access to information.

Brooks's work on insect-like robots in the late eighties created a paradigm shift in AI. He demonstrated that intelligent behavior, such as learning to walk and avoid obstacles, could emerge in these creatures without any explicit representation of the world!. The robots were built according to the subsumption architecture, in which simple, low-level parallel layers coupled the sensors of the robot with the actuators. Intelligent behavior emerged from the interaction of the robot with the environment. This revolutionary work created the field of behavior-based robotics. But, as in many fields, things slowed down when trying to scale up in complexity, moving from insect-level intelligence to human-level intelligence.

A.P.:   You also published an article in the *IEEE Spectrum* magazine that starts by saying:

Imagine a piece of technology that allows you to control an apparatus simply by thinking about it. Lots of people, it turns out, have dreamed of just such a system, which for decades has fired the imaginations of scientists, engineers, and science fiction authors. It's easy to see why: By transforming thought into action, a brain-machine interface could let paralyzed people control devices like wheelchairs, prosthetic limbs, or computers.[2]

However, although "transforming thought into action" is easy to say, turning it into reality is really complicated, don't you think?

J.C.:   Yes, it's complicated, but possible. There is a whole community of researchers worldwide working on the problem, and there are already many demonstrations of animals and humans volitionally controlling external devices (prostheses) with their brains.[3]

A.P.: Indeed, in this article you wrote that the holy grail of the brain-machine interface will be a system in which the brain can control and feel the machine, incorporated as part of one's own body, in a bidirectional interaction, is that right?

J.C.: Exactly. For example, in the case of a patient who wants to move a robotic arm to pick up a glass of water from the table, the first thing is to be able to "decode" the intention of movement. This entails using a mathematical algorithm to "translate" the motor intentions into control signals.

This is the motor part. But we would also like the patient to "feel" the prosthesis—what is the prosthesis holding or touching—and where is it located in space, to get a sense of proprioception. This information will have to be encoded somehow in the brain by artificial means. One possibility currently being explored by several groups is using intracortical electrical stimulation.

The goal is to have a bidirectional "port," that is, one from which we can "read" or "download" information from the brain concerning the action we want to perform and also "write" or "upload" sensory information about the state of the actuator used to perform the action. The idea is for this closed loop of action and perception between the brain and the machine to incorporate, using plasticity mechanisms, a representation of the actuator, the neuroprosthetic, in the brain, as an expansion of our body map.

A.P.: Jose, in the same article you also comment on an issue you describe as perennial among researchers of the interface connecting the brain and machines. It is related to the debate over invasive versus noninvasive.

J.C.: Yes, that's right—the use of invasive versus noninvasive brain-machine interface technologies is a subject that has generated considerable debate in the scientific community. Both types of technology have advantages and drawbacks, and my position in this respect is basically that the two should advance simultaneously, rather than our debating which is better. It's a meaningless debate, if you think about it. Wouldn't it be beneficial for patients to have several treatments for their disease? In the end, the only thing that matters is the opinion of two people: the patient and their physician. These two people are going to evaluate the risk-benefit ratio for each technology and decide which is the most appropriate.

A.P.: The central hypothesis of your work proposes that the brain's plasticity, through a brain-machine interface, enables a robotic device, a neuroprosthesis, to be controlled by thought alone. For example, for someone who has lost a limb, the capacity of the cerebral cortex will enable control of the prosthesis and the possibility of substituting a lost limb in an efficient way.

But what happens to all the memories stored in the brain for this limb, all these years of activity, for example, when someone is left paralyzed because of an accident? I imagine there must be a lot of memories stored away about this activity; about the limb which you had, but no longer have. What happens in this case?

J.C.: Very often, when the body undergoes a change, for example, the amputation of a limb, it simply adapts and comes to terms with the new situation. But in some cases the opposite, known as "maladaptive plasticity," occurs. One example is phantom limb syndrome, in which the brain of a patient with a lost limb refuses to accept that the limb no longer exists. The patient feels pain in a nonexistent arm, the phantom limb, just as when you sleep in a bad position and your limb goes numb. The arm hurts because of the bad position, but it isn't possible to change its position because there is no arm! It's a very interesting syndrome from a neurological point of view. One of the foremost experts, Vilayanur Ramachandran, at the University of California, San Diego, has obtained significant results in relieving the pain in many patients with very simple techniques using the nonamputated arm reflected in a mirror, as though it were the missing arm.

A.P.: What seems clear is that the cerebral cortex is very plastic. In fact, there are people who have lost their sight and in whom the parts of the brain that were dedicated to sight are now capable of devoting time to something else. The issue of plasticity is essential to your work, I suppose.

J.C.: Yes, neuroplasticity is key for brain-machine interfaces. We study how the brain adapts to the machine, which is interesting not only from a basic research point of view; the therapeutic application of neuronal plasticity in the field of the brain-machine interface is also very promising. An example that illustrates the power of neuroplasticity applied to assistive technology is the cochlear implant, which the brain's auditory system uses to learn to hear again.

A.P.: In a series of papers from your laboratory on work in rodents and primates, you talk about learning "neuroprosthetic skills."[4] Could you summarize what this kind of learning is like?

J.C.: In these studies, the aim was first to determine whether the brain can incorporate, through learning, a motor memory related to how to control an external device, a neuroprosthetic, and in this case, discover which brain circuits are associated with this learning.

We first focused on the cerebral cortex of macaque monkeys and later on (in collaboration with Rui Costa's team from the Champalimaud Centre for the Unknown, in Lisbon), studied deeper areas below the cortex of rats and mice, in the basal ganglia.

In these studies, we demonstrated that the brain's capacity for plasticity is greater than was thought, which allows it to adapt to tasks that at first sight aren't natural or biomimetic, such as moving your own natural arm, but are more abstract, for example, controlling the movement of a robotic arm by thought. In other words, the brain circuits forming links between the cortex and the basal ganglia—the built-in machinery for learning how to control our body—are also necessary to learn neuroprosthetic tasks, which by definition don't require physical movement of our own limbs.

A.P.:   In some of your recent research publications there are terms, such as "neural dust,"[5] that sound like science fiction but are becoming part of the domain of real science. Give me a brief outline of the neural dust project that you are immersed in.

J.C.:   "Neural dust" is a project we launched at Berkeley two years ago, together with my colleagues in the Department of Electrical Engineering and Computer Science, Michel Maharbiz, Elad Alon, and Jan Rabaey. It aims to solve one of the main bottlenecks in the field of the brain-machine interfaces and neuroprosthetics systems, that of a neural interface that is viable for a patient's lifetime.

Current implantable technology works for only a few years before the recorded signals start to degrade and end up being inseparable from the noise level. It is believed that one of the reasons for this degradation is the immunological reaction of the brain to the implant itself, which tends to be of considerable size and quite rigid. This reaction worsens as a result of the constant micromotion inside the cranial cavity, which, in the case of arrays anchored to the skull, creates constant friction between the electrode and the brain tissue. Another factor is the degradation of the implant materials themselves over time.

Our idea to alleviate all these problems is to substitute the implanted microeletrode array with tiny particles implanted in the brain that we call dust, no larger than a few dozens of cubic microns, so that they can act as a sensor with which the electrical discharge of the neurons can be measured. These particles are made of simple passive circuits mounted on a piezoelectric material that is energized using ultrasound. All of this is coated in a biocompatible polymer to keep the dust stable over time. The signals recorded by the dust are backscattered as ultrasonic echoes toward the "interrogator," a circuit located below the cranium that is in charge of emitting the ultrasonic pulses toward the dust and recording all the echoes and processing them in order to separate the signals obtained.

A.P.:   You are talking about scattering nanodevices, "dust," inside the brain. And you aren't talking about it just in theory, are you?

J.C.:   The project started out as a theoretical concept, and recently we have begun evaluation of the first prototypes in vivo. The number of neural dust particles to be implanted will depend on the application, but in principle, we think it will be in the region of hundreds to thousands. Another area in which we are thinking of using technology based on neural dust is in applications concerning the peripheral nervous system, an area that has recently gotten a lot of attention from pharmaceutical companies such as GlaxoSmithKline.

A.P.:   The scientific philosopher Javier Echeverria assures us in another dialogue in this book that "when talking of the emergence of a heightened consciousness, if there is not also a 'heightened unconsciousness,' then the result will not be comparable with what we see in the human being."[6]

Do you think an "artificial consciousness" should imitate human consciousness, or is there no reason why it should?

J.C.:   The unconscious is basically all our brain does that we are not "aware of," and this is a big part of the normal functioning of the brain. Some of it is made explicitly available to us through subjective experience, an illusion fabricated by the brain itself, which is what we typically refer to as consciousness.

Before tackling the issue of an "artificial consciousness," it would first be necessary to understand natural consciousness itself. The main problem with attempting such a quest is that its subjective nature makes it a "first-person view," which does not allow application of the scientific method (third-person view).

I am pretty sure, however, that neurotechnology and experimental paradigms based on brain-machine interfaces will play an important role in the quest for understanding how the brain makes the mind. This is because these paradigms allow causally defining the relationship between neural activity and behavior, providing new ways to study subjective experience by explicit manipulation of it. In fact, an important topic that is already being debated in neuroethics forums is how neurotechnology applied to mental health problems—for example, by means of implantable devices for electrical stimulation in the brain of people suffering depression, post-traumatic stress disorder (PTSD), and the like—will affect a person's "self."

A.P.: Thank you for your kindness and generosity. It's been a real pleasure.

J.C.: Thank you, Adolfo. The pleasure is mine.

## Notes

1. J. K. Chapin, K. A. Moxon, R. S. Markowitz, and M. A. Nicolelis, "Real-Time Control of a Robot Arm Using Simultaneously Recorded Neurons in the Motor Cortex," *Nature Neuroscience* 2, no. 7 (1999): 664–670, http://www.ncbi.nlm.nih.gov/pubmed/10404201.

2. Jose M. Carmena, "How to Control a Prosthesis with Your Mind," *IEEE Spectrum*, February 27, 2012, http://spectrum.ieee.org/biomedical/bionics/how-to-control-a-prosthesis-with-your-mind.

3. For a short review, see J. M. Carmena, "Advances in Neuroprosthetic Learning and Control," *PLoS Biology* 11, no. 5 (2013), http://journals.plos.org/plosbiology/article?id=10.1371/journal.pbio.1001561.

4. Karunesh Ganguly and Jose M. Carmena, :Emergence of a Stable Cortical Map for Neuroprosthetic Control," *PLoS Biology,* July 21, 2009, http://journals.plos.org/plosbiology/article?id=10.1371/journal.pbio.1000153; Karunesh Ganguly, Dragan F. Dimitrov, Jonathan D. Wallis, and J. M. Carmena, "Reversible Large-Scale Modification of Cortical Networks during Neuroprosthetic Control," *Nature Neuroscience* 14 (2011): 662–667, http://www.nature.com/neuro/journal/v14/n5/abs/nn.2797.html; Aaron C. Koralek, Xin Jin, John D. Long II, Rui M. Costa, and Jose M. Carmena, "Corticostriatal Plasticity Is Necessary for Learning Intentional Neuroprosthetic Skills," *Nature* 483 (March 2012): 331–335, http://www.nature.com/nature/journal/v483/n7389/full/nature10845.html; Aaron C. Koralek, Rui M. Costa, and Jose M. Carmena, "Temporally Precise Cell-Specific Coherence Develops in Corticostriatal Networks during Learning," *Neuron* 79, no. 5 (September 2013): 865–872, http://www.cell.com/neuron/abstract/S0896-6273%2813%2900563-1.

5. Dongjin Seo, Jose M. Carmena, Jan M. Rabaey, Elad Alon, and Michel M. Maharbiz, "Neural Dust: An Ultrasonic, Low Power Solution for Chronic Brain-Machine Interfaces," arXiv.org, July 8, 2013, http://arxiv.org/abs/1307.2196; Dongjin Seo, Ryan M. Neely, Konlin Shen, Utkarsh Singhal, Elad Alon, Jan M. Rabaey, Jose M. Carmena, and Michel M. Maharbiz, "Wireless Recording in the Peripheral Nervous System with Ultrasonic Neural Dust," *Neuron* 91, no. 3 (August 3, 2016): 529–539, doi: http://dx.doi.org/10.1016/j.neuron.2016.06.034.

6. Javier Echeverria participates in dialogue 28.

## 27 We Want Robots to See and Understand the World

Antonio Torralba and Adolfo Plasencia

Antonio Torralba. Photograph by Adolfo Plasencia.

*One of the reasons why there are not more robots in homes, ... is that they are not capable of understanding the world.*

*The true list of problems to solve before machines can do certain things by themselves is still undefined. In fact, it doesn't even exist.*

*Developing vision systems that can see but not learn does not make sense.*
—*Antonio Torralba*

Antonio Torralba is a Research Scientist and Professor of Electrical Engineering and Computer Science at MIT and a member of the MIT Computer

Science and Artificial Intelligence Lab (CSAIL). He holds a master's degree in telecommunications engineering from the Universitat Politècnica de Catalunya, or BarcelonaTech (UPC), Spain, and a doctorate in signal-image-speech from the Institut National Polytechnique, Grenoble, France. His research interests span computer and human vision, human visual perception, computer graphics, and machine learning.

Adolfo Plasencia:   Antonio, some people say that MIT is like the technological center of the universe; if someone wants to do cutting-edge technology and science, then the best place for it is MIT. Your research here focuses on machine vision, machine learning, and visual perception. You work in the area of vision and graphics, right?

Antonio Torralba:   Yes, I am part of the MIT Computer Vision Group. The Computer Vision Group is different from the MIT Computer Graphics Group, but lately these two areas have been closely connected; we are sort of neighbors. Let's say that if we were to divide it into districts, then I would live in the Vision and Graphics neighborhood.

A.P.:   We could say that you are like a digital-visual alchemist, yes? Scene setup and recognition of scene objects are usually separate elements, but you study them as a whole here. Your goal is initially to build recognition systems that can be much more efficient and robust. But how did you get to this point?

A.T.:   The study of vision allows you to focus on a more specific level than if you studied artificial intelligence (AI) in a more general way. But in fact the human visual center is a brain in itself. It has everything a broader knowledge system needs to solve. You can work on a focused problem and at the same time solve very general questions.

I became interested in the question of vision when I was a student at the Universitat Politècnica de Catalunya, or BarcelonaTech (UPC). Some people there were working on image processing. Image processing and vision are closely related but nevertheless very different things. Basically, in image processing, the input is one image and the output is another image that may have been treated in some way so as to highlight a certain type of information. In vision, the input is an image, but the output might simply be the interpretation of that image, and in this case, the objects that are present need semantic tags. Moreover, there may not even be an output image. It is a more closed system that processes the image in a certain way and produces, for example, information to be used to control a robot.

A.P.: First things first: you were already thinking about AI as a youngster in Madrid, your hometown. Nobody seemed to understand you. Would you explain AI now in a very different way from those days, when you dreamed of it and those around you scarcely understood a word?

A.T.: It is quite an interesting question. I think my explanation would be very similar, but there is something that has changed. People now listen to you, they are less skeptical. As regards AI, twenty or thirty years ago, it was common to think that a computer would never be intelligent. In fact, as obstacles were overcome, for example with Deep Blue winning in chess, people started to believe that it was possible to automate certain things.

A.P.: Yes, particularly lately, with Siri on everybody's iPhone.

A.T.: Yes, it is quite curious. People used to think some things were not possible and now they see them as absolutely normal. And they are not aware of the enormous complexity behind the systems they use.

A.P.: And they carry it in their pockets.

A.T.: Yes. For example, now, any digital camera detects faces and adjusts the focus automatically. This technology is only ten years old. Formerly, researchers themselves thought that to reliably detect faces, it would be necessary to overcome the problem of object recognition in general.

A.P.: Rodney Brooks told me that he is skeptical. Do you think your scientific generation will be able to answer the big questions on AI?

A.T.: I think some of those questions will be answered, but we will have to wait for answers to some of them, maybe for several generations. I do not think this is a field that will be clarified in the next ten or twenty years. It will take much longer.

A.P.: I referred to alchemy earlier, which is at the origin of chemistry, when everything was based more on beliefs about components than on genuine scientific knowledge. And today, in AI. ...

A.T.: Perhaps AI is at that point. We are playing, trying things out, learning about what works and what does not work. Collecting lots of data will take a long time, until we get to know the real fundamentals, to tell apart the things that work from those that do not work, and we are not there yet. In particular, in the field of vision, we're pretty far advanced. It is true, though, that now, in AI and computer vision, progress has been made for operating systems to function in such a way that people find them useful as engineering parts: digital cameras, phones, simple robots, and for production, and in many other applications that already include AI and vision algorithms to solve problems in ways that were previously unthinkable. In fact, there are already commercial products, such as Machine Vision, that use computer vision. But complex vision remains unresolved; it has to deal with the natural environment in which humans move.

A.P.:    And process their context, right?

A.T.:    Yes, and their context too. That environment is very complex. Applications are now starting to operate in that environment.

A.P.:    But AI self-expectations have not been fulfilled. Are people disappointed?

A.T.:    I understand that some people may not be satisfied with the speed at which certain areas have progressed. But there must be a reason for that. Basically, in AI, the problem was more complex than expected. For example, there is a very famous story about the onset of computer vision. Vision seems a very simple problem. After all, we all see without much of an effort. You open your eyes, you look at things, and you see them! You do not have to be resting. Anyway, it is a very automatic process. Therefore, the problem with vision seemed simple, so that is why an MIT professor suggested that some of his students, as a summer project, should make a vision system. It was not supposed to be an entire visual system but it was to include many of the elements that make up a complete system; and that was the students' summer assignment. It was in the 1960s. When they got to it, of course, they realized it was not so easy. For a start, they couldn't even capture and record images because in the 1960s there were no digital cameras like the ones we have today. They were very costly. They were large, expensive systems. A digital camera cost $30,000 or $40,000, and then where were you supposed to store the pictures? You had no memory. A single picture would take all of your computer memory. Maybe you could work with a few images, but not many. All in all, the prospects were not very good.

Some of the systems in today's digital cameras, like the face recognition system, were developed in 2000, and object recognition systems, which are now the state of the art in their field, are based on articles from the 1970s. But at that time some thought it wasn't an interesting line of work. In fact, there was heated controversy about it.

A.P.:    Controversy?

A.T.:    Yes, controversy. There were two major schools in computer vision. One, based on geometric methods, sought to shape the world with mathematics; the other one focused on learning techniques. At the time, learning techniques were not as fashionable as they are today, and these two communities were in conflict. It was very difficult to tell which was going to be right and bear fruit. Many of the fundamentals were laid out then that today are leading to more powerful systems that feed on both communities.

A.P.:   Antonio, you work in one area of computer science and AI that allows machines to interpret images in order to understand scenes and the meaning of objects based on their context. You want to create systems that can understand how objects relate to each other, just as you are trying to achieve in the project "Places-CNN."[1]

A.T.:   What we are studying is object recognition and understanding, that is, understanding scenes. And by scene I mean an environment, for example, a living room, which has lots of objects. Our goal is to develop a system that, by seeing images in that environment, is capable of understanding the objects in the scene, and then, perhaps, how they are interrelated.

A.P.:   So, for example, you are part of a scene, you are in a room, and you ask the robot, "Bring me the ball." It will do so as long as it knows what a ball is. It finds it in the scene and brings it to you. Is that what the dream is all about?

A.T.:   Yes. In fact, perhaps a good way to understand what we are trying to do is through applications, in particular robotics. One of the reasons why today there are no robots in homes, or driverless cars on the streets (although some experiments have been made), or why there are not more autonomous systems in the world, apart from mechanical aspects (motion has been almost resolved), is that they are not capable of understanding the world. So, when moving around, if you do not understand how the world is set up, it is very difficult to move without bumping into objects, or just to know where things are, or how to interact with objects. A robot would be helpful if you could say "Bring me the scissors" and it knows what scissors are, how to move in space until it finds them, where it has to go to find them because it remembers where it left them or, if it does not remember where it left them exactly, the places in which it is more likely to find them; and then, once it gets there, it can recognize the scissors, their three-dimensional structure, and is capable of grabbing them and bringing them back.

A.P.:   And it must also know that if something is made of glass it can break, or if something is made of wood. ... And even know the implications of the material the object is made of.

A.T.:   It must understand the material it is made of.

A.P.:   Because if the robot picks up a glass, if too much pressure is applied, it can break, right?

A.T.:   Yes. The right amount of pressure must be applied. For example, if it brings you a glass of water, the robot needs to know that a glass must be picked up in a particular way.

A.P.:   It cannot be held upside down.

A.T.:   And it has to move carefully, because the object has its own dynamics too. So as the robot moves, the water also moves inside the glass, and even if it has been positioned correctly, water may spill. The ability to understand all these different features of the world on the basis of visual information is what we are trying to tackle. Many groups are working on this issue.

A.P.:   Yes, and there are many things yet to be resolved.

A.T.:   Yes, there are many issues. It really is very difficult to deal with all this in a summer project, as they intended to do back in the 1960s. We focus on object recognition in particular, and not on face detection; the goal is for the object to be recognized and for the scene to be understood: to identify that a particular place is, for example, an office or a hallway. And to know the objects you expect to find in that space and what spatial distribution is typical of such an environment or what items can be found in that type of place. Using this information, we can better identify objects that are difficult to recognize if there is no information about the context. For example, a chair can always be identified as a chair; there is nothing special about it. It can be anywhere; it will always be a chair. But other objects are contextually defined—for example, a portion of something, or a blank sheet of paper, which is seen as a white rectangle. The definition—what the white rectangle really is—will very much depend on its location.

A.P.:   And on its contents, and whether it is valuable or not.

A.T.:   Sure. Then you must also understand the functions it has. But that rectangle ... if it is on the table then it is a sheet of paper, or if it is large and on the wall, maybe it is a whiteboard. The visual information received is ultimately very similar, but it is the rest of the scene that defines what it is, what the rectangle actually is.

A.P.:   There are many different rectangular objects.

A.T.:   Of course, and not of all them are sheets of paper. So in that sense, recognition, understanding the whole scene, is essential to giving meaning to objects. Objects are defined not only by their internal structure but also by the context in which they are found.

A.P.: And there are metaprocesses, too. There may be a display—all over the place nowadays—with a representation of the object, which is not the actual object.

A.T.: Exactly.

A.P.: So it gets tricky.

A.T.: Yes, it is rather complicated. In fact, when you work with many objects, the boundaries between them become blurred. There are objects that, as they are slowly changed, become a different object, and this happens with objects in real life, too. And that border, where you decide that an object starts to be another object, is not well defined yet. For example, a table has a very specific function, and a chair too. But there are many objects that are "in between": you can sit on them, but you could also use them as tables, and so it is not so clear, visually speaking, what the difference between them is.

A.P.: Let me refer to the tip you gave me earlier: "For machines to do certain things, some issues need to be resolved: this issue, this one and this one." So the list of issues to solve is huge. An interesting thing would be to have such a list of big issues, but who sets the priorities? Because that's yet another challenge.

A.T.: Yes. Yes, that is a complex question, because the list of problems in AI is not well defined. The greatest difficulty is not even knowing what the questions are. It is really a problem in research; you go forward blindly, not knowing what will end up working. You follow your own intuition and the things you think are going to pay off. But that list of questions, one that everyone agrees with, simply does not exist. There are questions people more or less agree with, but they are possibly just one piece in the jigsaw puzzle.

A.P.: And one of them is how to get machines to make decisions. In vision, technologies are needed for a machine to be autonomous and make decisions, am I right?

A.T.: Yes, you are. In fact, to overcome problems in object recognition in general, many of the challenges were already there in the 1980s, 1990s, or in 2000. One of the major problems has to do with the big data, a very trendy subject today.

A.P.: Yes, you were a pioneer in the use of big data for machine vision with a very attractive project, the "Visual Dictionary: Teaching Computers to Recognize Objects"—a very appealing title: "80 Million Tiny Images."[2] You used 7,527,697 images to create a huge mosaic of tiles whose colors matched

53,464 English nouns, arranged by their meaning. It is a project intended to teach computers to recognize objects. I am interested in the big data side to your project because, for this book, Ricardo Baeza-Yates, at Yahoo! Labs, told me, in a very simple way, what big data was to him. He said (and I'll paraphrase): "It's what we do: we work with data generated by seven hundred million people. We collect and manage data, and we draw conclusions from them."[3] That is a big data dimension, as happens with some aspects in your project, right?

A.T.:   Yes.

A.P.:   The order of magnitude of the information that you handle, in relation to objects or people, is huge. Computing enables this nowadays.

A.T.:   Yes. The concept is simple. But it is not just a matter of having a lot of data but rather of having enough data for certain patterns to emerge. Imagine you want to see a movie, and instead of the whole movie I just give you three frames (and there are twenty-four frames per second). You would miss the whole story. That's what I mean.

A.P.:   The story, the plot, the excitement. ...

A.T.:   Many things. Maybe you get to grasp what the film is about, but that's it, really. You miss the plot entirely. Then the three frames in the field of AI in particular would correspond to the type and extent of the data used for many years to train systems. The volume of data was so small that the system was confronted with a situation that made it impossible for it to understand the world. No chance. Only three frames! To put it in perspective with the film example, big data would mean having enough frames to really understand the movie. That's when you start getting the whole picture.

A.P.:   And even be moved by the film.

A.T.:   Indeed, to understand enough so that you do not miss anything. Then, having data about seven hundred million people is almost like watching the whole movie.

A.P.:   In society, or in a part of society. ...

A.T.:   Of course, it is already a very large proportion.

A.P.:   Coming back to your work in the vision "neighborhood." Can you already train computer systems? What is the state of art in your field in AI?

A.T.:   Today's computer systems must learn to see. Actually, the visual world is arbitrary. The fact that tables and chairs are the way they are is a random process in which human beings have designed objects with a sequence of functional constraints. But then there are a number of aspects and more open parameters that, perhaps, allow you to be more creative, right? The style of the chair, for example. That variety of visual appearances that objects possess is arbitrary. In fact, they change according to fashion. Therefore, developing vision systems that can see but do not learn makes no sense.

Progress has been made over the past twenty years in machine learning. The goal is, given a set of data, to be able to draw out essential properties that then allow you to generalize. And you have to learn to identify those features that give you the definition of the object. That is a very important area. In machine learning, having to deal with a huge quantity of data gives rise to a series of challenges. First technologically, because you have to deal with lots of information, store it, and gain quick access to it. To do that, new techniques must be developed that allow us to search for things very efficiently amid a large amount of visual data. This is already an area where there have been many developments, but we need people working on them to publish their latest advances. And then another major problem to be solved applies particularly to computer vision: the creation of large databases where images are provided with semantic information. For example, one can take a picture, but that picture is not something that we can directly inject into the computer for it to learn; it must be told what the different objects that make up the picture are: here is a table, here is a bottle. Accessing that information requires people to contribute to tagging such pictures on a large scale.

A.P.: That is a social aspect of research, right?

A.T.: Indeed. In the construction of these annotated databases, it is really difficult for a laboratory alone to build up a sufficient volume of data. So social media are being used intensively, so that people can also contribute to research in this field. There are already tools that allow people to annotate images. Then access to this information is given to researchers.

A.P.: From what you said earlier, I gather that Google's driverless car, for instance, would not be possible if there were not a gigantic visual database behind it, to be able to tell the car that is a street lamp, that is a dog, that is a pedestrian, and those lines on the ground represent a street crossing. You see a driverless car and you ask yourself how it does it, but we don't really think about the big data behind it, as you said. Would this be a good example for today?

A.T.: Yes. We must have not only visual knowledge about objects but also access to maps—for example, to Google Street View, and access to the three-dimensional structure of the scene. Just with the GPS coordinates of the vehicle, lots of semantic information is supplied about the car's environment.

A.P.: And the system must also say: a dog is crossing, a person is waiting. ...

A.T.: Of course, and then it has to be able to discriminate things that are not fixed.

A.P.: You mean they are not fixed in the territory.

A.T.: Exactly.

A.P.: Moving things, random things that may suddenly appear. ...

A.T.: Yes, and there are things that change over time.

A.P.: When a moving object appears, the system must know whether it is a dog, but it has to compare patterns of dog pictures to be able to do this.

A.T.: Exactly.

A.P.: And that requires a huge amount of information that must be handled by a car that looks so normal, doesn't it?

A.T.: Yes. It has to deal with an enormous amount of data because the list of objects it may come across is not a short one. In fact, there are many possible objects.

A.P.: Well, your job is huge, too.

A.T.: Yes, it is.

A.P.: Antonio, it has been a pleasure. Thank you for your time and your ideas.

A.T.: You are welcome. Thank you for the conversation.

## Notes

1. See MIT Computer Science and Artificial Intelligence Laboratory, "Places: The Scene Recognition Database," a new scene-centric database called PlacesPlaces, with 205 scene categories and 2.5 million images with a category label, using convolutional neural networks (CNNs) (http://places.csail.mit.edu).

2. See MIT, *Visual Dictionary*, "80 Million Tiny Images" (http://groups.csail.mit.edu/vision/TinyImages).

3. Ricardo Baeza-Yates takes part in dialogue 17.

# 28 Between Caves: From Plato to the Brain through the Internet

Javier Echeverria and Adolfo Plasencia

Javier Echeverria. Photograph by Adolfo Plasencia.

*Whether the universe is a hologram or not is something the scientists have to prove, but, philosophically speaking, I believe it fits perfectly.*

*Technologies are going to enter into our brains and our bodies. They are going to form part of us, in the most intimate sense of the word.*

*—Javier Echeverria*

Javier Echeverria is Research Professor at Ikerbasque (Basque Science Foundation) and on the Social Sciences and Communication Faculty of the Universidad del País Vasco. He holds degrees in philosophy and mathematics and a doctorate in philosophy, all from the Universidad Complutense de Madrid, and is a Docteur d'Etat ès Lettres et Sciences Humaines, awarded by the Université Paris-I (Sorbonne).

His awards include Anagrama de Ensayo (1995) and the Euskadi Research Award, granted by the Department of Education, Universities and Research of the Basque Government (1997).

Among his many publications are *Telépolis* (Destino, 1994), *Filosofía de la ciencia* (Akal, 1995), *Los señores del aire: Telépolis y el tercer entorno* (Destino, 1999); *Un mundo virtual* (Debolsillo, 2000), *Ciencia y valores* (Destino, 2002), *Entre cavernas: De Platón al cerebro pasando por Internet* (Triacastela, 2013), and "Innovation and Values: A European Perspective" (Reno, 2014).

Adolfo Plasencia:   Javier, thanks for agreeing to meet me to take part in this dialogue.

Javier Echeverria:   It's a pleasure to have you here with me today.

A.P.:   Javier, you are a philosopher, a scientist and a mathematician. You are also a specialist on Leibniz, who was an extraordinary mathematician, philosopher, and specialist in the law and who in 1702 provided us with the first modern version of the binary system: on a medallion he represented the fifteen whole numbers in binary language. The legend on his medallion of 1707 reads: "With only the number 1 (and its absence) everything can be expressed." In other words, using zeros and ones we can express everything intelligible. This is what the digital revolution seems to consist of: a "mathematicization" of the world into a binary system.

Do you believe this is so? That the digitalization of the world is truly a mathematicization of the intelligible world into zeros and ones? And that what is taking place is the coming about in our times of Liebniz's vision?

J.E.:   I did two dissertations, one for the Universidad Complutense and a second one for the Sorbonne, on Leibniz's *Characteristica geometrica*. The digital characteristic (*characteristica digitalis*), the binary number system, was invented by Leibniz.[1] As far as we know, he not only invented it but also realized its importance, and for that reason made that medallion for his benefactor, the duke. Leibniz commissioned a craftsman to make the medallion and presented it to the patron who paid and protected him. He had discovered the binary number system some years previously during his correspondence with Joachim Bouvet, french Jesuit missionarie who worked in China. They had sent him the *I Ching*, a system of symbols with various functions but which Leibniz realized had a strong formal relationship with his binary system.

A.P.:   An isomorphism. ...

J.E.:   Indeed, an isomorphism, and Leibniz realized that he had invented a universal language that would allow any number to be expressed in ones and zeros.[2] Hence, being a philosopher and mathematician, he predicted that "everything can be expressed through zeros and ones," and that is the idea of the medallion.

A.P.:   Really, with ones and their "absence."

J.E.:   True, with ones and their "absence," of course. In other cultures the zero is the absence of numbers, for example, in Mayan civilization. In the abacus, the zero is represented by the absence of a token in a row. Leibniz also saw that the zero is actually the absence of number, but that zeros and ones can be computed, the play between the presence and absence being the basis of the digital characteristic or the binary language. The binary system is the simplest possible: it has only two symbols.

A.P.:   Have you heard the anecdote about a Mesopotamian clay tablet that appeared in a wheat silo bearing the legend "five sacks of wheat minus five sacks of wheat equals: 'what you see!'"

J.E.:   Really? That's good. No, I didn't know about that.

A.P.:   They didn't have a zero.

J.E.:   The great mathematical invention is the zero. It was the Arabs who invented it, as far as we know. They introduced a special symbol to designate the absence of number, to designate nothing, which is how Leibniz interprets it. The play between everything and nothing—that is what allowed God to create the world, according to Leibniz. He put it on a medallion and then expressed it mathematically, as you say, up to fifteen in the binary system. He showed it as a sum of zeros and ones, and therefore invented what today we call the binary system. I was excited about this when computers first began. When I first heard about the existence of computers, I said to myself: "This is very important, the fact that the two states of the electron, charged and uncharged, positive and negative, will allow us to physically represent zeros and ones in the form of an electrical charge." That was a decisive step.

There had to be a theory of electricity to achieve that, and later another decisive moment was when the first digital computers were built. This became technology. First we have Leibniz, who realized the importance of binary, of digitization, then the physicists, who enabled play to take place between charges, and Kirchhoff's algebra, and it was shown that circuit algebra is equivalent to Kirchhoff's algebra, and of course Boole's. A mathematical structure has its isomorphism with a physical structure, and this allows us to establish a technological R&D development—but R&D in the "strong" sense of the term. Then came the machine, an artifact that has been improving ever since; this is what we call the computer, but its fundamental basis is symbolic, digital numerical. It is Leibniz's binary language. This is why Leibniz is recognized as one of the fathers of the modern computer, along with many others, such as Babbage and Turing.

A.P.:   So, digitalization is a mathematicization of everything intelligible?

J.E.:   On that, I have to say no. I'll give you a very clear example of how digitization does not mathematicize everything: so far it cannot mathematize dreams. I tried to do it many years ago in another book. Trying to mathematize dreams was one of the things I have done in my life, but there are enormous difficulties. We all dream practically every day, but not all thought processes are digitizable, at least not yet. This interests me a lot. Everything sensorially representable is digitizable, it's true. That is to say, not everything intelligible can be digitalized, but everything representable through the senses, through perceptions, is indeed digitizable.

Here's an example: the digitalization of the senses. Those of hearing, sight, sound, writing, mathematics, formulas, tables, databases are completely done. In contrast, the digitalization of smell, touch, or taste is not so fully developed. It has been done, but we are a long way from being able to implement it. This is what I was dealing within *Los señores del aire,* in the year 2000, and again in my latest book, *Entre cavernas,* to integrate the five senses,[3] especially the interaction between the five senses. An example that I always use is wine tasting. Digital wine-tasting sessions exist with "digital noses" that wine and cava producers use.

A.P.:   More examples. I don't know if you know, but currently one of the fields within IT security research is digital biometry. One part of digital biometry is based on the textures of skin and fingerprints.

J.E.:   First of all, one thing is digitalization, which is important, but I think that not everything is digitizable; secondly, digitalization requires computerization of the digital, i.e., it requires information technology languages that process the zeros and ones and computing power and needs hardware and software; and, thirdly, I see a third technological system—telecommunications that spring into action when those informaticized zeros and ones are transmissible through the networks, break out of their space, and go distances. Therefore, this combination of digitalization, informaticization, and telecommunications is what allows the Internet to exist, and this is what I call the "third environment."

A.P.: Ricardo Baeza-Yates, who is a computation scientist, told me that the digital revolution is not comparable to the invention of writing.[4] The MIT professor Hal Abelson told me that even if the digital revolution and the Internet are not comparable to the invention of writing, their impact would be comparable to the invention of Gutenberg's printing press.[5]

As a philosopher of science, which of the two hypotheses would you support? Or, to respond from a historical point of view, is all this too recent to make that comparison?

J.E.: It is true that we don't know how this huge digitization process will develop. But even so, hypotheses and hypothetical comparisons can be made. First, I would say that digitization is a kind of writing. Actually, it forms part of the invention of writing. The invention of writing, therefore, is more important than digitization, from my point of view.

If one compares it with changes in science I would say that digitization is a technological system. Digitization is extraordinary, but it does not explain anything. It does not provide reasons. Science is another type of knowledge in which one tries to explain the causes of phenomena. Digitization, as a system of representing phenomena, is the best that has been invented, but it does not explain anything.

A.P.: Then science comes first, as technology is an application of science?

J.E.: Absolutely. Leibniz comes first; the theory of electricity comes first. If there had been no theory of electricity, there would have been no Internet; if the binary language had not been invented and numbers had not been expressed in zeros and ones, there would also not have been any Internet. What I mean by that is that we are faced with a great revolution of enormous consequences, in particular social ones, and that affects everything, the economy, money, everything. But without writing, which came first, this could not have happened, and without science too. These are things that are linked together, but digitization, even though it is hugely important, is not so important. The invention of writing was much more important.

A.P.: And as to whether we have sufficient perspective? I ask you because the same must have happened to those who invented agriculture. As Avelino Corma says, the inventors of ammonia synthesis changed our relationship with the planet, but they didn't know that was going to happen. Perhaps things of huge import for the human condition and the history of humanity are taking place but we are too close in time to recognize them.

Perhaps they will be seen when more time has passed. Do you think this will happen?

J.E.: It is happening now. I believe the digital revolution has already changed the relationship of human beings to the planet (this is what I call the "third environment"). What has changed the relationship between human beings is the digitization implemented through IT and telecommunications, which has changed everything. This transforms human relationships. It is a revolution that transforms the social world, and this can be called whatever you like, the information society, the knowledge society, whatever, which did not exist before.

Information and knowledge have always existed in human history, but that social relationships could be based on the possibility of communicating with each other at a distance, transmitting files, images, and the like, instantly to any part of the world, and without a physical presence, has no precedent.

A.P.: Some humanists are totally against that and say that technology is only a tool, and therefore cannot affect the human condition. What do you think?

J.E.: Technology is much more than a tool. The instrumental condition—that's ingenious! No. We philosophers of science argue with sociologists who talk about it as a "toolbox." Science is much more than a toolbox. Naturally it has a toolbox, but mathematizing the world is no joke because we are transforming the world. Once you have mathematized it well you can create new kinds of worlds, new phenomena, new human relationships. Science, in this case technoscience—digitization—transforms the world, and may even involve an evolutionary leap forward. That is the current debate.

A.P.: In the dialogue with Alvaro Pascual-Leone, we were talking about what consciousness is, how it arises, and where it comes from. But there doesn't seem to be any definitive answer so far.[6]

J.E.: On the subject of consciousness, there is one thing that I would like to say, because it may be important. For me consciousness is not a very important thing for the human being. It is a highly developed cognitive function. It's debatable whether other mammals, for example, have a consciousness or not. That they have a brain is not in doubt, but whether they have a consciousness or not is debatable. It is the neuroscientists who have to clarify this.

I would like to say something that, philosophically, is very important, in my view. In the human being there is a consciousness, but there is also an unconsciousness. The human being is a hybridization of processes, some of which we are conscious of and many others of which—of importance to us—we are not. For example, we are not conscious of emotional processes. By that I mean that when talking of the emergence of a heightened consciousness, if there is not also a "heightened unconsciousness," then the result will not be comparable with what we see in the human being.

A.P.:   Michail Bletsas, director of computation at MIT's Media Lab, is considered to be a supporter of Marvin Minsky's AI line.[7] Michail is convinced that in the twenty-first century, it is very possible that a nonbiological intelligence will emerge, or at least one not based on *Homo sapiens*. How do you understand, from the philosophy of science point of view, the concept of artificial intelligence (AI)? Do you imagine the possibility of nonbiological intelligence?

J.E.:   I would prefer to talk about "intelligences" in the plural. There is not just one intelligence; there are various types of intelligence. Take, for example, emotional intelligence, which people have been theorizing about in recent years: it's not the same type of intelligence that Locke and Leibniz were contemplating at the end of the eighteenth century. So, first, I would say: "intelligences." Having said that, there is also AI. Of course there is, but it is true that AI has been developed by imitation, by trying to reproduce the processes of human intelligence. It seems to me to be a good thing that there may be others, that they are being investigated and that programs of AI based on, for example, other species and not on the human species are developed. I believe intelligence to be a question of degree. If, say, in ten years' time the neuroscientists tell me that ants are intelligent. ...

A.P.:   Are ants intelligent? Or rather is the anthill, which behaves like an intelligent body, intelligent? Because the termite colonies of Africa are like a joint intelligent entity. As far as we know there is insufficient DNA in each termite to be able to organize that immense complex. There must be something that establishes and maintains that order, that ensures the smooth running of the colony and so ensures survival.

J.E.:   It behaves like a collective intelligence. They have developed a coding system through which termites send messages to develop that so-called intelligence. Perhaps this will be explained one day. It would be a kind of nonhuman intelligence, but still biological, that would serve as a model for developing AI based on the anthill intelligence and not on that of the human being. It wouldn't be based on human intelligence but on another type of intelligence. Modeling it would be important. Therefore, there are intelligences and degrees of intelligence. And something else I think is important: intelligence is not only biological. Intelligence is biosocial. What I mean is that culture, science itself, technology have developed new kinds of intelligence.

A.P.:    The intelligence of the termite colony would be biosocial.

J.E.:    That's right—biosocial. Human beings have also developed an intelligence that is biosocial but is also scientific. For example, mathematics is much more developed in a society that has scientific knowledge than in one that does not have it.

A.P.:    Our dialogue includes in its title—"From Plato to the Brain through the Internet"—a very nice idea of yours, the cave of caves—Plato's cave and the technological cave that reproduces Plato's. It seems to be a conceptual fractal, doesn't it?

J.E.:    We should bear in mind one allegorical aspect of Plato's cave that critics often do not emphasize enough but I make use of in the book *Among Caves*: we live in a sensitive world. The sensitive world is a cave. We are prisoners of the sensitive world, chained to the material nature of our world, and cannot see beyond our world. So far we are not able to transcend it.

In Plato's allegory of the cave, what does Plato say happens? That there is a wall, a screen, and all the prisoners see the world there, perceiving it just as it appears to them. But what they see is really a projection that other little men are projecting onto the wall. These little men move objects, even gods, across the screen. This is what appears to us to be the real world.

Why do I say that the Internet is a cave? I put forward the metaphor of the cave because the little men are the IT programmers. They program and digitalize everything. They can even digitalize me, my image or my voice, my representation, and then project it perfectly to everyone. Behind that camera are the little men that have built it, and within it my voice and image are being digitized.

A.P.:    The elves of the digital....

J.E.:    Yes, the elves of the digital.... That in itself is updating Plato's metaphor, and it fits perfectly and is clearly applicable to the Internet. Today, when a youngster takes a selfie, he or she is digitizing him- or herself. On the one hand, he or she is in the first environment with his or her natural body, but they are at the same time the "elves" of themselves in their transit to the third environment, to their own digitalization.

That allegory of the cave when applied to the Internet is extremely rich and wonderful. I'm willing even to risk talking about not the Internet but "Intercaves." What I mean is that the human being goes from cave to cave, and there is nothing else; we are between caves. The Internet, then, is also a combination of caves, multiple interconnected caves, and if one can perfectly apply terms such as interconnection, internetworks, then why not "Intercaves"?

A.P.: That's something I have to ask you about. Not long ago in *Nature*, a group of scientists published an article with a very surprising hypothesis. They said in their research that recent simulations backed up the theory that the universe is a huge hologram, that is, there are features that suggest perhaps the universe could just be a huge representation, could be a giant hologram. What do you think of that hypothesis?

Could the universe be a hologram?

J.E.: That's something I affirmed in passing because I'm no specialist in cosmology, but I do believe that cosmology and black holes are also kinds of "caves."

A.P.: Plato's. ...

J.E.: Yes, Plato's cave. The Big Bang itself. Logically the hypothesis has arisen that behind the Big Bang there are other caves, other worlds, and this is a very important philosophical topic. This is also in my book *Among Caves*.

A.P.: String theory, the multiple dimensions of the universe, all that. ...

J.E.: That's right. It's a scientific hypothesis.

A.P.: Nine dimensions of the universe, eleven dimensions of the universe. ...

J.E.: But the philosophers have already said that before, as always. William James, the North American pragmatist and philosopher, said it. He talked about "multiverses," which is now a technical term on the Internet. James should be read. I quoted him in *Among Caves*. James based his theories on Leibniz, of course. Leibniz spoke of possible worlds. He was clear that there is a plurality of worlds, which was a topic of debate in the seventeenth century. So I would say: there is no single universe; there are many, multiverses, when the term "universes" is well understood. As for proving empirically that the universe is a hologram, that is something for the scientists to show, but, philosophically speaking, I believe it fits perfectly.

A.P.:   It will be just as for the brain: one has to wait for the empiricists to do their work.

J.E.:   We philosophers put forward very audacious conjectures based on concepts, and several of them are picked up by scientists and others are not. I am pleased that this metaphor, William James's conceptual proposal of multiverses, has been taken up by physicists, who are now investigating it.

A.P.:   Nicholas Carr wrote an article titled, "Is Google Making Us Stupid?," and later, in 2011, a book, which became a finalist for the Pulitzer Prize, the title of which asked *What Is Internet Doing with Our Brains?*[8]

Javier, does technology modify the human condition? Do you believe it is causing changes in it?

J.E.:   It is already changing things, and the changes are enormous. This will produce qualitative changes in the functioning of the human brain and in brain-organ, brain-body relationships. What I mean is that the bodies of our successors will be techno-bodies (a term I invented); there will be a "techno-brain," a "techno-liver," some "techno-kidneys," a techno-medico-technological prosthesis that can detect at any given moment a kidney stone and quickly operate on it.

This hypothesis that you've mentioned is quite probable, but there is another hypothesis that I am quite clear about since I learned something about the theory of innovation and read authors such as Rogers, Joseph Schumpeter, and many others, and that is, innovation is always destructive. What I mean is that all these innovations like the Internet, mobile telephony, all digital technologies are destructive. They are creative but induce creative destruction, which is Schumpeter's maxim (of Nietzschean origin). So what will happen? The logical thing is that as the third environment develops and advances (something that is going to happen both in the external world and in the internal world, as in our bodies themselves), a series of cultural forms will be destroyed.

The more power develops in the third environment and enters in bionic form into our body, the more the forms, abilities, and skills that we have or had will disappear. The typical example is the book. The book is not going to disappear, but it is going to be like papyrus. Papyruses have not disappeared, but they are now museum pieces.

A.P.: Do you believe that the humanities will have to seriously rethink in order to analyze these, let's say "eternal," schemas, faced with what is happening now? The idea we have in the humanities of what the human condition consists of, of what its components are?

Has this got to be reviewed?

J.E.: Yes. A very worrying thing is that these technologies of information, communication digitalization, the digital world are technologies of control. This has been recognized by many authors such as Ignacio Ramonet, who said so years ago, and he is right.

A.P.: And then there's that of the NSA and Edward Snowden....

J.E.: There are many examples of control. At the moment in which our brain starts to become bionic, starts becoming a techno-brain (in my terminology), the power of the "lords of the air" to control the human brain is going to increase. Then the humanists will have a real challenge.

Humanists in general have been completely reticent and opposed to the process of technologization, but there is no turning back, and the challenge for the humanities is going to be a very serious one; the human condition is going to come into question. However, I insist I am not talking about posthumanism.

A.P.: And on that, the humanities can no longer claim ignorance.

J.E.: The challenge that this digital revolution poses to the humanities is huge. Some of us have made certain hypotheses on this point, but now many more people are going to have to take part in the debate in a serious fashion, because it is happening now. We are not talking about new technologies but of now well-established technologies that are going to enter our brains and our bodies. They are going to form part of us, in the most intimate sense of the word.

A.P.: Many thanks, Javier.

J.E.: Thanks to you.

## Notes

1. See Javier Echeverria, "'Characteristica Digitalis' y escritura electrónica," *Debats* 69 (2000): 76–81, http://dialnet.unirioja.es/servlet/articulo?codigo=226219.

2. Joachim Bouvet was a seventeenth-century French-born Jesuit missionary to China, where he taught the emperor mathematics and astronomy and wrote mathematical treatises.

3. Javier Echeverria, *Los señores del aire: Telepolis y el tercer entorno* [The lords of the air: And the third environment telepolis] (Destino, 1999); *Entre cavernas* (*Among Caves*) (Madrid: Triacastela, 2013).

4. Ricardo Baeza-Yates participates in dialogue 17.

5. Hal Abelson participates in dialogue 14.

6. Alvaro Pascual-Leone participates in dialogue 24.

7. Michail Bletsas participates in dialogue 16.

8. Nicholas Carr, "Is Google Making Us Stupid?," *Atlantic Monthly,* July–August 2008; idem, *The Shallows: What the Internet Is Doing to Our Brains* (New York: Norton, 2010).

## 29 There Will Be No End of Work

Paul Osterman and Adolfo Plasencia

Paul Osterman. Photograph by Adolfo Plasencia.

*There will always be work to do and there will always be a demand for things, goods, and so forth, so we need people to build them.*

*The question of whether the United States is a model to imitate is complicated because, on the one hand, we have a lot of innovation, which is good, but on the other hand, we have levels of inequality that are much higher than in Europe. We have many more people working for minimum wage, or not working at all. So I believe that one has to be careful about the balance between the advantages of our model and its disadvantages.*
—*Paul Osterman*

Paul Osterman is the Nanyang Technological University Professor of Human Resources and Management at the MIT Sloan School of Management;

Co-Director, MIT Sloan Institute for Work and Employment Research; and a faculty member of the Department of Urban Planning.

His research concerns changes in work organization within companies, career patterns and processes within firms, economic development, urban poverty, and public policy surrounding skills training and employment programs. He has been a senior administrator of job training programs for the Commonwealth of Massachusetts and has consulted widely to government agencies, foundations, community groups, firms, and public interest organizations.

Among his publications are (with Beth Shulman) *Good Jobs America: Making Work Better for Everyone* (Russell Sage Foundation, 2011), *The Truth About Middle Managers: Who They Are, How They Work, How They Matter* (Harvard Business School Press, 2009), *Gathering Power: The Future of Progressive Politics in America* (Beacon Press, 2003), *Securing Prosperity: The American Labor Market: How It Has Changed and What to Do about It* (Princeton University Press, 1999), and *Working in America: A Blueprint for the New Labor Market* (MIT Press, 2001).

Adolfo Plasencia:   Thanks, Paul, for making time for me in your busy agenda. We are at the MIT Sloan School of Management, the prestigious and universally recognized business school.

Paul Osterman:   Thanks for coming. I'm very happy to be here with you.

A.P.:   Paul, you're a professor here at Sloan, and your field includes human resources, working practices, and labor markets, particularly in the United States. As the United States has become a model for the entire world, your experience and successes here can serve as an example for everyone.

I want to ask you something that may be obvious to you. MIT is a magnet for people and talent from all over the world. Many of the best young minds of the world come here. With those minds you can work marvels, isn't that so?

P.O.:   Absolutely. We have incredible students here, but we only admit individuals on the basis of their capacity, and therefore, we have people here who are very willing to work and who come from all over the world to do so.

A.P.:   I know that Sloan has tools for analyzing and observing the global economy and business models that would perhaps form a powerful "macroscope," to use the terminology of Joël de Rosnay. Using your own pupils, professors, and colleagues throughout the world as sensors, that macrosope is capable of analyzing all the decisive ideas on what is happening in the world economy in real time.

In Spain, where I come from, and throughout southern Europe there has been quite a serious problem in the traditional sectors for a number of years. Industrial manufacturing has largely moved to areas of the world with a considerable difference in salaries and in manufacturing circumstances. Does that contextual labor problem also exist here in the businesses of the United States? Or is it less important and urgent to deal with this problem than in Europe?

P.O.:   We have the same problem. For example, in the automobile industry, we have lost jobs; in the textile and clothing industries we are also losing jobs. But our strategy is always to look for new fields, new frontiers, so as to improve our technology, to innovate. Everyone faces the same problem of countries with much lower salaries. Yes, it's the same.

A.P.:   That problem in Spanish is called "*deslocalización*," or offshoring. I wonder whether in the future companies, instead of moving to Asia, will go to Africa and Latin America.

Can what has happened to us happen in Asia? After all, labor costs are rising quickly in China.

P.O.:   Yes, in the long term, of course. What is happening now is nothing new. If you look at history and the economy over the last one hundred or two hundred years, you can see the same process. And, within—I don't know—fifty or so years, Asia will face the same problem. But that's not a bad thing, it's normal, and the response, as I've just said, is to look for new technologies, new products, and everyone will end up better off through that process.

A.P.:   I was, some time ago, in a company that made cell phones in Spain. I asked the manufacturing manager about offshoring, and he said, "Here we don't have that problem because the salary only affects 4 percent of the cost of cell phones." Do you think he was right?

P.O.:   In theory, he's right, but the problem is what percentage of real costs comes from salaries. It depends. If they have fixed costs that do not allow them to maneuver and the salaries are a high percentage of the variable costs, then there can be a problem.

A.P.:   That head of manufacturing also said that in Spain there was an abundance of engineers and the living conditions were good, and for those reasons it was a good place for his company. However, we have seen that nonetheless, companies are leaving Europe. Look at what happened to

Nokia. And I don't know if that logic about the engineer supply is still valid or changing.

P.O.:   I don't know much about the careers of engineers in Europe, but the important thing, I think, is to build a ladder for their careers. We have to show them that they have a future within the companies, that they can climb the rungs of responsibility and technical challenges. If they can see they will have opportunities in the future, they are going to stay in Europe.

A.P.:   There's an old book by Jeremy Rifkin with the blunt title, *The End of Work*. I know that Jeremy Rifkin is not taken very seriously by the scientific sectors of economic science, but his books are so widely distributed in Europe that they affect opinions. Rifkin said that we are moving toward a world without work, among other things as a consequence of technological development. It's been some years, maybe twenty years, since that book was published. *The End of Work* made a fortune and has often been referred to in recent years. However, it's quite clear that the end of work has not arrived. Now, in the United States, the employment figures would suggest that he is wrong.

What trends in employment do you see in the United States for the near future?

P.O.:   In principle, I don't agree with Rifkin. There will always be work to do and there will always be a demand for things, goods, and so forth, so we need people to build and manage all that.

In the United States, we need only consider medical services, for example. There is a lot of work that has nothing to do with manufacturing but with the necessities of life, the requirements of many people. In addition, we still have a lot of work to do in the manufacturing sector, in factories or the high-technology sector, but also in sectors with less technology, so we will have to continue making progress in the process of innovation. That's the key: to improve the products and the manufacturing process. If we can achieve that, we are going to have long-term work.

A.P.:   The United States, in terms of innovation and experiments in working practices, usually does make progress, doesn't it? I would like to ask you if those changes of model are a good mirror for us in Europe to look into.

P.O.:   The question of whether the United States is a good model to imitate is complicated because, on the one hand, we have a lot of innovation, which is good, but on the other hand, we have levels of inequality that are much higher than in Europe. We have many more people working for minimum wage, or not working at all. So I believe that one has to be careful about the balance between the advantages of our model and its disadvantages. Europe has to choose which aspects of our model should be copied and which aspects should be rejected. I don't think it would be a good idea to copy all aspects of our model. You have to pick and choose.

A.P.:   I come from the Valencia region in the east of Spain. Our region has a very special business fabric. Its economy represents more or less 10 percent of the Spanish economy. But within its business ecosystem, an enormous proportion of companies, more than 90 percent, are small and medium-sized enterprises.

Should these small companies recycle their human resources culture as if they were large companies? Should they grow, if possible?

P.O.:   Yes. I'm not an expert on the economy of human resources in Spain, but what I do know is that small companies in Spain are a little behind the times in terms of their human resources strategy. There has to be more training; you have to build ladders, or careers within companies; you have to have cooperation between small companies in order to join forces. So yes, I think family businesses in Spain must change.

A.P.:   In Spain, another surprising thing is that we are the European country with the most university graduates. However, it turns out that often university graduates choose their careers based on a question of social opinion. Often that opinion is stronger than the supply-and-demand factors, and many choose careers that have no future prospects and end up doing jobs at a much lower intellectual level than they are trained for. I read that the same is true of the United States.

Do you think that something can be done about this? At MIT you have pupils from all over the world, so perhaps you have some opinion on this.

P.O.:   Yes, it's a problem in many countries all over the world. In Asia, for example, in a country like India, they have many law graduates who are out of work. I suppose in Spain it's the same. There are many people with qualifications in law but without work in the legal profession.

When there is an excess of university graduates in a society, the response is to increase wages in other fields in which more people are needed and to reduce the wages in surplus fields to balance the labor market. This works in the long term. The question is whether the government can help in this process through incentives or whatever so as to push people into fields where there may be more demand, and also whether public policy can ease the costs and pain of adjustment.

A.P.:   Yes, because in Spain, we have arrived at the paradox that a plumber earns three times as much as an engineer or university graduate.

P.O.:   That can happen.

A.P.:   And more than an economist. That's a problem, isn't it?

P.O.:   Yes. Although in the United States—I don't know about Spain—plumbers earn more per hour, but over the whole year they don't, because they are quite often out of work, and you have to think in terms of annual salary, not just hourly wages.

A.P.:   Some time ago, the *New York Times* published on its front page that the Chinese computation scientist Chen Yi, who is almost a national hero in his country for the success of his chip creations for cell phones, had been accused of fraud. There was also the case of the South Korean scientist Hwang Woo-suk, who was lauded for his pioneering successes in cloned embryo stem cells but then fell into disgrace for fraud. That case had an enormous worldwide repercussion, especially in Europe, as his work had been published by *Science* and other prestigious journals. There have been more cases of scientific misconduct, such as Haruko Obokata of the Riken Center for Developmental Biology in Kobe and his later fight with *Nature* in 2014, and the Schön scandal on progress in semiconductors in Germany. All of these scandals and frauds have become symbols of the imperative to obtain scientific and economic results in the face of the hard economic worldwide competition that the explosion of the Asian economy has produced.[1]

Do you believe that high-responsibility professions or jobs exert too much pressure on those who occupy those positions because of the huge competition and the constant economic need to be first?

P.O.:   Yes, there is a problem with the level of competition and pressure to achieve results, but I'm not a scientist; what I do know is the peer review system, which involves evaluating the work done by your professional colleagues.

A.P.:   The only evaluation by "equals."…

P.O.:   Those that evaluate your work or research are other scientists, and in most cases it works. However, despite that, there are sometimes mistakes. We in the United States also have cases of fraud, but most of the work of scientists is, I believe, honest and clean.

A.P.:   In your students' case, I imagine that to set up a company, to direct a large company, a world company, there is also a lot of pressure from the competition. In your business school, do you teach them how to handle that pressure?

P.O.:   Yes. At the Sloan School of Management we run ethics courses. That word is in the course title. At Sloan, in recent years, we have increased the number of ethics courses. They are to explain and teach the students that they have to be totally honest in their work. Yes, yes, yes. That's really important.

A.P.:   That's why I asked you, because you have been a professor of business ethics and the "fairer" use of the economy, haven't you?

P.O.:   True. I agree that this has to be taught. I believe there is no trade-off between ethics and good results. And yes, you can have both in business.

A.P.:   We're now in the middle of the second decade of the twenty-first century. The Social Internet is in full expansion. The second generation, it is said, will be the generation of the collaborative Internet and open knowledge.

Are we going to work in this second part of the decade in a more collaborative way, with the creation and management of more open knowledge?

Or will the most decisive information and knowledge continue to be encoded, closed and treasured by a small, privileged minority, especially in the world of business?

P.O.:   I don't know the answer to that. In my own work, and in that of everyone I work with, the system is increasingly open, and there is more and more cooperation between us. There are, however, pressures to "close" these issues, pressure that comes from the companies that have more money to gain if the systems remain closed, and pressure from the government too, which is afraid, for instance, of terrorism. All that may end up closing the system even more.

There are, nevertheless, many advantages to an open system, advantages that we must reveal and fight for. I think these advantages are very great and widespread. So there will always be pressure to open the system as well, and I wish it would open up, but I don't know what's going to happen. I can guess, but I really don't know.

A.P.:   I asked you because I collaborated once with the manager of the Ford Factory in Almussafes, Valencia, which is one of the largest, if not Ford's largest, in the world. I helped him prepare some strategic presentations because, as it turned out, every quarter the directors of many factories of companies in Europe that strongly compete with each other get together to tell each other what they are doing. This surprises people who don't know much about the economy. I suppose it's more common than we think, isn't it?

P.O.: Yes, it's quite normal. Information on competitors' trends is now available, and there is also a trend to collaborate, as competing companies often purchase from each other.

Take the automobile industry as an example. Here in the United States, Ford, General Motors, and Chrysler work together, for example, on devising systems that are cleaner and use less fuel and oil. So yes, there is a balance between competition and collaboration. This balance is something very important in our world.

A.P.: Paul, the issue of individual quality at work has always been at the center of your research and thought. You published a now well-known book with the lawyer Beth Shulman, the title of which is very illustrative on all this and a veritable declaration of intentions: *Good Jobs America: Making Work Better for Everyone* Moreover, the *New York Times* echoed with an eye-catching article titled "The Challenge of Creating Good Jobs."[2]

Today, I want to ask you whether you still maintain the thesis of that book. Would you change anything, bearing in mind the huge changes in the market and working practices that have taken place as a result of the growing automation and robotization of many jobs that digitization, among other things, is causing in most areas of human activity?

P.O.: The challenge of creating good work remains central. The key is to deploy a range of tools. Regulation has a role to play, as does political pressure. The recent success of movements to pressure firms such as Wal-Mart and McDonald's to raise their wages is encouraging, as are recent developments such as the $15 per hour living wage recently passed in Los Angeles. But we also need to work constructively with employers and convince them that treating their workforce better is profitable and leads to higher quality. This in turn requires creative public policy to enhance workforce skills and to support those employers who choose to follow the "high road."

A.P.: Last, I want to ask you about a complex but very contemporary issue. What do you think of the idea by Tim O'Reilly, referred to in a conference at MIT Media Lab, of the Mechanical Turk, which was generally associated with repetitive work and automatons? Tim used that name from an Amazon service in his Media Lab conference to take up the Mechanical Turk concept as a metaphor of "the man within the machine." According to

O'Reilly, it is also the way—metaphorically speaking, of course—the Amazon software developers "live" within their systems, to explain the immense cybernetics that allows the internal machinery of the Amazon giant to function.

It turns out that that same name is also a service, the Amazon Mturk, through which Amazon offers people the opportunity "to earn money from their homes," and through which they can establish their own working hours and pay scales. In it, each task is called a HIT (human intelligence task).[3]

Do you think that these new ways of working are temporary and the result of successive waves of new technologies? Or are they signs that we have to rethink and adopt a radically different approach to how we understand the relationship between people and what we have called for a long time now their "work" or their "job"?

What do you think?

P.O.:   There is no question that the organization of work is changing, and the example you give of the Mechanical Turk illustrates this. There are two points to be made here. First, while important, these developments can be exaggerated. The strong majority of employees still work in traditional settings. But, second, these changes require that we rethink the laws and regulations with which we seek to maintain decent employment standards. Many people are in settings in which the traditional definition of employee is no longer applicable. Employment law needs to be updated.

A.P.:   Thank you, Paul, for agreeing to meet me for this dialogue.

P.O.:   Thanks to you.

### Notes

1. Sources for some of the scandals mentioned by Adolfo Plasencia in this paragraph include the following: David Barboza, "In a Scientist's Fall, China Feels Robbed of Glory," *New York Times,* May 15, 2006, http://nyti.ms/1D6zkzH; David Brown, "Korean Researcher Is Said to Admit Stem Cell Fakery," *Washington Post,* December 16, 2005, http://www.washingtonpost.com/wp-dyn/content/article/2005/12/15/AR2005121502243.html; "Special Online Collection: Hwang et al. Controversy: Committee Report, Response, and Background," *Science* (online), December 1, 2006, and after, http://www.sciencemag.org/site/feature/misc/webfeat/hwang2005; Jeff Akst, "The Top Science Scandals of 2014," *The Scientist,* December 25, 2014, http://bit.ly/1zrAFgR; Tracy Vence, "STAP Papers Retracted," *The Scientist,* July 2, 2014, http://www.the-scientist.com/?articles.view/articleNo/40408/title/STAP-Papers-Retracted; Alison Abbott, "Japanese Scientist Resigns as 'STAP' Stem-Cell Method

Fails," *Nature*, December 19, 2014, http://www.nature.com/news/japanese-scientis t-resigns-as-stap-stem-cell-method-fails-1.16631.

2. Paul Osterman and Beth Shulman, *Good Jobs America: Making Work Better for Everyone* (New York: Russell Sage Foundation, 2011); Steven Greenhouse, "The Challenge of Creating Good Jobs," *New York Times*, September 7, 2011, http:// nyti.ms/1Q2LPkS.

3. From Wikipedia: "The Amazon Mechanical Turk (MTurk) is a crowdsourcing Internet marketplace that enables individuals and businesses (known as Requesters) to coordinate the use of human intelligence to perform tasks that computers are currently unable to do. It is one of the sites of Amazon Web Services" (http://en .wikipedia.org/wiki/Amazon_Mechanical_Turk).

# 30   A Smart Mob Is Not Necessarily a Wise Mob

Howard Rheingold and Adolfo Plasencia

Howard Rheingold. Photograph by Adolfo Plasencia.

*I believe the term artificial intelligence has actually been used to describe machine intelligence, and in this case the intelligence doesn't reside in machines; the machines are a mediator for people's actions.*

*Current tools are really tools for communication, not for thinking.*
—*Howard Rheingold*

Howard Rheingold, credited with inventing the phrase "virtual community," is an adjunct faculty member of the Transformative Leadership Studies program at the California Institute of Integral Studies, San

Francisco, and a critic, writer, teacher, and expert on the cultural, social, and political implications of modern communication media and virtual communities.

He served as editor of *The Whole Earth Review* and editor in chief of *The Millennium Whole Earth Catalog,* and as the first executive editor of *HotWired.* Subsequently he founded the website Electric Minds and started a consultancy for virtual community building.

He is the author of *Smart Mobs: The Next Social Revolution* (Basic Books, 2002), *The Virtual Community: Homesteading on the Electronic Frontier* (MIT Press, 1993), and *Net Smart: How to Thrive Online* (MIT Press, 2012).

Adolfo Plasencia:   Howard, thanks for accepting this conversation with me, here in Barcelona.

Howard Rheingold:   It's a pleasure.

A.P.:   Why don't we start with the concept of your most famous books?—a concept that has generated much comment owing to the numerous possible interpretations of the title, *Smart Mobs,* particularly in this age of social media and the almost universal use of cell phones, which are now, with their ubiquitous connection to the Internet, called "smart phones." The smart phone issue appears to be an application directly related to your book.

How would you most directly define "smart mobs"?

Do you believe that distributed intelligence, in the form of, for example, being contacted by cell phone, generates an emerging collective intelligence?

H.R.:   I'll give you a definition of smart mobs. Smart mobs are groups of people who emerge as such, when technology allows them to collaborate and organize their activities in a collective way, through an interaction that we could call something like "intelligent."

A smart mob is not necessarily a wise mob. In English, I think that "smart mobs" has a somewhat different connotation from that given by the Spanish translation of "mobs," in the sense that "mobs" also has a bit of a sinister connotation. It has numerous meanings, one of them being "collective intelligence." Another is that a group is capable of organizing its activities in ways in which it would be unable to organize itself without mobile and Internet technologies. And there is also a kind of description (concept) (so they say) that adding computers to things turns them into smart things or homes into smart homes.

People aren't smarter because their use of a cell phone somehow makes them better or more intelligent but because this technological device and the technologies linked to it give people the chance to engage in a world, in a setting, different from which until now had only been possible to access through Internet, using computers connected to a socket on the wall.

So there are different connotations, which obviously don't fully describe the phenomenon. I explain this quite widely in my book. I speak about the use of mobile communications, I speak about the emergence of collective intelligences, and I also think there is a difference between people who use technologies to coordinate their activities and people who, in addition, understand how they work. And this is the reason why I wrote the book: because there is a phenomenon that is emerging and people who participate in it don't necessarily understand it or understand all aspects of it.

We're in the early stages of the emergence of a new medium that combines the personal computer with the Internet and the cell phone (telephone + Internet + wireless + ubiquitous chips and a ubiquitous connection. too).

A.P.:   Let's talk about artificial intelligence (AI) and machines.

Do you think that AI now exists in digital machines?

Is it going to be possible to use genuinely intelligent machines soon?

H.R.:   I believe the term artificial intelligence has actually been used to describe machine intelligence, and in this case the intelligence doesn't reside in machines; the machines are a mediator for people's actions. What simply happens is that technologies such as the Internet or cell phones allow people to carry out collective actions, even in the virtual world or the physical world, which we wouldn't be able to do collectively without the technologies.

A.P.:   Let's talk now about processes, the ones you have written about in your books, both in the pioneering *The Virtual Community: Homesteading on the Electronic Frontier* (1993) and in *Smart Mobs: The Next Social Revolution* (2002), as well as in the most recent *Net Smart: How to Thrive Online* (2012).

Do you think these processes in relation to smart mobs are in some way related to Hayek's concepts of spontaneous evolution, or biological theories on self-organization and morphogenesis, which is what Alan Turing talks about in his latest works? Because some scientists insinuate that in a large enough mass with special biological activity, there might be a spontaneous generation of intelligence, precisely as a result of surpassing a certain threshold or quantity of critical mass of the activity.

In summary, what do you think about the idea of a new intelligence arising as a result of an accumulation of many individual interacting intelligences? Hayek and Turing have spoken about these types of processes applied to intelligence.

H.R.:   I'm not too sure that this is related to the ideas revolving around the intelligence of machines. Current tools are really tools for communication, not for thinking. I insist: they are not a tool for thinking. Actually, I think it doesn't matter whether a computer has the capacity to emulate AI or not. The important thing is that it allows people to act collectively, and I suppose that wanting to call it an emerging phenomenon is a philosophical question, in the same way that we call the sum and accumulation of many people's actions civilization.

A.P.:   The next question I have for you concerns what happened in the Madrid bombings—known as 11-M, for March 11—the series of terrorist attacks on four commuter trains in Madrid carried out by jihadist terrorists, as revealed by the police and judicial investigation.[1] I recall that in the prologue of your book, *Multitudes inteligentes* (smart mobs), you talk about the smart mobs in Spain following the Madrid bombings; the social catharsis that affected many people, prompted by the impact of the events of March 11; and the twisted response given by the authorities about their origin.[2] There are some who say that the popular response wasn't self-organized, although we were all out there in the streets.

Do you believe it was a totally spontaneous phenomenon? That it modified itself on its own? Because in the text messages (SMS) it said: "The Government is lying. Pass it on." In other words, there was information and there was an order. So did the mob carry out an order or was it something that was, by itself, self-regulated?

H.R.:   That doesn't really matter.

A.P.:   I mention it because of the "pass it on" thing that people put at the end of their short texts so that the messages could be passed on from one person to another, to more and more people at full speed.

H.R.:   I'm not sure I understand it. It's quite a common and colloquial sentence in any language for people organizing a political action. And I cannot claim to have any experience in Spanish politics, but what I do find quite logical is that people join in with a message of this kind.

A.P.:   Nicholas Negroponte explained wireless technology in its Wi-Fi mode in his article "Being Wireless" by using the metaphor of water lilies, the horizontal flowers floating on the surface of the water, and frogs, didn't he?[3] Information jumps from one side to another, from one water lily to another, in a horizontal system with no hierarchy.

Do smart mobs have no hierarchy?

H.R.: The smart mobs phenomenon is even more horizontal, more lateral.

A.P.: In other words, there isn't any hierarchy?

H.R.: They are lateral communications. Communication is passed on laterally from person to person. In that sense, yes, I suppose that Negroponte's metaphor is a good one too.

A.P.: Howard, you also said in the prologue to your book about the 11-M events that, in a certain way, television died with the text messages in Spain. I don't know if you think that new things always kill off something that is old.

What could die, apart from television, if smart mobs become universal?

H.R.: In fact I didn't say that; perhaps it came up in the translation. However, it's not a case of putting the blame on television. Instead, people who didn't have access to television, to the radio, or to newspapers were able to use communication technologies for disseminating their message from person to person, and actually it has nothing to do with television, apart from the fact that it is an alternative, a means for spreading your message.

Of course, radio hasn't killed off movies, movies haven't killed off theater, television will not kill off movies, and the Internet will not kill off television. I think it was Marshal McLuhan who said that every time a new medium comes out, it simply alters other media, but this is not necessarily related to doing away with it or substituting for it. Many people simply used text messaging because in a delocalized environment they didn't really have access to television. It isn't that it killed off television. What was altered instead was the relation of these people with the television media because what they used on that occasion was a new medium.

A.P.: You said in one of your essays some time ago that the Internet was born out of love and not through a lust for money. How can we combine a free and open space without controls and the economic business space in the mobile cyberspace?

H.R.: I don't see why there should be any reason for the two not to coexist. People use language for commercial ends and after that they use it for poetry, and they are not mutually exclusive.

A.P.: In the third part of your most recent book, which is titled "Participation Power," you speak about "participation skills." As you usually do when speaking about this participation, you were referring to that carried out using a cell phone, a smart phone. Time has obviously proved you right.

Do you think this phenomenon of delocalized or mobile connection (of smart mobs) will continue to be decisive for humanity? Do you believe it's going to be the next social revolution?

H.R.:   Well, yes, because there are more people who have cell phones than computers. In just the first quarter of the previous year, 267 million smart phones were sold. Telephones are giving us more and more access to Internet; people carry theirs around with them and it is a part of their lives. Personal computers and laptops haven't been so readily available. So there are more and more people in more places who are able to do the same things they used to do with personal computers and the Internet. This is important: many, many more people in many more places.

A.P.:   Thanks very much, Howard, for your time and opinions and for this conversation.

H.R.:   You're welcome.

## Notes

1. The 2004 Madrid train bombings are known in Spain as 11-M. A controversy arose between Spain's two main political parties over the government's handling of the attacks.

2. The reference is to Howard Rheingold, *Multitudes inteligentes: La próxima revolución social (smart mobs)* (Gedisa, 2009).

3. Nicholas Negroponte, "Being Wireless," *Wired,* October 1, 2002, http://archive.wired.com/wired/archive/10.10/wireless.html.

# 31   Measuring the Intelligence of Everything

José Hernández-Orallo and Adolfo Plasencia

José Hernández-Orallo. Photograph courtesy of J.H.-O.

*Any nonanthropocentric view of intelligence must be based on the theory of evolution or on algorithmic information theory, or both.*

*A fully fledged unification (of the definition and evaluation of cognitive abilities), also covering machines and based on first principles, is what we call universal psychometrics, the measurement of cognitive traits for any kind of subject, including hybrids and collectives.*
*—José Hernández-Orallo*

José Hernández-Orallo is a Reader in the Department of Information Systems and Computation, Universitat Politècnica de València, (UPV), Spain. He holds a master's degree in computer science from UPV and a doctorate in logic with a Doctoral Extraordinary Prize from the University of Valencia.

His research in machine learning and data mining focuses on the construction of declarative models using logic and functional logic programs and on the analysis of classifier performance in a wide range of contexts, with a more sustainable view of the data-to-knowledge process.

His research on machine-intelligence evaluation stems from the construction of the first intelligence tests derived from algorithmic information theory. He is the author of *The Measure of All Minds: Evaluating Natural and Artificial Intelligence.*[1]

Adolfo Plasencia:    José, I'm delighted you have agreed to hold this dialogue with me today.

José Hernández-Orallo:    It's a pleasure. Thanks for coming.

A.P.:    Answers to the questions of what is human intelligence and how does it work will probably differ, depending on the view each discipline takes when tackling this complex theme. Neuroscience or neurophysiology will not have the same view as the sciences of computation, philosophy, or artificial intelligence (AI).

J.H.-O.:    The answer of what intelligence is should be the same whether the subject is a human, another animal, or a machine. How *human* intelligence *works* is a different, more specific question. Another different, related question is why and how intelligence appeared.

All the disciplines you have mentioned, including psychology and evolutionary biology—many of them usually grouped together under the umbrella of cognitive science—have a say, and can focus on some issues of intelligence, its purpose, and even its interpretation.

One key question for me, however, is which of these disciplines is most explanatory and least anthrophocentric about human intelligence. If one day human intelligence is fully understood, how will this understanding be explained to lay people? I don't think science has the duty of rendering all the mysteries of nature intelligible to everyone outside the specialists of a discipline, but the understanding of how human intelligence works is a sine qua non to understanding ourselves, without which all other understanding is lame.

## 1.   On what is measurable in intelligence, and what its ingredients are

A.P.:    A while ago you and your colleague, David L. Dowe, published a scientific work titled "Measuring Cognitive Abilities of Machines, Humans and Non-Human Animals in a Unified Way: Towards Universal Psychometrics."[2] You expounded on the results of your research into the ways of measuring intelligence. The scientific press spoke of your articles as proposing a "universal measurer" that could be used for measuring the level of intelligence of both humans and machines or robots, as well as other types of "bodies." One can imagine how many question marks were raised just by the mention of "cognitive abilities" in the title! So I have various questions to ask you about that.

a.   What conceptual vision of intelligence do you apply in your research?
b.   What distinction, if any, do you make between human biological intelligence and what Michail Bletsas calls "human nonbiological intelligence"?[3]
c.   How would you define the cognitive abilities of a machine, and how do they differ from human cognitive abilities?
d.   Is it possible to compare different types of intelligence?

J.H.-O.:   The principles of our approach rely on algorithmic information theory, a mathematical theory that combines information theory with computation. We claim that intelligence, and other cognitive abilities, can be defined in these terms. In fact, I believe that any nonanthropocentric view of intelligence must be based on the theory of evolution or on algorithmic information theory, or both.

I also believe that many of the problems of the current science of intelligence stem from the different conceptions of intelligence for biological and nonbiological systems. Cognitive science has contributed significantly to all these disciplines. However, we still evaluate humans, nonhuman animals, and machines by applying very different principles. In other words, the definition and evaluation of cognitive abilities have not yet been unified, despite the efforts of comparative cognition embracing humans and nonhuman animals. A fully fledged unification, also covering machines and based on first principles, is what we call universal psychometrics, the measurement of cognitive traits of any kind of subject, including hybrids and collectives.

I think it is possible to formally define a common series of cognitive abilities and personality traits from very first principles, and to derive tasks and their degree of difficulty from them. We could then use them to evaluate humans, other animals, and machines, with the appropriate adaptation of the interface, in order to compare their psychometric profiles. Some of our previous work has shown that this is a promising path to follow, providing some working examples for some kinds of tasks. Of course, there are many difficulties; measurement is always more difficult the more general we want our instruments to be. But this is a path that needs to be explored.

In the end, we do not make any distinction between biological and non-biological intelligence, or between the definition of cognitive abilities for one or another. In fact, only with the same definitions—the same rule—is it possible to compare different systems in a meaningful way.

## 2. On how to universally measure intelligence

A.P.:   Having explained the vision of the intelligence that you propose to measure:

Can you explain to me how your way of measuring intelligence works?
Can we say that there is a universal "parameter" of measurable intelligence?

J.H.-O.:   To solve a problem—hiding from a lion, finding a mating partner, lighting a fire, or proving a theorem—we need an appropriate behavior, a policy, an algorithm. If these policies are not innate, they must be acquired in some way: transmitted, imitated, or discovered. Cognitive abilities make the acquisition, integration, and application of these policies possible.

With this view, the evaluation of cognitive abilities relies on the construction of tasks, usually interactive, for which a policy is required. The difficulty of the task is determined according to the computational effort that is required to acquire and deploy the policy. In a way, this is a very similar approach to psychometrics, but the policies are formalized and their difficulty is estimated without the need for a population (such as *Homo sapiens*). Instead, the difficulty of a task depends on the computational resources involved in the policy, including its length and execution time.

Also, through the analysis of the properties of the task and most especially the policy that solves it, we can understand what kind of ability we are measuring, from first principles.

Intelligence is one of these abilities and is measurable, as all others are. Of course, intelligence plays a central role, especially if it is defined as the ability to acquire, integrate, and apply all kinds of successful policies. The rest is not very different from the way humans, and most especially other animals, are evaluated. For instance, rewards and penalties are used, as in the evaluation of animals and small children.

## 3. On the Turing test

A.P.:   The Turing test, although having come in for some criticism of late—some say it has been superseded—has been and still is a classic test from the conceptual point of view for finding out about a machine, a computer, and so forth.

It consists—and please correct me if I'm wrong—of comparing a machine or intelligent body in its logical behavior through language with that of a human being.

What opinion do you have today of the Turing test?

J.H.-O.:   Turing's imitation game is a beautiful and insightful philosophical exercise. It was introduced as a powerful resource to support a series of arguments in favor of the possibility of machine intelligence.

The Turing test as is known today is not appropriate as an effective intelligence test. In 2000 I wrote a paper titled "Beyond the Turing Test" proposing an alternative pathway for machine intelligence evaluation.[4] Since then there have been several other papers, even workshops, with the same title. I'm inclined to think that this shows that AI evaluation is still stigmatized by the Turing test.

### 4.   On compared intelligences and the IQ (Intelligence Quotient)

A.P.:   Another of your articles, published in *The Intelligence Journal*, is titled "IQ Tests Are Not for Machines, Yet."[5]

How can one define the IQ of a machine, if it can indeed be done, and what does it have in common with, or how is it different from, the human IQ?

J.H.-O.:   IQ tests are normed, which means that their values are normalized to a human population. It does not make sense to talk about the IQ of machines, basically because there is no normative population of machines. Another thing is whether the items that compose IQ tests are useful or not for constructing tests to evaluate machine intelligence. Unfortunately, the items found on IQ tests are specialized for humans, and their validity and reliability are dubious for other kinds of subjects. It is highly questionable whether they can be either a necessary or a sufficient condition when applied to machines.

That said, at the moment, the philosophy behind IQ tests and human psychometrics is decades ahead of the current mainstream evaluation approaches in the area of AI, based on task-oriented benchmarks or on variants of the Turing test. The choice is not between human tests or AI tests but rather the construction of principled alternatives for measuring the abilities that IQ tests and other tests seek to measure by deriving items using algorithmic information theory.

## 5. On the AI agents of software

A.P.:   At the Artificial General Intelligence 2011 conference, held in Mountain View, California, you presented a paper titled "Comparing Humans and AI Agents."[6] AI agents are a fundamental piece of the giant cybernetics that circulates and acts on the networks and the Internet today. They are able to deal with thousands of transactions per second in the stock market (this happens every day in high-frequency trading), something quite impossible for a human being. They are also able to negotiate between many offers, comparing and choosing the most profitable. Once again, I have various questions to ask you.

Is this not very close to machines (with software) making decisions?
Do human beings come out badly in some aspects when compared with AI
     agents?

J.H.-O.:   Some machines can do many things much better than humans. We can compare a machine that has been specialized for a task with a human (such as chess or Jeopardy!), and see that the machine beats the human. Also, in particular domains, computers can make much better decisions than humans.

The problem is that no machine today can deploy a wide, previously unforeseen, range of policies better than humans.

In that particular paper we made an effort to derive a broad task class spanning a range of policies to compare a general but very simple learning agent, such as Q-learning, with humans. Since humans did not outperform Q-learning (despite being much more intelligent), the experiment simply falsified that the task class could be considered sufficiently diverse.

But the point was precisely this. By comparing AI algorithms and humans, we can discard those tests that are not sufficiently comprehensive. This refutation power of any measurement device when applied to more situations is one of the motivations behind universal psychometrics. A thermometer can work very well in the Caribbean but may provide inaccurate measurements, or even break, when brought to Antarctica.

## 6. On whether the human condition is mathematizable

A.P.:   In science it is usually said that that which cannot be measured, cannot be improved. In your research, you and your colleagues usually "mathematize" things concerned with intelligence (of humans, machines, and other "bodies"), using mathematical formulas that may astound me, but you're not the only ones to do so:

An article in PNAS by the scientists Rutledge and colleagues and titled "A Computational and Neural Model of Momentary Subjective Well-being" includes this "Happiness Formula":

$$\text{Happiness}(t) = w_0 + w_1 \sum_{j=1}^{t} \gamma^{t-j} CR_j + w_2 \sum_{j=1}^{t} \gamma^{t-j} EV_j + w_3 \sum_{j=1}^{t} \gamma^{t-j} RPE_j,$$

where $t$ is the trial number, $w_0$ is a constant term, other weights $w$ capture the influence of different event types, $0 \leq \gamma \leq 1$ is a forgetting factor that makes events in more recent trials more influential than those in earlier trials, $CR_j$ is the $CR$ if chosen instead of a gamble on trial $j$, $EV_j$ is the $EV$ of a gamble (average reward for the gamble) if chosen on trial $j$, and $RPE_j$ is the $RPE$ on trial $j$ contingent on choice of the gamble. If the $CR$ was chosen, then $EVj=0$ and $RPE_j=0$; if the gamble was chose, then $CR_j = 0$. Parameters were fit to happiness ratings in individual subjects. We found that $CR$, $EV$, and $RPE$ weights were on average positive [all $t(25) > 4.6$, $P < 0.0001$] with $EV$ weights lower than $RPE$ weights [$t(25) = 4.3$, $P < 0.001$]. The forgetting factor $\gamma$ was $0.61 \pm 0.30$ (mean $\pm$ SD). This model explained moment-to-moment fluctuations in happiness well with $r^2 = 0.47 \pm 0.21$ (mean $\pm$ SD) and, when judged according to complexity, explained this reactive happiness better than a range of alternative models, including models without exponential constraints, parameters for unchosen option, and utility-based models.[7]

The Spanish writer and scientist Javier Sampedro ironically commented on this, saying:

The formula is fed by numbers that measure the activity of the dependent circuits of dopamine—the crucial neurotransmitter of the pleasure circuits in the brain- as well as the influence of environmental factors on mood, their proximity in time, the amount of reward for winning some kind of gamble and other experimental parameters. In the end a number is produced which tells you whether you are very happy, happy or unhappy.

The ironic Sampedro ended his reflection by saying, "Little hot air and sound mathematics: who could ask for more."[8]

Anyway, the authors of the article do indeed seem to have mathematized happiness!

You and your colleagues are conducting research into the intersection of computation, formal logic, and the development of intelligent bodies through software, and I'm interested in your opinion, from the research on intelligence in which you use mathematics and the formal logic that that brings with it as your main tool …

In your paper "Measuring Universal Intelligence: Towards an Anytime Intelligence Test,"[9] you published this definition of a "universal" intelligence test, expressed precisely as an algorithm with the heading "Universal Intelligence Test considering time … etc." (p. 1596), that is, through "sound mathematics":

**Definition 18** *(Anytime universal intelligence test taking time into account).*
We define $\Upsilon^v(\pi,U,H,\Theta)$ as the result of the following algorithm, which can be stopped anytime:

1.  ALGORITHM: Anytime Universal Intelligence Test

2.  INPUTS: $\pi$ (an agent), $U$ (a universal machine), $H$ (a complexity function),
        $\Theta$ (test time, not as a parameter if the test is stopped anytime)

3.  OUTPUTS: a real number (approximation of the agent's intelligence)

4.  BEGIN

5.      $\Upsilon \leftarrow 0$                       (initial intelligence)

6.      $\tau \leftarrow 1$ microsecond         (or any other small time value)

7.      $\xi \leftarrow 1$                       (initial complexity)

8.      $S_{used} \leftarrow \varnothing$            (set of used environments, initially empty)

9.      WHILE (TotalElapsedTime < $\Theta$) DO

10.     REPEAT

11.         $\mu \leftarrow \mathrm{Choose}(U,\xi,H,S_{used})$  (get a balanced, reward-sensitive
                    environment with $\xi - 1 \leq H \leq \xi$ not already
                    in $S_{used}$)

12.         IF (NOT FOUND) THEN   (all of them have been used already)

13.             $\xi \leftarrow \xi + 1$         (we increment complexity artificially)

14.         ELSE

15.             BREAK REPEAT     (we can exit loop and go on)

16.         END IF

17.     END REPEAT

18.     Reward $\leftarrow V_\mu^\pi \parallel \tau$     (average reward until time-out $\tau$ stops)

19.     $\Upsilon \leftarrow \Upsilon + \mathrm{Reward}$      (adds the reward)

20.     $\xi \leftarrow \xi + \xi \cdot \mathrm{Reward}/2$   (updates the level according to reward)

21.     $\tau \leftarrow \tau + \tau/2$        (increases time)

22.     $S_{used} \leftarrow S_{used} \cup \{\mu\}$     (updates set of used environments)

23.   END WHILE

24.   $\Upsilon \leftarrow \Upsilon / |S_{used}|$      (averages accumulated rewards)

25.   RETURN $\Upsilon$

26.   END ALGORITHM

It's clear, then, that for you and David L. Dowe, you do think that sound and rigorous mathematics is sufficient to define and measure intelligence. Bearing that reflection in mind, and from the point of view of your discipline, can you explain to me your opinion on this, and obviously I would also like you to comment on what position you believe this leaves philosophy and humanists in when you tell them that the human condition (the intelligence that forms part of it) can be "mathematized," something that many of them, as you know, still reject.

J.H.-O.: Formalization in science is very important, and mathematics is an excellent way to formalize a concept and operate with it in an unambiguous, reliable way. However, not every scientific theory needs to be mathematical. For instance, Darwin did not need too much formalization. His evidence and language were sufficiently precise for his theory to be falsifiable.

The above formula for happiness is simply a composite indicator with a weighting of three sums with some coefficients. It is just a way of giving more or less relevance (in an unambiguous way) to some other indicators for their contribution to a joint measure. What I mean is that this is not particular to science: any sports competition can in the very same way derive scores from different parts of the game in accordance with several rules. At school we are taught mathematics for things like this, and many philosophers also use mathematics.

Algorithms are another way of formalizing ideas and are more appropriate for the expression of procedures. As John Mayfield says in his recent book, "It is just not possible to write a simple formula for a soufflé."[10] For that we need a recipe, a procedure, an algorithm. An intelligence test is a measuring procedure.

Philosophers do not complain about the use of mathematics but about its overuse. For example, for the sake of a beautiful formula, some scientists oversimplify a phenomenon. But even if this is the case, philosophers can always ask questions about what has been left out of a formal definition. These uncovered phenomena can again be addressed by scientists. This is a very healthy way of making science advance.

But even if a theory is philosophically unpleasant, an indicator of happiness, intelligence, or neuroticism is meaningful or not according to its predictive or explanatory power. The question is whether any indicator, alone or together with others, can predict some other human phenomena more reliably than other indicators or competing theories.

For instance, IQ scores do not only measure how ingenious one is at finding a solution to a problem; they also have a very strong predictive power for many other aspects of life, such as education level, socioeconomic state, fertility, health, and even myopia, to name a few.

## 7. On the relationship between intelligence and humor

A.P.:   Together with anticipation, humor is considered to be one of the most advanced human forms of intelligence.

Scott Adams, with his habitual irony, published a blog post recently with the title "The Illusion of Intelligence," which is pure provocation.[11]

Apparently it is a humorous piece, but it is much deeper than it appears. He says that it is clear why computers are still not intelligent: "So what parts of [human] intelligence are computers failing to duplicate? Answer: The parts that only *look* like intelligence to humans but are in fact just illusions."

The provocative sarcasm of Adams (even though he says at the beginning that he is not speaking in jest) has helped me formulate another question.

Do you think that there may be parts of human intelligence that cannot be replicated in any way, and that to create an intelligence one would have to take a different path to imitating human intelligence?

J.H.-O.:   The interesting point of Adams's post is that he turns the AI effect upside down. The AI effect refers to the phenomenon that whenever AI solves a problem, its mystery disappears, and we no longer consider it part of intelligence. Adams says that whatever remains to be captured by AI is not intelligence but an illusion. Basically, I disagree with both views. The AI effect happens because we construct AI artifacts that automate one specific task, such as driving a car, which, if done by humans, would require intelligence. However, we find a way of solving the specific problem without intelligence. The key issue is that in order to find solutions, that is, policies, for a wide, unexpected, range of tasks, the task-specific approach is doomed to failure. This is why a new approach to AI is based on systems that are able to learn to do new tasks, to acquire the policies by interacting with a world, instead of being programmed to do so. With this more modern view of AI, based on machine learning, both the AI effect and the intelligence illusion disappear.

Regarding the replicability of human intelligence, I have a rather mechanist position here: I do think that a brain can be replicated completely.

I have not seen any piece of evidence that suggests this should not be the case.

Actually, replicating the brain is a possible way toward AI, and many scientists are taking this path to reproduce intelligence. The result of this path can lead to the understanding of how human intelligence works, and has many therapeutic applications. Also, from this understanding, future technology may create variants and even migrate the mind to other milieus.

However, it is for me more fulfilling if we can understand what intelligence is in general, in a computational way. In the long term, using an approach that does not depend on human intelligence, we can also adjust the hardware to the best possible way of achieving some cognitive abilities without going through low-level neuron emulations.

What I'm not sure of is which pathway will prove to be faster. Perhaps it just depends on how much money is invested in each of them, or a hybrid of both, which seems to be the one providing the most promising results.

### 8. Are there universal ingredients in what we call intelligence?

A.P.: Returning to your research, I would now like to ask you a question that draws on the title of a scientific work. How "universal" can an intelligence test be?

J.H.-O.: We don't know the answer yet, but my opinion is that we must be able to create tests that are valid for humans, nonhuman animals, and many kinds of machines. However, there is a general rule about measurement: the wider the set of objects one wants to measure, the more difficult it is to get a reliable measurement. Also, there is another finding when one analyzes this question: universal tests must be adaptive.

In any case, in light of the cosmic variety of animals and machines one can think of, we cannot persist much longer with thousands of specific tests depending on the kind of subject. This would be like having hundreds of thermometers depending on the day or place they are to be used.

A.P.: Michail Bletsas says in another dialogue in this book that a nonbiological intelligence is sure to arise in this century. The astrophysicist Sara Seager also says in another dialogue that "beyond the Solar System, it is most likely that intelligence acts in 'nonbiological' ways."[12]

I don't know whether, on looking to the future in your field, universal psychometrics, you also see these ideas as probable.

How do you see the conundrum of intelligence when you look into the future of your discipline?

J.H.-O.:   Universal psychometrics is agnostic about this, but if these forms of intelligence exist, then the notions and tests of intelligence and other abilities should be applicable to them, provided we can find an appropriate interface.

About the timing on Earth, I'm inclined to think that Michail is right, although I would require a more precise statement about the expected psychometric profile of such nonbiological intelligence. That is another reason why we need more precise measurement instruments. Any serious prediction must clearly state the instrument that is going to be used and which level is predicted for that instrument. Otherwise we cannot even make a falsifiable guess. In any case, I prefer to paraphrase Steven Pinker and say that human-level AI is still the standard fifteen to twenty-five years away, just as it always has been.[13]

Finally, about extraterrestrial intelligence, my understanding is that once a high degree of intelligence appears biologically in any corner of the universe, the exploration of nonbiological milieus and the improvement of intelligence will take place no later than a few million years, which is an instant in cosmological terms. However, it is difficult to tell how empowering or self-destructive intelligence and especially super-intelligence can be, in order to say whether the current moment of Earth is unstable and as singular as we may think. Intelligence is unpredictable, and I guess that this length of a civilization is the most uncertain factor in the Drake equation.

A.P.:   The future of science is usually full of surprises and illusions, but recently in relation to AI, there are also those who see several threats, among them none other than Stephen Hawking, Elon Musk, and Bill Gates, who have even signed a manifesto.[14] You will surely have heard of the polemic on the ethics of robots and the possible (probable for some) threat from AI, personified in the form of "lethal autonomous weapons."[15]

Do you think that AI can become a threat to humanity, as several leading personalities in the technological world claim?

J.H.-O.:   *Homo homini lupus est*—a man is a wolf to man. By that I mean that many, if not all, of the AI-related risks that we will face in the following decades are political; their roots lie in humanity rather than in technological or scientific progress. Machine intelligence will be dangerous if it is concentrated and controlled by a few hands or by a few companies. For instance, there is the possibility that the current oligopoly of money, knowledge, and resources is transformed into an oligopoly of intelligence. This can have a tremendous impact.

Today, inequalities in the world are extreme, despite intelligence being well distributed. We may imagine how much the problem can be aggravated if this intelligence distribution is altered significantly. Indeed, the race has already started and some companies are hiring the smartest guys on the planet to develop superintelligent machines. It is certainly too much responsibility in a few hands, even if I prefer these hands to those of many political and any religious leaders.

I might be wrong, as this enters the realm of futurology and the ethically abominable for some, but I think that rights must be linked to the manifestation of certain cognitive abilities. This is vindicated by the Great Ape Project and the Nonhuman Rights Project.[16] I do not see why we should not have the same view for intelligent machines in the future. But of course, this inclusive view would require a complete overhauling of what democracy is, new regulations about the growth of artificial populations, the concentration of intelligence, or universal access to cognitive enhancement, and many, many other things we can hardly foresee.

As a result, the recent emphasis on ethical AI and its risks is welcome, including both a technical and political analysis; it is never too soon to think about all this. I would just ask that, if possible, we were less anthropocentric. Humans are not that special.

A.P.:    Thank you very much, José.

J.H.-O.:    Many thanks; it's a pleasure for me.

## Notes

1. He is the author of *The Measure of All Minds. Evaluating Natural and Artificial Intelligence* (Cambridge: Cambridge University Press, 2017). http://www.cambridge.org/us/academic/subjects/computer-science/artificial-intelligence-and-natural-language-processing/measure-all-minds-evaluating-natural-and-artificial-intelligence.

2. José Hernández-Orallo, David L. Dowe, and Victoria Hernández-Loreda, "Universal Psychometrics: Measuring Cognitive Abilities in the Machine Kingdom," *Cognitive Systems Research* 27 (2014): 50–74, http://www.sciencedirect.com/science/article/pii/S1389041713000338.

3. Michail Bletsas participates in dialogue 16.

4. José Hernández-Orallo, "Beyond the Turing Test," *Journal of Logic, Language and Information* 9, no. 4 (October 2000): 447–466, http://link.springer.com/article/10.1023/A:1008367325700.

5. David L. Dowe and José Hernández-Orallo, "IQ Tests Are Not for Machines, Yet," *Intelligence*, 2012, http://www.sciencedirect.com/science/article/pii/S0160289611001619.

6. Javier Insa-Cabrera, David L. Dowe, Sergio España-Cubillo, M. Victoria Hernández-Lloreda, and José Hernández-Orallo, "Comparing Humans and AI Agents," *Lecture Notes in Computer Science* 6830 (2011): 122–132, http://link.springer.com/chapter/10.1007%2F978-3-642-22887-2_13.

7. Robb B. Rutledge, Nikolina Skandali, Peter Dayan, and Raymond J. Dolan, "A Computational and Neural Model of Momentary Subjective Well-Being," *PNAS* 111, no. 33 (2014): 12252–12257, http://www.pnas.org/content/111/33/12252.abstract.

8. Javier Sampedro, "La fórmula de la felicidad," *El País*, "Cultura," http://cultura.elpais.com/cultura/2014/08/06/actualidad/1407351595_968289.html.

9. José Hernández-Orallo and David L. Dowe, "Measuring Universal Intelligence: Toward an Anytime Intelligence Test," *Artificial Intelligence* 174, no. 118 (December 2010): 1508–1539, http://www.sciencedirect.com/science/article/pii/S0004370210001554.

10. John Mayfield, *The Engine of Complexity: Evolution as Computation* (New York: Columbia University Press, 2013), 33.

11. Scott Adams, "The Illusion of Intelligence," blog post, *Scott Adams' Blog*, July 28, 2014, http://blog.dilbert.com/post/103051144811/the-illusion-of-intelligence.

12. Michail Bletsas participates in dialogue 16, and Sara Seager participates in dialogue 3.

13. Steven Pinker, "Thinking Does Not Imply Subjugating," in "2015: What Do You Think about Machines That Think?," an online discussion, *Edge,* 2015, http://edge.org/response-detail/26243.

14. The opinions of these eminent scientists are rendered in "Stephen Hawking: 'Transcendence Looks at the Implications of Artificial Intelligence—but Are We Taking AI Seriously Enough?'" *The Independent*, May 1, 2014, http://ind.pn/1i3aWlU; "Elon Musk: Artificial Intelligence Is Our Biggest Existential Threat," *The Guardian*, October 27, 2014, https://www.theguardian.com/technology/2014/oct/27/elon-musk-artificial-intelligence-ai-biggest-existential-threat; and Aaron Mamiit, "Bill Gates, like Stephen Hawking and Elon Musk, Worries about Artificial Intelligence Being a Threat," *Tech Times,* January 29, 2015, http://www.techtimes.com/articles/29436/20150129/bill-gates-like-stephen-hawking-and-elon-musk-worries-about-artificial-intelligence-being-a-threat.htm.

15. "Lethal Autonomous Weapons Systems: Future Challenges," *CSS Analyses in Security Policy* 164 (November 2014), http://www.css.ethz.ch/publications/pdfs/CSSAnalyse164-EN.pdf. The Center for Security Studies (CSS) is located at ETH

Zurich. See also Matthias Englert, Sandra Siebert, and Martin Ziegler, "Logical Limitations to Machine Ethics with Consequences to Lethal Autonomous Weapons," Technische Universität Darmstadt, http://arxiv.org/pdf/1411.2842v1.pdf.

16. The Great Ape Project, founded in 1993, advocates a "Declaration of the Rights of Great Apes" that would afford nonhuman apes legal protection. The Nonhuman Rights Project seeks to change the status of at least some animals from possession to person. The organization has filed lawsuits on behalf on nonhuman animals held in captivity.

# 32   Touching the Soul of Michelangelo

## Gianluigi Colalucci and Adolfo Plasencia

*I believe that art, throughout the path that humanity has taken, is its highest, its most spiritual expression. Humanity has made extraordinary technical scientific progress, but if there isn't a spiritual element to accompany man on that path, it's an incomplete process.*

*In Europe and the East, the concept of "original" is not the same. An Eastern temple can be constantly remade and continue to be regarded as original. For us in the West, something that has been made and is then modified is no longer original.*

*An artist should also wait until his or her doubts are resolved. When those doubts are cleared up, he or she can begin to work.*

*—Gianluigi Colalucci*

Gianluigi Colalucci is technical director of the restoration of medieval frescoes of the Monumental Cemetery of Pisa, consultant for the restoration of paintings in the Vatican Museums, and former Chief Restorer of the Vatican Museums. He received a diploma in restoration from the Central Restoration Institute in Rome and undertook further professional work in the Restoration Laboratory of the National Gallery of Sicily. From 1979 to 1995 he was Chief Restorer of the Vatican Laboratory for the Restoration of Paintings, Papal Monuments, Museums, and Galleries. He spent fourteen years restoring Michelangelo's frescoes in the Sistine Chapel and has also restored works by such artists as Raphael, Titian, Leonardo, and Giotto. He has served as adviser to the Prado Museum of Madrid, and also consultant for the restoration of the ceiling of the Church de los Santos Juanes in Valencia, both of them, in Spain.

Among his publications are (with F. Mancineli and others) *Michelangelo: La Cappella Sistina. Rapparto sul restauro del Guidizio Universale,* and (with F. Mancineli and N. Gabrielli) *Michelangelo, La Cappella Sistina. Rapparto sul restauro della Volta.*

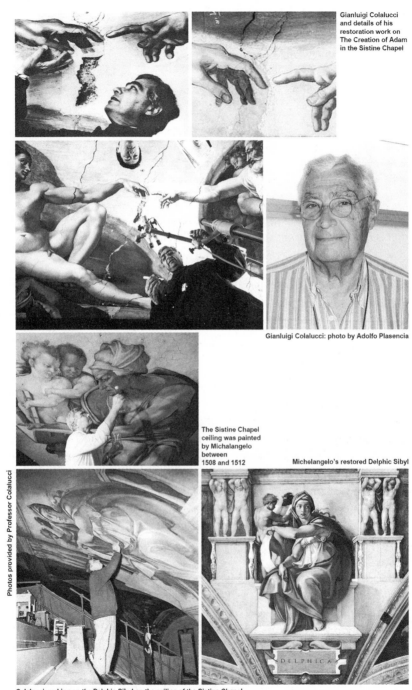

Gianluigi Colalucci and details of his restoration work on The Creation of Adam in the Sistine Chapel

Gianluigi Colalucci: photo by Adolfo Plasencia

The Sistine Chapel ceiling was painted by Michalangelo between 1508 and 1512

Michelangelo's restored Delphic Sibyl

Photos provided by Professor Colalucci

DELPHICA

Colalucci working on the Delphic Sibyl on the ceiling of the Sistine Chapel

Gianluici Colalucci. Photograph by Adolfo Plasencia, and courtesy of G.C.

He is a member of the International Institute of Conservation, Cultural Art Association Educatrice Museum of Roma, and was named a Knight of the Equestrian Order of San Gregorio Magno by Pope Paul VI.

In addition to restoration of the *Volta* and *Last Judgment* frescoes in the Sistine Chapel and other works under the Vatican's purview, he has restored frescoes by Raphael in the Room of the Oath of Leo III in the Vatican, and frescoes of the Scrovegni Chapel by Giotto in Padova; Caravaggio's *St. Jerome* canvas in Rome; three Titian frescoes in Scuola del Santo in Padova; frescoes by Andrea Mantegna in the Basilica del Santo, S. Antonio and S. Bernardino, also in Padova; the *Venus and Cupid* table by Luca Cranach; and the *San Girolamo* table by Leonardo da Vinci.

Adolfo Plasencia:   Professor Colalucci, thank you very much for receiving me.

Gianluigi Colalucci:   It's a pleasure.

A.P.:   When you were named Doctor Honoris Causa by the Universidad Politécnica de Valencia (UPV) , the president at that time, Justo Nieto, began his speech by saying:

Art and artists have managed to make the path taken by humanity more legitimate. From more than twenty thousand years ago, from Altamira to Leonardo, from Michelangelo to Picasso, the anguish, contradictions, and rebelliousness of those giants has left us with a heritage that we have to look after, as, through it, we can constantly recognize ourselves.[1]

Professor Colalucci, do you think that art and artists have really managed to make the path taken by humanity more legitimate?

G.C.:   Yes, surely. I believe that art, throughout the path that humanity has taken, is its highest, its most spiritual expression. Humanity has made extraordinary technical scientific progress, but if there isn't a spiritual element to accompany man on that path, it's an incomplete process. Science and technology alone are not enough. Art is absolutely indispensable for humankind, although it is true that humans themselves have created it.

A.P.:   In the speech following those words, the resident referred to you as an artist-scientist. I suppose you agree with that definition. What alchemy provides the balance for these two facets of creators? And how are the two, art and science, balanced in your research and work?

G.C.:   I don't think of myself as an artist. I think a restorer must be a person who understands art, who knows about art. The artist is a creator. The restorer is not a creator. He or she is a person who has the basic technical and scientific baggage to preserve art and on occasion to restore it. There are two "souls" in the restorer: the art-loving soul, which knows how to interpret it, and the technical soul, which knows when and what has to be done technically to conserve art.

A.P.:  Let's talk now about your personal universe. You have touched the surface, and I imagine sometimes the inside, of the most highly valued artwork in the world, especially the work of Michelangelo, but also the works of Leonardo da Vinci, Giotto, Raphael, Andrea Mantegna, Caravaggio, Titian ... with your own hands, and I want to ask you:

What feeling does this universal artwork arouse in you on touching it? Do you feel different from the rest of us, who can only look but not touch?

Do you have the feeling that under that so extraordinary surface, there is an "inside," an inside that contains the soul of its creators?

G.C.:  Touch in restoration is very important, but it is a question mainly of technique. When one cleans a picture, touch is fundamental because, with its help, one understands what is happening on that surface. For example, and to continue talking about Michelangelo, the frescoes had a very rough surface before they were cleaned because it was a surface where there were materials that did not belong to Michelangelo's paints, materials that had been deposited and were gradually consolidated to form a rough surface and, I would go as far as to say, a surface that was dirty to the touch. When the frescoes were freed of those foreign substances, the paintings of Michelangelo shone through. We could then see that they had an enamel surface, like that of porcelain, something that is the consequence of the artist's technique, showing us that he had used mortar and also how he had painted them. With the passage of time, mortar oxidizes, consolidates, and becomes very compact, and this surface, if one looks at it side on, is very flat and, like plaster, very smooth.

And one realizes that this is what the artist wanted because he is someone working to achieve that material. He is an artist who, also through this material, is sending his message, speaking his language. Obviously, material is only material, and before being worked it is not a work of art. How to transform the material into a work of art is the ultimate mystery of art. How do we know when a material can be converted into a work of art? We can only know it there, where it has taken place.

A.P.:  Is there something spiritual inside, behind the artistic surface?

G.C.: Inside, yes, because it is that which the artist transmits. It's similar in music. Music needs a piano to make you feel what the player wants you to feel. The quality of the piano, which is a technical means, will also have an influence. It may be in the hands of Michelangelo or it could also be in the hands of an amateur who only paints on Sundays. The colors are the same, but the results may be very different. This forms part of all that we call art. I continue to believe it is an absolute mystery.

A.P.: Professor Colalucci, you have also been called a "caresser" of art. When one can approach the most sublime works of art without encountering any barriers at all, as in your case, do you "caress" those works with your hands or also, close up, with your eyes?

G.C.: Although we see with our eyes, we could say that the hand also sees, because observation can then be accompanied by a reasoning that tries to understand the details and feel of a color, of a figure, and so on, something that is much fuller.

A.P.: Normally, over the years, over the centuries, the accumulated smoke from the candles darkens the original colors of frescoes and old paints, as has happened in the Sistine Chapel.

What do you feel when you begin to clean that darkened surface, and suddenly the original colors begin to shine through just as they were when the artist saw them when he had finished his work?

Do you think that the feeling this new sheen to the colors produces has the same effect as that felt by the artist? Do you think that is possible, or does the time difference mean that we see them in a different way?

G.C.: Materially speaking, the colors there are the originals, they are the same material. I think the difference is us. It is in us. If I had been one of those present at the unveiling of *The Last Judgment*, I would have been a man of the Cinquecento with no knowledge of art coming after that period because the fresco was, in that era, one of the latest works of art of that moment.

A.P.: Clearly, because at that time, in that moment, the work of Picasso, Goya, Kandinsky did not yet exist.

G.C.: There literally didn't exist anything of what you say. So much so that this *Last Judgment* by Michelangelo has served all the later generations of painters to make their own art, as it provided a great impetus for a grand transformation over the following centuries.

When we see this *Final Judgment* today, we also see in our mind's eye all the other works of art that other artists have painted—Picasso, Goya, up to the artists of today, with their so-called performances—and all those things that provide us with a diverse culture. We look at that painting in a certain way that probably brings to mind a whole series of experiences that we have had.

From this point of view, we can say that the painting is diverse, not materially but in the way we look at it today. Equally, the men of the sixteenth and seventeenth centuries saw it in accordance with their own time and experience. There's something I would like to add. The true work of art is that which in every century gives language to the people of that century, and that means it is a work that is capable of outlasting the centuries and becoming universal.

A.P.:   Professor Colalucci, as you know, there has been great controversy over how to preserve the great works of artistic and cultural heritage, but is this only a question of applying scientific criteria, or are there differences in interpretations and criteria depending on the dominant cultural-artistic currents or guidelines of restoration schools?

Is this a strictly scientific issue or can it be culturally interpreted in a different way?

G.C.:   Conservation—conserving things as they are—is a substantially technical matter, concerned with issues of environment, microenvironment, microclimate, and the like, whereas restoration isn't. Restoration is now a concept of state philosophy, and schools of restoration change, so there may be contrasts. Today, there is generally a uniform method of restoration throughout the world, although it's not only the material that is involved but the final intention. For example, in Europe and the East, in Asia, the concept of "original" is not the same. An Eastern temple can be constantly remade and continue to be regarded as original. For us in the West, something that has been made and is then modified is no longer original.

A.P.:   I said that because here, in this region, during the Baroque era, certain ecclesiastical powers considered, for example, Gothic art as being old, poor, and outdated, so they destroyed many elegant Gothic towers and replaced them with ornate Baroque towers, as you well know. These cultural interpretations often depend on power, on who is governing at a particular moment.

G.C.:   It depends on the concept that a society has of a work of art. We were talking about the past. In the past, they didn't have the problem we have today. The works that we have decided to maintain for history today, the past did not worry about. The rooms in the Vatican where Raphael's paintings are displayed were before that covered with frescoes by Piero della

Francesca. These were destroyed with hardly a second glance so that new ones could be painted. Piero della Francesca was no longer liked, and the artist who seemed more interesting at that time was Raphael. That is linked to the concept of each age, and today we are linked to "conserving."

A.P.: Returning to your own field, restoration, and your work and research, what has most impressed us is your relationship with the work of Michelangelo Buonarroti, who, like Leonardo da Vinci, is one of the preeminent figures in universal art. Let's talk now about the Sistine Chapel. You were one of the restorers of that chapel and also a restorer of that majestic fresco within it, the *Last Judgment*. You told me that you spent a long time there, many a night alone in front of those paintings, in front of those giants that Michelangelo created.

I'd like to ask you: what did you feel there, alone in front of Michelangelo's giants?

Did you feel small? When alone, did they humble you? Does the scene make you feel afraid? Does it make the spectator's soul float in the air?

What emotions are aroused in that solitude?

G.C.: The initial feeling is one of realizing that you are a privileged person to be there where others are not. I believe that when the restorer works on a work of art for a certain period of time, he or she becomes the patron of that work of art and has no need to collect paintings, because for that brief period of time spent on restoring them, the works actually belong to them. I devote myself to this job, which is a privilege, because I have always thought that I wanted to make a direct contribution to the world of art. It is morally uplifting. It's as if a string quartet plays for you alone, and not for the three hundred other people who are also listening to it with you. You are there alone, and it plays just for you. It's extraordinary.

Do I feel small faced with these paintings? Always. It's good for a restorer to feel small in front of an artist, even before those of a lesser stature than Michelangelo. I should point out that a restorer serves the artist, and this must be so.

A.P.: Do these paintings make one feel small, more humble?

G.C.: Yes, it's necessary to feel humble. You don't have to be in front of the giants of art: the restorer always has to be humble.

A.P.: Now let's employ a metaphor using a detail I know from the *Last Judgment* fresco. There is, in one of the figures of the *Last Judgment*, a very long leg, the complicated profile of which Michelangelo resolved in a single gesture, as if he had no doubts about doing it.[2]

Before you gave your speech, there was talk about the relationship between the greatness of artists such as Michelangelo and the anguish, doubt, contradictions, and rebelliousness of those artists—Michelangelo left Florence to escape the pressure of Pope Julius II, who had commissioned a tomb from him, for which he sculpted another great work, the famous *Moses*. He fled so as to be able to work in peace.

You've had the opportunity to see many details, like that leg, close up. That will have let you analyze and reflect on them.

From what you have seen, can you imagine the doubts, anguish, and contradictions that the artist must have felt when he made the paintings? The way that leg is painted would seem to indicate the opposite of doubt.

G.C.:   When one is very close to the work, one sees the most "material" part of its execution, because in the work, its execution is the development of a creative person, the creation the result of undertaking it. Making a fresco that has to be placed on a wall, for example, involves the development of technique. Yes, when one is very close, in front of it, one sees the enormous effort of the execution technique that the artist has had to make to transfer his or her ideas there, so that its design functions safely. But later, when the restoration is being done, one has to be like a lithographer, one has to be well positioned before the whole. The effort involved in the execution is also hard work. When one does restoration work, one sees all the traces, all the fingerprints of the effort of the artist.

A.P.:   In these great works, the artists prioritize certainty rather than doubt, don't they? They give the idea of having chosen one final "path" from the various possible ones, don't you think?

G.C.:   If I've understood the question correctly, I would say that all that forms part of the technique. First of all, in the field of restoration, before making any decision, we have to develop all the necessary techniques to know exactly what should be done. Once we are technically prepared and have decided what to do, there's not going to be much doubt, because if one begins to work while having doubts, things don't work out well. An artist should also wait until his or her doubts are resolved. When those doubts are cleared up, he or she can begin to work.

A.P.:   Do you think that in art, out of all the possible paths, there is always one that is better?

G.C.:   No. Or maybe yes, in the case of artists like Michelangelo, who had a great influence on many other artists who came along later. But in art it is difficult to find the best path, one that is higher or lower. Think about music: who can choose between Mozart and Bach? They are different and both extraordinary.

A.P.: Let's talk now about the present and what is happening now in the world. This is a period of crises and change. There is in this book a conversation with Yung Ho Chang, a professor of architecture at MIT.[3] He told me two things that are of interest here. The first is that the era of grand egos in architecture has finished; and the second is that he is convinced that cutting-edge architecture can be done through knowledge derived from the past. His dialogue is titled "Going Forward in Architecture by Looking Back." Perhaps the market, the large companies, the short-sightedness of present life make us forget that wisdom.

Do you think, Professor Colalucci, that something similar has happened in the world of art and that we have put the accumulated wisdom of the past, many centuries of art, aside?

G.C.: The question is very complicated and needs thinking about. What I understand is that at the moment, a revolution is taking place, and we are abandoning the past. There is a transition, not only of century but also of millennium. It's inevitable that, after hundreds of years, someone will arrive here, see this, and say no, I don't know what this is. For me it's difficult to know what we should do and what will be done. As far as the future is concerned, any future that I've imagined has turned out to be ridiculous once I've seen what really happened. So I don't know. However, it's usual now to run away from anything that seems too old. It's as if you put your foot in an ants' nest and everybody begins to flee. It's true, for example, that today in contemporary art things are presented that are quite strange, quite unexpected.

A.P.: The philosopher of science Javier Echeverria, whom I also have a conversation with in this book, told me that he believes that inventing something is not an instantaneous event, a momentary stroke of genius, but a process, and, as Picasso said, "Inspiration exists, but it has to find you working."[4]

G.C.: Yes, it's true. Creation is a process, something that comes from a brain in which many things have entered beforehand.

A.P.: I'd like to mention something else that is going on now. The restoration work on the frescos in the Sistine Chapel took fourteen years to complete and was financed by Nippon Television, the biggest private TV broadcaster in Japan. In exchange, they have the exclusive rights to the reproduction of images of the different stages of the restoration over all of those years.

The Sistine Chapel belongs to all humanity and forms part of our common heritage, but in exchange for finance, the said company possesses the image reproduction rights for decades to come.

Do you think this is a good method and compatible with heritage? Or do you think that everything that involves recovering universal heritage should be paid for out of public funds? Finally, what was the relationship like between you scientists and the company in the case of the Sistine Chapel?

G.C.:   Yes. But we have to remember that, apart from the Vatican, which is a very small state, we often do not have the capability of financing so many great works of art. We restorers earn the same whether we are restoring Michelangelo's frescoes or restoring a painting that was in the warehouse. For us the stipend is always the same. Nippon TV of Tokyo made this offer and in exchange was granted the rights to the photographs taken during the process. That's normal. World heritage is so immense, so extensive, and so in need of intervention that the money of a single state is insufficient, and so, if there is a sponsor, they can intervene. They shouldn't ask for too much, but the sponsor can be granted things in exchange. Private money for restoration is very important.

A.P.:   Does a company have to be sensitive in order to help to maintain the common heritage of all, as in the case we are speaking about? A company has to be sensitive to art and culture, doesn't it?

G.C.:   There are companies that dedicate certain funds to culture. Companies have to be careful not to dirty the planet, and it would also be a good thing if a part of what they earned was dedicated to culture.

A.P.:   I'd like to ask you one more question, and you can decide for yourself whether to answer it or not.

Today, a different concept of the global world is beginning to take hold in which many people from other continents also consider any important work of world heritage as their own. We can see this in the destruction of the giant Buddhas of Bamiyan in Afghanistan, which, in 2001, after having survived 1,500 years, were destroyed by the Taliban Islamist government.

Parallel to this is another debate as to whether, for example, the friezes of the Parthenon in the British Museum in London should be returned to Athens, the city they came from.[5] Do you think this type of art should be left where it is, that it is better to leave things of the past as they are, and to conserve them there as best as possible? Or do you agree with the people who think that they should be returned to their original location. What's your opinion?

G.C.: To that last question, I'll have to give a rather inarticulate response.

He who loves art, feels it to be his own, whether African, Japanese or Valencian. When I am in Valencia in front of a work of art, I don't ask myself if it is Valencian, European, or Japanese because it belongs to me. I've been in Japan in a Zen temple, but I feel it's mine. I didn't feel at all foreign there. And the same is true anywhere, for any part of the world. If I go to London and see the Parthenon sculptures, I don't think about where they came from. There's no need for them to be there or anywhere else for me to feel they are mine. However, returning things to their former site forms part of the negative side of our society. The more global society becomes, the more small societies close themselves off and withdraw into themselves. There are examples in Spain and in Italy too.

I believe it's nonsense because I cannot live in a global world and keep closing myself off more and more. If the culture of a small center expands, all feel it, even those far away. That is more important than the money to be gained from tourism, but in the end, it boils down to the same thing, economics. I consider a work of art as being for everyone. In all lands, all works of art belong to everyone, to the human condition.

A.P.: Thank you very much. It's been a privilege talking to you, Professor Colalucci.

G.C.: You are welcome. Thanks to you. It's been a pleasure.

## Notes

1. Words by Justo Nieto, the president, at that time, of the Universidad Politécnica de Valencia (UPV) in Investiture as Doctor Honoris Causa by the UPV, of Professor Colalucci in *Doctores Honoris Causa por la Universidad Politécnica de Valencia,* ed. Universidad Politécnica de Valencia (Valencia: UPV, 1998), 260, https://www.upv.es/organizacion/la-institucion/honoris-causa/gianluigi-colalucci/index-es.html.

2. Profile of a leg from a figure in the fresco painted by Michelangelo in the *Last Judgment*: https://www.upv.es/organizacion/la-institucion/honoris-causa/gianluigi-colalucci/images/colalucci_cuadro6.jpg.

3. Yung Ho Chang participates in dialogue 10.

4. Javier Echeverria participates in dialogue 28.

5. The British Museum, "The Parthenon Sculptures," press release, n.d., http://www.britishmuseum.org/about_us/news_and_press/statements/parthenon_sculptures.aspx.

# IV  Epilogue

As George Steiner says in his *Grammars of Creation*, "More than *Homo sapiens*, we are *Homo quaerens*, the animal that asks and asks,"[1] maybe an animal that pushes the limits of language and images. He goes on to ask, "Has any painter invented a new color?"[2] Steiner's point is that no artist, however gifted, is an absolute creator who conjures new worlds out of nothing. We start with what has been given and all that has been created before. As humans we are constrained by the limits of our own senses in our ability to perceive the world. Steiner suggests that the humility that accompanies this realization is essential to the project of art and to the creation of new knowledge more generally.

Certainly this is true in science. Isaac Newton's words about standing on the shoulders of giants are by now a cliché: in science, all new knowledge builds on a long legacy of hard-won discovery. Are artistic creation and scientific discovery two sides of the same coin or different facets of the same transparent whole? It is hard to deny that the genius and talent of individuals have contributed to scientific achievement, as they have to major achievements in art. But Steiner recognizes a difference between the overlapping magisteria. Scientific development draws on a tacit, collective ecosystem of knowledge with its own distinctive weather. In this ecosystem, disruptive "flashes of lightning" can spark insight and clear the way to new knowledge in specific areas. Does the same happen with artistic creation? And if not, why not?

One answer may be that the scientific method is distinct in its skeptical approach. All scientific knowledge is temporary and contingent. No solution can be permanent because nothing can be definitively proved for all time and all circumstances.

Still, art and science share many fascinations. They include mathematics, which, as Goethe pointed out, has "the completely false reputation of yielding infallible conclusions."[3] Science uses mathematics as its natural language, and art has had its own passionate (if turbulent) relationship with the apparent harmony and symmetry that geometry provides. The same quest for truth might also be said to inspire art and science, although the paths are different.

The scientific method as we understand it today owes much to Leonardo da Vinci, one of the greatest artists of the Western tradition. Is it still possible to be an artist and a scientist in the way that Leonardo embodied, or have we traveled too far in each realm, with too many divergences along the way? Have we developed two entirely distinct methods of approaching the unknown that are nonoverlapping magisteria, as some have claimed?

The manuscript edition of Leonardo's *"A Treatise on Painting,"*[4] which he wrote at the beginning of the sixteenth century, begins with an inquiry on whether painting is or is not science. Leonardo's manuscript states that "no human investigation can be called real science if it cannot be demonstrated mathematically." He goes on to say, "And you will tell me that sciences have their beginning and end in the mind, they participate in the truth; that I will not concede to you; I reject it for many reasons; the first because, in such discourses of the mind, experience is not gained without some certainty being produced." How interesting it would have been to hear Albert Einstein defend his known counterarguments on this before Leonardo. Impossible! But possible to imagine, as an artist would say. Five centuries after Leonardo penned those arguments, the projects of science and art continue to operate as distinct pursuits, but with crisscrossing paths. One does not prevail over the other; often they are complementary and necessary to one another even as they attack, annoy, and delight each other.

In the following pages there are reflections of an artistic-scientific nature on this timeless conflict that remains seductive and bloodless, even when it is ill-humored—a multifaceted reflection on the intersections and overlaps between the arts and sciences. We will encounter thoughts on geometry and its absence, the metaphysics of art in the work of Mark Rothko, the perhaps impossible solution to equations involving truth and beauty, and the eternal dilemma of order versus chaos. The following reflections also try to clarify whether it is possible for art to change the past, and conclude with an exploration of science, art, and scientific reductionism.

## Notes

1. George Steiner, *Grammars of Creation* (London: Faber and Faber, 2010), 22.

2. Steiner, *Grammars of Creation,* 25.

3. A. N. Mitra, "Mathematics: The Language of Science. Introduction," arXiv.org, last revised February 28, 2012, http://arxiv.org/abs/1111.6560v3.

4. Leonardo da Vinci, *Treatise on Painting,* 1452–1519, trans. and annotated by A. Philip McMahon, with an introduction by Ludwig H. Heydenreich (Princeton, NJ: Princeton University Press, 1956).

## 33 Geometry of a Multidimensional Universe: Weightless Art and the Painting of the Void

José María Yturralde and Adolfo Plasencia

*For me, geometry is a kind of mental field that permeates everything and is the basis and sustenance of my approach to reality.*

*The response of art to reality ends up being a kind of 'enlightenment' that weighs us down, fills us up in a very special way, and escapes all scientific analysis.*

*Both artistic abstraction and quantum physics do not require "representation." It is as absurd to ask oneself what the White on White painting represents as it is to ask oneself, in quantum mechanics, what the electron looks like.*

—*José María Yturralde*

José María Yturralde is a Professor of Fine Arts, Facultad de Bellas Artes de Valencia, UPV, and Full Academician of the Real Academia de Bellas Artes de San Carlos, Valencia. He received his doctorate in fine arts from the Universitat Politècnica de València (UPV). In 1968, while a scholar at the Calculus Centre of Madrid University, he started his first computer works, which were followed by his first exhibition of computable forms in 1969.

His works have been exhibited in numerous galleries, both as solo exhibitions and in group shows, often as an invited artist. In 1996, during a one-month stay at UNAM, Mexico, he lectured and conducted various actions around *Flying Structures*. For the First European Nature Triennial in Dragsholm Slot Odsherred, Denmark, in 1998, he installed the pieces *Frame the Forest, Hyperweb,* and *Flying Structure.* Recent solo exhibitions include those at Javier López & Fer Frances, Madrid, in 2016; the Centro de Arte Contemporáneo, Málaga, in 2015; *Horizontes,* Galería Miguel Marcos, in 2014; *The Absent Space,* Galeria Mário Sequeira, Braga, in 2010; and *New Paintings,* Gering & López Gallery, New York, in 2008.

376 José María Yturralde and Adolfo Plasencia

José María Yturralde. Photograph by Adolfo Plasencia.

Awards have included the Ibizagràfic Award (1972, 1976), the "B.G. Salvi" and "Premio Europa" awards, Ancona, Italy (1972); a grant from the Juan March Foundation (1974) to support a stay at MIT, where he also became Research Fellow at the Center for Advanced Visual Studies; a Fulbright-Hays grant to attend the Salzburg Seminar in American Studies (1978); and the Alfonso Roig Award, Diputación de Valencia (1995).

He is author of the books *Estructuras 1968–1972: Series Triangular-Cuadrado-Cubos-Prismas, La cuarta dimension: Ensayo metodológico para la proyección geométrica de estructuras N-Dimensionales,* and co-author of *Hypergraphics: Visualizing Complex Relationship in Art, Science and Technology.*

Adolfo Plasencia: José María, thank you for receiving me here in your studio.

José María Yturralde: Thanks for coming.

## 1. MIT, art, and science

A.P.: One of your formative stages as an artist was when you attended MIT. Arriving at MIT, at the Center for Advanced Visual Arts and the surrounding campus, was for you like arriving at a huge source of discovery, invention, science and technology, mathematics, geometry, but also a place with machines, computers, improbable experiments, all in a physical and practical manner that you could take advantage of and use. I think that left its mark on you and guided your intellectual and artistic development.

Was the MIT you found there decisive for you?

Did you find there the meeting point of technology, art, and science?

J.M.Y.: When I went as a Research Fellow to the Center for Advanced Visual Studies at MIT, I felt immensely privileged and a combination of feelings and emotions that the well-known poem by Gustavo Adolfo Bécquer expresses and which I can't resist quoting:

Hoy la tierra y los cielos me sonríen,
hoy llega al fondo de mi alma el sol,
hoy la he visto …, la he visto y me ha mirado …,
¡Hoy creo en Dios!
(Today the earth and sky are smiling on me,
Today the sun has touched my soul,
Today I have seen her and she has looked at me.
Today, I believe in God!)

A.P.: There's a good metaphor for that mood you wish to express in the comment of an MIT student's hack (*Fire Hydrant Water Fountain*) on display in the Stata Center building, next to the Forbes Café.[1] It says: "Getting an Education from MIT is like taking a drink from a Fire Hose."

J.M.Y.: Really? I didn't know that. When I was there, the new Frank Gehry CSAIL building had not yet been built. Anyway, it wasn't actually there where I discovered that meeting point that you mentioned between art and technology. It started with my participation in the creation, in 1967, in

Valencia of the Antes del Arte group, which proposed reestablishing and updating a dialogue between the state of science at that moment and the basic "artistic theory" of those days.

We didn't then consider sciences as an abstract notion but as an arsenal of new data, methodologies, and strategies, essential to the understanding of human communication and most of all art. We hoped to overcome certain anachronisms and several of the more irrational aspects of creativity. We understood that intuition, the emotions, sensitivity, and the ineffable nature of art had to begin where the proven and experienced ended.

They were the processes involved before considering a work of art as Art, those that were of interest to us, at the same time as we tried to assimilate the very latest technology available to us to exemplify those ideas. I built various mobile, luminous artifacts using lasers and also industrial materials, electric motors, and other things.

My stay at MIT was a natural move to make. Of course, being in the most important scientific-technological university in the world was a tremendous leap for me, opening up so many possibilities for understanding the universe of knowledge.

It was perhaps the most important stage of my life as far as learning is concerned.

A.P.:  Do you think that the current interaction between people in the digital world and the Internet world, now deployed all over the world, is different from those past times, before digitalization?

J.M.Y.:  The Internet, as everybody knows, extends our access enormously and at the speed of light to a multitude of options and choices. Knowledge is just one of them. The vast amount of information now available presents a problem for the mind, but a very welcome one!

## 2.  Geometry and its absence

A.P.:  When we analyze the development of your work, it's clear there are various grand stages. At MIT you developed your "sky art," crossing tornadoes with the first lasers that science and technology at MIT provided you with, then you went on to ethereal flying structures, like that huge and brilliant octahedron that was floating for a number of days over the Venice Biennial in defiance of gravity, then to impossible geometries, and finally you have arrived at your present work, in which you oscillate between the painting of emptiness and that of horizons and eclipses, where there are no clear forms and where geometry and form are apparent more from their absence than by their presence.[2]

What do you remember of that development?

J.M.Y.:  I've never abandoned those notions related to the shapes of what we call geometry, which for me is a kind of mental field that permeates everything and is the basis and sustenance of my approach to reality. For example, space. The idea of space has changed so much for humanity, and for me also, over the last century.

In my work, I have never tried to illustrate an idea but rather to flow with it through painting in order to understand, or feel, access to that reality with love, passion, and "sentient" (emotional) knowledge, as Zubiri would say.

That idea of space (and geometry is an extraordinary conceptual tool for understanding it) has been and continues to be central to my artistic thinking, as the means to approach such classical questions as "Who are we?," "Where are we?," "Where did we come from?," and "Where are we going?" Painting and its basic expressive elements of form, light, color, space, time, surface, texture, auditory sensations, and so forth make up a language with which I try to gain access to an understanding of what those questions mean to me.

But there's a lot more. The more we question, the more complex things become, as illustrated by the famous response of Augustine of Hippo to the question, what is time?

"If you don't ask me, I know. If you ask me, I don't know."

I'd say that's a very artistic answer. The response of art to reality is of a similar kind. It ends up being a kind of "enlightenment" that weighs us down, fills us up in a very special way, and escapes all scientific analysis. It is not as if we haven't tried. In the 1960s at the Madrid Centro de Cálculo we tried to apply mathematical formulas to attain the maximum degree of aestheticism in a work of art. We based our research on, among other things, the ideas of the law of the mathematician Birkhoff, of *Gestalt* theory, whose formula contemplates:

$$M \text{ (degree of aestheticism)} = \frac{O \text{ (order)}}{C \text{ (complexity)}}$$

Or his laws of *prägnanz*:

"One form will be all the more *prägnanz*, the greater the number of axes it possesses."

From there, I moved on to *Impossible Structures*, where in some ways I was trying to break our perceptive experience by creating a certain sensorial versus meaning conflict, so as to catch a glimpse of a new way of conceiving space as a multidimensional body.

The flying structures[3] that I began at MIT in 1975 followed that attempt to establish several "attractors" of the attention, this time in the sky, that would lead us on to intuit different ways of contemplating space-time, the surrounding reality flowing with other more complex and exciting realities. Some of these structures, capable of flying, were a reflection on different hyperpolyhedrons.

Later the geometries of fractals and the ideas of quantum physics led me to "dilute" the classical forms of polygons and polyhedrons to arrive at the notion of emptiness, the infinite, the "transinfinite," the origin and end, the possible starting point of everything. Or, as the writer Marguerite Yourcenar asked herself:

"Quoi? L'Éternité."[4]

Astronomy has always interested me as the sublime backdrop to immense, ancestral knowledge. We human beings look to the stars when we ask ourselves who we are. There, possibly, we will find the answer. But it is not the questions or the answers that most interest me in this process of my own life but the possibility of flowing with nature by accepting that we are part of it. For me, perhaps, the most complex and intense possibility is becoming aware of the universe.

### 3.   On the metaphysics of art and Mark Rothko

A.P.:   You've just said that painting is your "specific medium." I know that one of the most important references of contemporary art at present is the painting of Mark Rothko. There's a sentence in a famous letter to the *New York Times,* written by Adolph Gottlieb and Rothko in 1943, that reads: "There is no such thing as good painting about nothing."[5]

Rothko also said, "Pictures must be miraculous."[6]

What attracts you to the painting and philosophy of Mark Rothko?

J.M.Y.:   Mark Rothko produced essential painting, of a superior order—even sacred, if I may say so.

It is for me a source of continuous discovery, a kind of door that, on opening, offers us the mystery of everything, that imagined unified theory that scientists look for with such determination. His paintings are "attractors," places of meditation, culminating in the Rothko Chapel in

Houston, perhaps the most ambitious undertaking of the artist.[7] It was officially opened in 1971, just one year after the death of the painter, so he never lived to see it completed. He didn't think that certain works should end up in museums. His own work he preferred to see in relatively small places, without interferences, such as old churches that could be reused as places of art. The combination of those works, situated in an octagonal space, envelops the spectator in an atmosphere of mystery and serious, profound emotion. For me the high degree of spirituality, devotion, and personal enlightenment is similar to that experienced in another great space of a similar kind, the Ryōan-ji in Kyoto, a Zen garden of the fifteenth century that provokes similar aesthetic and spiritual tensions. It is a not a very big space. Its rectangular shape contains fifteen rocks surrounded by a small circle of moss on a fine gravel floor that is raked every day. From any point on the rectangular perimeter there is always one rock that remains hidden. The important thing is the intense, almost physical shock you feel on entering it. In both cases, you feel a metaphysical vibration, feel immersed in the transcendental, in Rothko's paintings and in the Ryōan-ji.

Art "happens," as James Abbott McNeill Whistler said. The miracle happens and as such lends itself poorly to description or interpretation. It's emotional knowledge, living poetry, therefore. The cultured and precise philosophy of the painter, although providing us with the keys to understanding, in no way gets near to real pure contemplation faced with his work.

The North American painting of the era in which Mark Rothko lived, especially around World War II, was influenced by Oriental art and spirituality, the idea that emptiness is not empty but the origin of everything.

## 4.  Beauty ≠ truth

A.P.:  José María, I would like you to reflect on the relationship between beauty and truth. I am sure you will remember Keats's famous aphorism, "Beauty is truth, truth beauty,—that is all / Ye know on earth and all ye need to know."[8]

Some artists, you among them, also defend minimalism, the idea expressed in the famous phrase that Mies van der Rohe adopted from the Robert Browning poem, "Andrea del Sarto" ("The Faultless Painter"): "Less is more." However, not everyone agrees with that.

There's an article by the English science writer Philip Ball, on whose web page one can read: "Beauty is truth? There's a false equation."[9] Ball provides weighty arguments to back up the equation that crowns his essay: "Beauty ≠ truth."

In the text he quotes the debate between Niels Bohr and Einstein:

That was Einstein all over. As the Danish physicist Niels Bohr commented at the time, [Einstein] was a little too fond of telling God what to do. But this wasn't sheer arrogance, nor parental pride in his theory. The reason Einstein felt general relativity must be right is that it was too beautiful a theory to be wrong.

Ball attempts to refute John Keats's statement and, by doing so, involves science and is helped by the growing complexity of the theories of quantum physics and cosmological physics. According to Ball, "The gravest offenders in this attempted redefinition of beauty are, of course, the physicists." For him, the beauty of art and that of science must be on separate pages. He states that science and art should not try to come up with a "united search for beauty." He concludes his argument with the sentence: "Beauty, unlike truth or nature, is something we make ourselves."

What do you think of Ball's refutation of Keats's statement?

Do you think that the quoted equation that Ball proposes makes sense? And given that the art you have made has always had some relationship with science, I think it is only logical to ask for your opinion on whether art and science should seek a united search for beauty or not.

What do you think of the separation he proposes?

J.M.Y.:  Truth has been represented in the history of art fundamentally as a beautiful female nude. There are many works in the history of painting to prove it. The famous painting of Sandro Botticelli, *The Calumny of Apelles*, is just one example. It is allegorical: a female figure pointing heavenward represents naked and beautiful Truth. She approaches the action, where there appears a judge with the ears of an ass, King Midas, who is positioned between Suspicion and Ignorance. Other figures represent Envy, Anger, Calumny, and so on. Next to the figure of Penitence is Truth, who will triumph in that description represented in the painting.

In the history of art, we can also find the myth of a woman accused of a crime that she didn't commit, whose defending counsel undresses her in front of the court. The court declares her innocent, so impressed are the judges by her beauty. Beauty and truth are considered in the Greek aesthetic ideal as one and are also linked to other qualities, such as the just and good.

The astronomer Johannes Kepler came up with a "beautiful" but erroneous theory on the marvelous proportion between the heavenly spheres, the movements of the planets and their circular orbits contained in the five platonic solids. The model of the Solar System he proposed in his work at the end of the sixteenth century, *Mysterium Cosmographicum,* I admire for its beauty, not for its scientific truth. It inspired me to make a series of variant sculptures from that model. Later Kepler corrected himself in his work *Astronomia Nova* after realizing that the planets' orbits describe ellipses. In a way, this great genius of astronomy accepted defeat by showing that those orbits were not actually circles, which were considered to be the simplest and most perfect figures.

To me as a painter, Kepler was fascinating because of his expressive and poetic force, which I feel can be found in the depth of his scientific work.

The idea of beauty has varied enormously for humans over the ages. The consideration of what is beautiful and also, I'm sure, of truth depends on different epochs and cultures.

However, contrary to what Ball says, there are very serious attempts to understand the unique constants of assimilating beauty as a single common denominator for all civilizations and for all time.

For me that equation, beauty ≠ truth, is a little off the mark. I experience art deep down inside, and to paraphrase the painter Barnett Newman, in the same way that birds know nothing of ornithology, artists should not speculate too much on aesthetics.

## 5.  Order versus chaos

A.P.:  Saul Bellow, winner of the Nobel Prize in Literature, said in an interview:

The first time I opened my eyes to the world, I had the impression that it was enslaved to the ideas of order, which are useless in every sense. More basic than all of this is the singularity of the point of view.[10]

José María, making use of Bellow's statement, could you explain to me from which point of view you tackle the chaos/order dichotomy?

Are order and chaos reconcilable in a single work of art?

J.M.Y.:  The artistic generation prior to the one I belong to were abstract expressionists, followers of expressive immediacy, Jackson Pollock being one of its leading lights.

I tried to assimilate the concept of entropy, the second law of thermodynamics, which indicates the degree of molecular disorder of a system, as it can be applied to the theory of information, art—or at least it seems so to me—together with the principles of uncertainty, chance theory, turbulence, and fluid mechanics.

However, as I advanced, I came to understand that Douglas Hofstadter was right when he said: "It turns out that an eerie type of chaos can lurk just behind a façade of order and yet, deep inside the chaos lurks an even eerier type of order."[11] Order and disorder, regularity and irregularity complement and contradict each other or come together in harmony in a work of art. I fully agree with what Hofstadter said because of what I have experienced as a painter in terms of the structural variables in an artistic composition, in which geometric formations, such as Sirpinsky curves or the concept of Benoit Mandelbrot's fractals, intervene, where chaos becomes orderly geometry.

We managed to incorporate these notions into the project of some paintings, like those of Pollock and some other artists of apparently radical compositional attitude, such as Piet Mondrian, where it is clear that chance and chaos partly determine the paintings of the former and the cosmos, proportional harmony, is what we perceive in the latter. Or if we consider the case in the sense of order of the famous painting of Rafael, *The School of Athens*, with its projective regularities.

My undertakings as a painter have always been conscious of that dichotomy, and I have used it to try to reach new degrees of aesthetic and expressive complexity, in the same way as chiaroscuro or complementary colors are used.

## 6. Time: Changing the past

A.P.:   When you finish a work, a painting, does that piece become something of the past?

Have you an inner need to return to a painting and eliminate it in order to paint it again, or do you only paint the present?

Is there regret in painting?

J.M.Y.:   Yes and no. The work remains in the past. Like everything that is not the fleeting "eternal present." A painting continues to be from the past, a landmark, a link in the chain of production that structures the future of other pieces, making up the present, but that remains stored in time. Even to remind myself what shouldn't be done, these pieces must remain in some way in the memory of the present activity. In the end, the time flow of each and every work of an artist determines the discourse of his or her passage through existence as a creator.

You can always return to a painting, but you can never eliminate it. This is something that I believe has never happened to me. One returns to develop, to perfect it. The difficult thing is to know when a piece of work is finished. For me, it's something that is never entirely finished. Rather, it's left to continue its own "life."

Just consider the origins, the beginnings of art. We continue, for example, to contemplate exceptional works, such as the cave paintings in Chauvet in France from 35,000 years ago, which at that time were contemporary art. They were present, they had a meaning conferred on them by their creators, and even now they transmit emotions and we admire them, although knowing that we cannot understand them very well, through the eyes of modern times.

### 7.   On the void

A.P.:   José María, your relationship with emptiness is a long one. I remember hearing a journalist at a press conference remark "the painter of the void," almost certainly because your exhibitions in Barcelona and Zaragoza in 2003 were called *On the Sublime Void* and your 2009 exhibition at the Galería Rafael Ortiz in Seville was called *The Absent Space*.

In art, as you know, there have been revolutionaries who have investigated different forms of artistic "void" (conceptually speaking): the void in painting (Kazimir Malevich's *White on White*), the entirely black canvas (Ad Reinhardt's black paintings), the absence of color (Li Yuan-Chi's *Monochromatic White Painting*, Robert Rauschenberg's *Erased de Kooning Drawing*) and the void in music (John Cage and his composition *4′33″*): silence as the absence of the music that the listener expects to hear.

Why does emptiness interest you? What is it for you?

In art, is nothing the same as emptiness?

And can you paint the void? And if so, how do you do it?

J.M.Y.:   The idea of understanding, or rather practicing, the notion of emptiness occurred to me as a result of the constant that I employ, not as neglect but of "less is more," which has been naturally projected right from my very first works.

The Zen gardens were, without my being aware of it at that time, my first real contact with Oriental art, which I came to truly admire. The Spanish painter Zobel was a great expert on Oriental art and I learned a lot from him, including the notion of the void.

As a minimalist practice, the principles of Zen aesthetics attracted me immediately. Some of those principles are asymmetry, austerity, liveliness, spontaneity, essentialness, meditation, serenity … and the sum of them all, emptiness. The object in this type of art is not understood as something finite, enclosed in its own limits, but as something that is gradually constituting itself all the time, completing itself, losing itself and recovering itself within the rhythms and vibrations of a universe in endless production.

Of course, I'm talking in metaphorical, poetic terms. But I prefer painting it than talking about it. I'm not developing artistic formulas that are more or less exotic, imprecise or scientific, and so on. I speak from the emotional experience of a painter whose reality moves him and teaches him in parallel with other forms of knowledge, including the symbolic and the spiritual.

"Nothing" and "emptiness" are terms I and other artists use indistinctly for the same thing, or we differentiate between them, in some cases, to understand each other better. Emptiness is everything; nothing is nothing.

Returning to my art masters, I think that for Malevich, the *Black Square* painting represented the supremacy of pure sensitivity in the visual arts. It gravitated on white, which means nothing. It was a concept that reflected to a certain extent its time, when mechanistic ideas still prevailed, with the notion that solid and indestructible particles were moving in the void.

Cage, closer to our times, was strongly influenced by Zen. His inspirational character was Buddha. He studied Oriental thought on the void, on silence, with D. T. Suzuki. He expressed that in the piece you mention: *4′ 33″*, in a very radical way. He is also considered a precursor to assuming whatever noise as a part of music. In the case of this piece, for example, the noise was that of the audience, or the sounds coming from outside the hall. I am a whole-hearted admirer of Cage. I met him at MIT and, although rather slow-wittedly, I did manage to learn something from his teachings.

A.P.: José María, let's talk now of the void according to current quantum physics:

The energy of point zero is in physics the lowest energy that a quantum mechanics physical system can possess, and is the energy of the fundamental state of the system. The concept of zero point energy was put forward by

Albert Einstein and Otto Stern in 1913 and was at first called "residual energy."[12] In quantum field theory, it is a synonym for the energy of the void, or dark energy, an amount of energy that is associated with the emptiness of the void space. In cosmology, the energy of the void is taken as the basis for the cosmological constant.

All of this that appears so abstract we can perhaps express by saying that for the present quantum physicist, the void is not nothing, for in the void, the aforementioned residual energy exists.

However, these things may be closer than we think. On the experimental level, zero point energy creates the Casimir effect (or Casimir-Polder force). This is the best-known mechanical effect of the fluctuations in the void and has something to do with how well or not the micromachines that we use every day function, for example the accelerometers of the iPhone that allow the content of the screen to change between vertical and horizontal when the smart phone is turned.

Perhaps, José María, a good metaphor for these phenomena would be some of your paintings that a critic classified as the "painting of the void."

Maybe the void of your paintings in reality is like a quantum void because within their vibrating colors there is perhaps a dark energy like that the present physicists are talking about, a probable cause of the emotion that one feels on contemplating them closely.

What do you think about this metaphor as far as your paintings are concerned?

J.M.Y.:   Your metaphor seems suitable and very interesting to me. The idea of the void was for me a natural continuation of the notion of space, basic for a painter who, in preferring classical mechanics, considered the void as something in which solid and indestructible particles move.

The space of a Renaissance painting is a scene that the painter constructed with his perspectives, with one or two points of view and geometric compositions, including canonical ones, following preestablished orders, almost compulsory, of the objects, landscapes, individuals who peopled that scene, subject to proportions, such as the golden ratio, for example.

The idea of space changed radically with Einstein's theories, which associated gravitational fields and spatial geometry. This change became even more pronounced with the combination of the theory of relativity and quantum fields, where the subatomic particles are no longer so "solid and indestructible." They are considered as mere condensations of the field and dilute in some way on interacting with the space surrounding them.

Space is no longer a passive void. This void is now recognized as a dynamic "entity" crossed by forces of great importance.

In art, these considerations all filter through to artists. They are notions of our times that we must incorporate into the pictorial space, as we too try to capture reality. Picasso, a great innovator of spatial relationships with cubism, never dared to make the leap from figuration to abstraction (what we call abstract art). In parallel, something similar took place with Einstein, who never accepted the theories of quantum mechanics of Bohr and Heisenberg. Both artistic abstraction and quantum physics do not require "representation." It is as absurd to ask oneself what the *White on White* painting represents as it is to ask oneself, in quantum mechanics, what the electron looks like.

Among my proposals is, I suppose, the exploration of the void as an approach to essential reality, perhaps a mystic reality like that found in the Oriental thought we were talking about and that considers the ultimate reality to be emptiness. They call it so because, for that vision, it is the essence of all forms and the source of all life.

I approach multispatial theories, quantum physics, superstring theories, multiverse theory, and so on, with awe and respect. Even the mysticism that we mentioned and that I assume, I try to understand. I study them as basic concepts, as well as their projection in my painting, faced with the fascination for the void, for space-time as cardinal expressive elements, obviously together with others, such as form, light, and color in plastic arts, and most important of all, their relationship and transcendental significance for the flow of humanity.

## 8.  Science, art, and scientific reductionism

A.P.:  Jonah Lehrer in his book, *Proust Was a Neuroscientist,* offers various statements on the relationship between art and science, which I would like to ask you about:[13]

• "What science forgets is that this isn't how we experience the world. (We feel like the ghost, not like the machine.) It is ironic but true: the one reality science cannot reduce is the only reality we will ever know. This is why we need art. By expressing our actual experience, the artist reminds us that our science is incomplete, that no map of matter will ever explain the immateriality of our consciousness."

- "Like a work of art, we exceed our materials. Science needs art to frame the mystery, but art needs science so that not everything is a mystery. Neither truth alone is our solution, for our reality exists in plural"
- "There is a presence in what is missing. That presence is our own."

Moreover, Lehrer says that the artist describes what the scientist cannot describe and that a comprehensive description of the functioning of the brain demands both cultures: that of art and that of science.

He also says that art is not reducible to physics and to time, that it is impossible to understand art without taking into consideration its relationship with science, and therefore, in our times, art is "a counterweight to the glories and excesses of scientific reductionism."

What do you think of that?

J.M.Y.:   I don't believe that science overlooks the ultimate reality, which, clearly, cannot be reached. I prefer to think that the tools for finding truth through the knowledge and technology that science uses are not the most appropriate ones for the task. However, part of science, especially stemming from the findings of quantum physics and the contradictions and new enigmas posed regarding the experimental observation of noted phenomena, is narrowing the gap between science and certain conclusions from the world of philosophy and even the spiritual.

Indefinite yet existential, art provides us with experiences that are in the end little understood by science. For me, however, art has always been a form of knowledge.

I think it's clear that there is a new attitude on the part of science, or of certain scientists who are trying to explore apparently observable fact, where consciousness may affect physical phenomena.

I fully agree that art needs science and science needs art. Both fields I understand as something more than disciplines. They are a coherent part of our cognitive and emotional capabilities. Art and science display, in every case, parallelisms, concomitances, and convergences, but also divergences and ruptures. All of them, individually and collectively, I see as expressive elements that constitute a fascinating "hyperpicture," perhaps overflowing with the multiverses that we form part of.

Science itself, with all its realism, also seems to me, without trying to mystify it, a miracle, as is art, as is everything. If not, how else can it be understood when science seems to be getting closer to a total understanding of the universe and its immutable laws? These inexorably have to change or evolve. Now we find we have learned to know that we know "less," at the same time as we confess to knowing only 5 percent of the total energy and matter of the universe that is observable with current technology.

Obviously, I'm speaking in terms of "beliefs." It couldn't be otherwise from my position as painter and with a very different education from that of a scientist. Finally, I do agree with what Lehrer says about our plural identity. Possibly, truth should not strictly be the final aim of thought.

A.P.:    Thank you, José María. It's been great talking to you.

J.M.Y.:    My pleasure and a privilege for me.

## Notes

1. MIT Hacks: "Hack: an Inventive, Anonymous Prank." *Fire Hydrant Water Fountain* (https://www.flickr.com/photos/adolfoplasencia/2573546585).

2. Some of the works mentioned by Adolfo Plasencia are *Sky Art*, MIT, 1982 (http://www.yturralde.org/Paginas/Etapas/et08/et0865-es.html), *Flying Structures: Structure Wheel, Octahedron,* Venice Biennale, 1978 (http://www.yturralde.org/Paginas/Etapas/et08/et0811-en.html), *Impossible Figures* (http://www.yturralde.org/Paginas/Etapas/et04/index-en.html), and an installation view, *Selected Works,* Sandra Gering Inc., 2008 (http://www.sandrageringinc.com/exhibitions/2008-01-17_jose-m-yturralde/#/images/7).

3. José María Yturralde, *Estructuras Volantes: Homenaje a A. Graham Bell,* 1979. http://www.yturralde.org/Paginas/Etapas/et08/et0869-es.html.

4. Marguerite Yourcenar, *Le Labyrinthe du monde,* vol. 3: *Quoi? L'Éternité* (Paris: Gallimard, 1988).

5. "There is no such thing as good painting about nothing": the letter was published on June 7, 1943, and addressed the themes of myth and primitivism in art, themes later explored in detail by the New York school.

6. Quoted in E. B. Breslin, *Mark Rothko: A Biography* (Chicago: University of Chicago Press, 1998), 240.

7. Pat Dowell, "Meditation and modern art meet in Rothko Chapel," NPR, March 1, 2011, http://www.npr.org/2011/03/01/134160717/meditation-and-modern-art-meet-in-rothko-chapel.

8. John Keats, "Ode on a Grecian Urn," written in May 1819 and published anonymously in the January 1820 issue of *Annals of the Fine Arts*.

9. Philip Ball, "Beauty ≠ Truth," *Aeon,* May 19, 2014, https://aeon.co/essays/beauty-is-truth-there-s-a-false-equation.

10. Norman Manea, "Interview with Saul Bellow: When Writing, My Job Is to Be Myself," *La Maleta de Portbou* 11 (May/June 2015), http://www.lamaletadeportbou.com/articulo/saul-bellow.

11. Douglas Hofstadter, quoted in David Parrish, *Nothing I See Means Anything: Quantum Questions, Quantum Answers* (Boulder: First Sentient Publications, 2006), 72.

12. Martin J. Klein, A. J. Kox, Jürgen Renn, and Robert Schulman, eds. *The Collected Papers of Albert Einstein, Volume 4: The Swiss Years: Writings 1912–1914* (Princeton: Princeton University Press, 1995), 270.

13. Jonah Lehrer, *Proust Was a Neuroscientist* (New York: Houghton Mifflin, Harcourt, 2007), xii, xii, 15.

# Name Index

# Subject Index